Thermal Physics

Clear and reader-friendly, this is an ideal textbook for students seeking an up-to-date introduction to thermal physics.

Written by an experienced teacher and extensively class-tested, *Thermal Physics* provides a comprehensive grounding in thermodynamics, statistical mechanics, and kinetic theory. A key feature of this text is its readily accessible introductory chapters, which begin with a review of fundamental ideas. Entropy, conceived microscopically and statistically, and the Second Law of Thermodynamics are introduced early in the book. Throughout, new topics are built on a conceptual foundation of four linked elements: entropy and the Second Law, the canonical probability distribution, the partition function, and the chemical potential. As well as providing a solid preparation in the basics of the subject, the text goes on to explain exciting recent developments such as Bose–Einstein condensation and critical phenomena. Key equations are highlighted throughout, and each chapter contains a summary of essential ideas and an extensive set of problems of varying degrees of difficulty. A solutions manual is available for instructors.

Suitable for both undergraduates and graduates in physics and astronomy.

Born in 1936 and educated at Harvard and Princeton Universities, Ralph Baierlein is currently Charlotte Ayres Professor of Physics at Wesleyan University, Middletown, Connecticut. He is a fellow of the American Physical Society and in 1998 received a Distinguished Service Citation from the American Association of Physics Teachers. He is also author of other university textbooks including *Atoms and Information Theory, Newtonian Dynamics,* and *Newton to Einstein: The Trail of Light*.

Thermal Physics

RALPH BAIERLEIN

Wesleyan University

CAMBRIDGE
UNIVERSITY PRESS

PUBLISHED BY THE PRESS SYNDICATE OF THE UNIVERSITY OF CAMBRIDGE
The Pitt Building, Trumpington Street, Cambridge, United Kingdom

CAMBRIDGE UNIVERSITY PRESS
The Edinburgh Building, Cambridge CB2 2RU, UK http://www.cup.cam.ac.uk
40 West 20th Street, New York, NY 10011-4211, USA http://www.cup.org
10 Stamford Road, Oakleigh, Melbourne 3166, Australia

First published 1999

Printed in the United Kingdom at the University Press, Cambridge

Typeset in Times 10/13pt, in 3B2   [KT]

A catalogue record for this book is available from the British Library

Library of Congress Cataloguing in Publication data

Baierlein, Ralph.
    Thermal physics / Ralph Baierlein.
        p.   cm.
    ISBN 0 521 59082 5
    1. Thermodynamics.   2. Entropy.   3. Statistical mechanics.
    I. Title.
    QC311.B293   1999
    536'. 7–dc21   98-38617 CIP

ISBN 0 521 59082 5 hardback

Contents

Preface   *xi*

1   Background

1.1   Heating and temperature   *1*
1.2   Some dilute gas relationships   *4*
1.3   The First Law of Thermodynamics   *8*
1.4   Heat capacity   *11*
1.5   An adiabatic process   *13*
1.6   The meaning of words   *16*
1.7   Essentials   *18*
       Further reading   *21*
       Problems   *21*

2   The Second Law of Thermodynamics

2.1   Multiplicity   *24*
2.2   The Second Law of Thermodynamics   *28*
2.3   The power of the Second Law   *29*
2.4   Connecting multiplicity and energy transfer by heating   *31*
2.5   Some examples   *35*
2.6   Generalization   *39*
2.7   Entropy and disorder   *44*
2.8   Essentials   *45*
       Further reading   *46*
       Problems   *47*

3   Entropy and Efficiency

3.1   The most important thermodynamic cycle: the Carnot cycle   *51*
3.2   Maximum efficiency   *55*
3.3   A practical consequence   *59*
3.4   Rapid change   *60*

3.5 The simplified Otto cycle   *62*
3.6 More about reversibility   *67*
3.7 Essentials   *69*
     Further reading   *70*
     Problems   *71*

4   Entropy in Quantum Theory

4.1 The density of states   *75*
4.2 The quantum version of multiplicity   *80*
4.3 A general definition of temperature   *80*
4.4 Essentials   *86*
     Problems   *87*

5   The Canonical Probability Distribution

5.1 Probabilities   *89*
5.2 Probabilities when the temperature is fixed   *91*
5.3 An example: spin $\frac{1}{2}\hbar$ paramagnetism   *94*
5.4 The partition function technique   *96*
5.5 The energy range δE *99*
5.6 The ideal gas, treated semi-classically   *101*
5.7 Theoretical threads   *109*
5.8 Essentials   *109*
     Further reading   *111*
     Problems   *112*

6   Photons and Phonons

6.1 The big picture   *116*
6.2 Electromagnetic waves and photons   *118*
6.3 Radiative flux   *123*
6.4 Entropy and evolution (optional)   *128*
6.5 Sound waves and phonons   *130*
6.6 Essentials   *139*
     Further reading   *141*
     Problems   *141*

7   The Chemical Potential

7.1 Discovering the chemical potential   *148*
7.2 Minimum free energy   *155*

7.3   A lemma for computing μ *156*
7.4   Adsorption   *157*
7.5   Essentials   *160*
        Further reading   *161*
        Problems   *162*

8   The Quantum Ideal Gas

8.1   Coping with many particles all at once   *166*
8.2   Occupation numbers   *168*
8.3   Estimating the occupation numbers   *170*
8.4   Limits: classical and semi-classical   *173*
8.5   The nearly classical ideal gas (optional)   *175*
8.6   Essentials   *178*
        Further reading   *179*
        Problems   *180*

9   Fermions and Bosons at Low Temperature

9.1   Fermions at low temperature   *182*
9.2   Pauli paramagnetism (optional)   *192*
9.3   White dwarf stars (optional)   *194*
9.4   Bose–Einstein condensation: theory   *199*
9.5   Bose–Einstein condensation: experiments   *205*
9.6   A graphical comparison   *209*
9.7   Essentials   *212*
        Further reading   *214*
        Problems   *215*

10   The Free Energies

10.1 Generalities about an open system   *222*
10.2 Helmholtz free energy   *225*
10.3 More on understanding the chemical potential   *226*
10.4 Gibbs free energy   *230*
10.5 The minimum property   *233*
10.6 Why the phrase "free energy"?   *234*
10.7 Miscellany   *236*
10.8 Essentials   *238*
        Further reading   *239*
        Problems   *240*

11 Chemical Equilibrium

11.1   The kinetic view   *244*
11.2   A consequence of minimum free energy   *246*
11.3   The diatomic molecule   *250*
11.4   Thermal ionization   *257*
11.5   Another facet of chemical equilibrium   *260*
11.6   Creation and annihilation   *262*
11.7   Essentials   *264*
        Further reading   *266*
        Problems   *266*

12 Phase Equilibrium

12.1   Phase diagram   *270*
12.2   Latent heat   *273*
12.3   Conditions for coexistence   *276*
12.4   Gibbs–Duhem relation   *279*
12.5   Clausius–Clapeyron equation   *280*
12.6   Cooling by adiabatic compression (optional)   *282*
12.7   Gibbs' phase rule (optional)   *290*
12.8   Isotherms   *291*
12.9   Van der Waals equation of state   *293*
12.10 Essentials   *300*
        Further reading   *301*
        Problems   *301*

13 The Classical Limit

13.1   Classical phase space   *306*
13.2   The Maxwellian gas   *309*
13.3   The equipartition theorem   *314*
13.4   Heat capacity of diatomic molecules   *318*
13.5   Essentials   *320*
        Further reading   *322*
        Problems   *322*

14 Approaching Zero

14.1   Entropy and probability   *327*
14.2   Entropy in paramagnetism   *329*
14.3   Cooling by adiabatic demagnetization   *331*
14.4   The Third Law of Thermodynamics   *337*

14.5   Some other consequences of the Third Law   *341*
14.6   Negative absolute temperatures   *343*
14.7   Temperature recapitulated   *347*
14.8   Why heating increases the entropy. Or does it?   *349*
14.9   Essentials   *351*
        Further reading   *352*
        Problems   *353*

15   Transport Processes

15.1   Mean free path   *356*
15.2   Random walk   *360*
15.3   Momentum transport: viscosity   *362*
15.4   Pipe flow   *366*
15.5   Energy transport: thermal conduction   *367*
15.6   Time-dependent thermal conduction   *369*
15.7   Thermal evolution: an example   *372*
15.8   Refinements   *375*
15.9   Essentials   *377*
        Further reading   *378*
        Problems   *378*

16   Critical Phenomena

16.1   Experiments   *382*
16.2   Critical exponents   *388*
16.3   Ising model   *389*
16.4   Mean field theory   *392*
16.5   Renormalization group   *397*
16.6   First-order versus continuous   *407*
16.7   Universality   *409*
16.8   Essentials   *414*
        Further reading   *415*
        Problems   *415*

Epilogue   *419*

Appendix A   Physical and Mathematical Data   *420*
Appendix B   Examples of Estimating Occupation Numbers   *426*
Appendix C   The Framework of Probability Theory   *428*
Appendix D   Qualitative Perspectives on the van der Waals Equation   *435*

Index   *438*

Preface

Several aims guided me while I wrote. My first goal was to build from the familiar to the abstract and still get to entropy, conceived microscopically, in the second chapter. I sought to keep the book crisp and lean: derivations were to be succinct and simple; topics were to be those essential for physics and astronomy. From the professor's perspective, a semester is a short time, and few undergraduate curricula can devote more than a semester to thermal physics.

Modularity was another aim. Instructors' tastes vary greatly, and so I sought maximal flexibility in what to teach and when to cover it. The book's logical structure is displayed in figure P1. Chapters 1 to 3 develop topics that appear in the typical fat textbook for introductory physics but are rarely assimilated by students in that course, if the instructor even gets to the topics. Thus the book presumes only an elementary knowledge of classical mechanics and some rudimentary ideas from quantum theory, primarily the de Broglie relationship $p = h/\lambda$ and the idea of energy eigenstates.

A benefit of modularity is that one can study chapter 13—the classical theory—any time after chapter 5. I placed the classical theory so far back in the book because I think students should get to use the quantum machinery of chapters 4 and 5 on some important physics before they face the development of more formalism. But students need not go through chapters 10, 11, and 12 before they do the classical theory. Chapter 13 is relatively easy, and so it is a good break after an intense chapter (such as chapter 6 or 9). In my own teaching, I tuck in chapter 13 after chapter 9.

The book's conceptual core consists of four linked elements: entropy and the Second Law of Thermodynamics, the canonical probability distribution, the partition function, and the chemical potential. You may welcome the conceptual economy. All too easily, thermal physics seems to require a radically new tool for every new topic. My aim is to use the four elements again and again, so that my students become comfortable with them and even moderately proficient.

A note about teaching strategy may be welcome. My students come to thermal physics without knowing that the density in an isothermal atmosphere drops off exponentially. Therefore, I assign problem 7.1 (the first problem in chapter 7) early enough so that my students have done the problem before I start to talk about chapter 7 in class. Thus the students know what should emerge from the statistical calculation in section 7.1, the calculation that "discovers" the chemical potential.

With gratitude, I acknowledge the advice and expertise of colleagues. At Wesleyan University, my thanks go to William Herbst, Lutz Hüwel, Richard Lindquist, Stewart Novick, and Brian Stewart. Faculty around the globe read the first draft and offered me

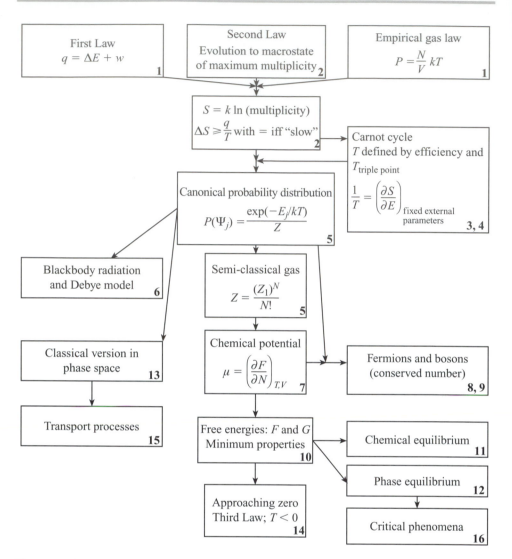

Figure P1 The logical structure of this book. The number in a box's lower right corner denotes the chapter that develops the topic. This flow chart is, of course, a bare-bones outline; looking at the table of contents or leafing through the pages will reveal the subtopics.

suggestions; I send thanks to Professors Michael Broide, Richard J. Cook, Hans Kolbenstvedt, Daniel Mustaki, Steven T. Ratliff, and Yako Yafet.

My heartfelt thanks go to Janet Morgan, who rescued me when my computer lost its ability to print $\hbar$, choked over fractional exponents, and otherwise subjected me to the trials of high-tech writing. I thank Vacek Miglus for setting up demonstrations and for making a video of the blackboards after class.

Comments by students were a great help, and so I send my appreciation to Andrew Billeb and Kristin Burgess.

A sabbatical at Northern Arizona University enabled me to produce the second draft. My thanks go to the Department of Physics and Astronomy and to its chair, Barry Lutz, for a congenial and productive stay.

Professor Carl Wieman and Michael Matthews provided the cover illustration: the classic "photo" of Bose–Einstein condensation in gaseous rubidium. I thank them here and provide a citation in the references for chapter 9.

As I have done with gratitude in other prefaces, I thank my wife Jean for her good advice and steadfast support. At the Press, Rufus Neal and Rebecca Mikulin were encouraging, effective, and a pleasure to work with. To them, I express my warm appreciation.

Let me end the list here and note that, despite all the comments and advice, I remain responsible for errors and infelicities.

For instructors, a solutions manual is available. Each problem has been worked out (by me) in a format appropriate for posting or distribution to students.

Middletown, Connecticut

1 Background

1.1 Heating and temperature

1.2 Some dilute gas relationships

1.3 The First Law of Thermodynamics

1.4 Heat capacity

1.5 An adiabatic process

1.6 The meaning of words

1.7 Essentials

Chapter 1 is meant as a review, for the most part. Indeed, if you have taken a good general physics course, then much of chapters 2 and 3 will be review also. Thermal physics has some subtle aspects, however, so it is best that we recapitulate basic ideas, definitions, and relationships. We begin in section 1.1 with the ideas of heating something and of temperature.

1.1 Heating and temperature

Suppose you want to fry two eggs, sunny-side up. You turn on the electric range and put the copper-bottomed frying pan on the metal coil, which soon glows an orangish red. The eggs begin to sizzle. From a physicist's point of view, energy is being transferred by conduction from the red-hot coil through the copper-bottomed pan and into the eggs. In a microscopic description of the process, one would say that, at the surface of contact between iron coil and copper pan, the intense jiggling of the iron atoms causes the adjacent copper atoms to vibrate more rapidly about their equilibrium sites and to pass such an increase in microscopic kinetic energy along through the thickness of the pan and finally into the eggs.

Meanwhile, your English muffin is in the toaster oven. Near the oven's roof, two metal rods glow red-hot, but there is no direct contact between them and the muffin. Rather, the hot metal radiates electromagnetic waves (of a wide spectrum of frequencies but primarily in the infrared region); those waves travel 10 centimeters through air to the muffin; and the muffin absorbs the electromagnetic waves and acquires their energy. The muffin is being heated by radiation. Now the microscopic view is this: the electrons and nuclei in the red-hot metal, being in erratic accelerated motion, emit photons of a broad spectrum of frequencies and polarizations; the muffin absorbs the photons.

Two more examples will suffice. In some early studies of paramagnetic salts at low temperature, gamma rays were used to heat the samples by irradiation. At a club picnic, the cans of soda and beer cool quite conveniently when immersed in a tub of water and crushed ice.

What are the common characteristics of these diverse means of heating and cooling? The following provides a partial list.

1. There is net transfer of energy (to or from the system, be it frying pan or muffin or soda).
2. The amount of energy transferred may be controlled and known at the macroscopic level but not at the microscopic level.
3. The transfer of energy does *not* require any change in the system's external parameters.

The phrase "external parameters" is new and needs explanation, best given in the context of physics applications rather than a kitchen or picnic. If steam is confined to a hollow cylinder fitted with a movable piston, then the volume V of the container is an external parameter for the gas. For a piece of soft iron wrapped with many turns of current-carrying wire, the magnetic field (produced by the electric current) is an external parameter. For a crystal of barium titanate between the plates of a capacitor, the electric field produced by the charges on the plates is an external parameter. In general, any macroscopic environmental parameter that appears in the microscopic mechanical expression for the energy of an atom or electron is an *external parameter*. If you are familiar with quantum mechanics, then a more precise definition of an external parameter is this: any macroscopic environmental parameter that appears in the Schrödinger equation for an atom, electron, or entire physical system is an *external parameter*.

In a fundamental way, one distinguishes two modes of energy transfer to a physical system:

1. by heating (or cooling);
2. by changing one or more external parameters.

To be sure, both kinds of transfer may occur simultaneously (for example, if one irradiates a sample at the same time that one changes the external magnetic field), but the distinction remains absolutely vital.

Energy transfer produced by a change in external parameters is called *work*.

Again, if you are familiar with quantum mechanics, you may wonder, how would heating be described in the Schrödinger equation? Consider the muffin that is being toasted. The Schrödinger equation for the muffin must contain terms that describe the interaction of organic molecules with the incident electromagnetic waves. But those terms fluctuate rapidly and irregularly with time; at most one knows some average value, perhaps a root mean square value for the electromagnetic fields of the waves. Although it may be well-defined at the macroscopic level, energy transfer by heating is inherently irregular and messy at the microscopic level. Later, in chapter 14, this insight will prove to be crucial.

Whenever two objects can exchange energy by heating (or cooling), one says that they are in *thermal contact*. For heating by conduction, literal contact is required. For heating by radiation, only a path for the electromagnetic radiation to get from one object to the other is required.

Elementary physics often speaks of three ways of heating: conduction, convection, and radiation. You may wonder, why is convection not mentioned here? Convection is basically energy transport by the flow of some material, perhaps hot air, water, or liquid sodium. Such "transport" is distinct from the "transfer" of energy to a physical system from its environment. For our purposes, only conduction and radiation are relevant.

To summarize: think of "heating" as a process of energy transfer, a process accomplished by conduction or radiation.

Temperature

We return to the kitchen. In colloquial language, the red-hot coil on the stove is hotter than was the copper-bottomed pan while it hung on the pot rack. In turn, the eggs, as they came out of the refrigerator, were colder than the pan was. Figure 1.1 illustrates the relationships. We can order objects in a sequence that tells us which will gain energy (and which will lose energy) by heating when we place them in thermal contact. Of two objects, the object that loses energy is the "hotter" one; the object that gains energy is the "colder" one. *Temperature* is hotness measured on some definite scale. That is, the goal of the "temperature" notion is to order objects in a sequence according to their "hotness" and to assign to each object a number—its temperature— that will facilitate comparisons of "hotness." A *thermometer* is any instrument that measures the degree of hotness in a calibrated way.

Over the centuries, many ways have been found to achieve the ordering. The length of a fine mercury column in glass constitutes a familiar thermometer, as does the length of an alcohol column (dyed red) in glass. In a professional laboratory, tempera- ture might be measured by the electrical resistance of a commercial carbon resistor, by the vapor pressure of liquid helium, by the voltage produced in a copper-constantan thermocouple, by the magnetic susceptibility of a paramagnetic salt such as cerium magnesium nitrate, or by the spectral distribution of the energy of electromagnetic waves, to name only five diverse methods. Calibration to an internationally adopted temperature scale is an item that we take up in section 4.3.

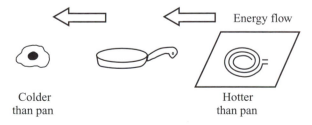

Figure 1.1 Ordering objects in a sequence according to energy transfer by heating. The broad arrows indicate the direction of energy flow when the objects are placed in thermal contact.

Return for a moment to the club picnic mentioned earlier. If you put a can of warm soda into the tub of water and crushed ice, the soda cools, that is to say, energy passes by conduction from the soda through the can's aluminum wall and into the ice water. In the course of an hour or so, the process pretty much runs its course: energy transfer ceases, and the soda now has the same temperature as the ice water. One says that the soda has come to "thermal equilibrium" with the ice water. More generally, the phrase *thermal equilibrium* means that a system has settled down to the point where its *macroscopic* properties are constant in time. Surely the microscopic motion of individual atoms remains, and tiny fluctuations persist, but no macroscopic change with time is discernible.

We will have more to say about temperature later in this chapter and in other chapters. The essence of the temperature notion, however, is contained in the first paragraph of this subsection. The paragraph is so short that one can easily underestimate its importance; I encourage you, before you go on, to read it again.

1.2   Some dilute gas relationships

As you read this, the air around you constitutes a dilute gas. The molecules of diatomic nitrogen and oxygen are in irregular motion. The molecules collide with one another as well as with the walls of the room, but most of the time they are out of the range of one another's forces, so that—in some computations—we may neglect those intermolecular forces. Whenever we do indeed neglect the intermolecular forces, we will speak of an *ideal gas*.

We will need several relationships that pertain to a dilute gas such as air under typical room conditions. They are presented here.

Pressure according to kinetic theory

Consider an ideal gas consisting of only one molecular species, say, pure diatomic nitrogen. There are N such molecules in a total volume V. In their collisions with the container walls, the molecules exert a pressure. How does that pressure depend on the typical speed of the molecules?

Because pressure is force (exerted perpendicular to the surface) per unit area, our first step is to compute the force exerted by the molecules on a patch of wall area A. To split the problem into manageable pieces, we write the word equation

$$\begin{pmatrix} \text{force on area } A \\ \text{due to molecules} \end{pmatrix} = \frac{\begin{pmatrix} \text{momentum transferred} \\ \text{to wall per collision} \end{pmatrix} \begin{pmatrix} \text{number of collisions} \\ \text{in time } \Delta t \end{pmatrix}}{\Delta t},$$

(1.1)

where Δt is a short time interval. The reasoning is based on Newton's second law of motion,

$$F = \frac{\Delta \mathbf{p}}{\Delta t},$$

where $\mathbf{p}$ denotes momentum and where we read from right to left to compute the force produced by molecular collisions.

Figure 1.2 shows the wall area and a molecule traveling obliquely toward the wall. When the molecule strikes the wall, its initial x-component of momentum mv_x will first be reduced to zero and will then be changed to $-mv_x$ in the opposite direction. Thus, for such a molecule, we have

$$\begin{pmatrix} \text{momentum transferred} \\ \text{to wall per collision} \end{pmatrix} = 2mv_x \qquad (1.2)$$

because only the x-component of momentum changes. The letter m denotes the molecule's rest mass.

Only the velocity component v_x transports molecules toward the wall; it carries them a distance $v_x \Delta t$ toward the wall in time Δt. To hit the wall in that time interval, a molecule must be within the distance $v_x \Delta t$ to start with. Consequently, we write

$$\begin{pmatrix} \text{number of collisions} \\ \text{in time } \Delta t \end{pmatrix} = (v_x \Delta t A)\tfrac{1}{2}\begin{pmatrix} \text{total number of molecules} \\ \text{per unit volume} \end{pmatrix}. \qquad (1.3)$$

As figure 1.2 illustrates, the factor $v_x \Delta t A$ is the slant volume of perpendicular length $v_x \Delta t$ and cross-sectional area A in which a molecule with $v_x > 0$ can be and still hit the wall area A within time Δt. The *number density* of molecules (regardless of their velocity) is N/V, but only half the molecules have $v_x > 0$ and hence travel toward the wall (albeit obliquely so). Therefore, if all molecules with $v_x > 0$ had the same value for v_x, then multiplication of $v_x \Delta t A$ by $\tfrac{1}{2}(N/V)$ would give the number of collisions in time Δt. In a moment, we will correct for that temporary assumption.

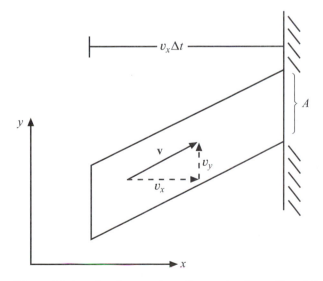

Figure 1.2 A molecule near the wall area A and traveling obliquely toward it.

Insert into equation (1.1) the two factors that we have worked out and then divide by the area A to find the provisional expression

$$\text{pressure} = \frac{2mv_x(v_x\Delta t A)\frac{1}{2}(N/V)}{\Delta t A}. \tag{1.4}$$

One step remains to be taken. The product v_x^2 appears, and we must average over its possible values. We can usefully relate that average to $\langle v^2 \rangle$, the average of the square of the speed. (Angular brackets, $\langle\ \rangle$, denote an average. Another, analogous meaning will be explained later, when it is first used.) Because v^2 equals the sum of the squares of the Cartesian components of velocity, the same equality is true for their averages:

$$\langle v^2 \rangle = \langle v_x^2 \rangle + \langle v_y^2 \rangle + \langle v_z^2 \rangle$$

$$= 3\langle v_x^2 \rangle; \tag{1.5}$$

the second line follows because the averages of the squared Cartesian components must be equal. Thus, denoting the pressure by P, we emerge with the relationship

$$P = \frac{1}{3}\frac{N}{V}m\langle v^2 \rangle$$

$$= \frac{2}{3} \times \frac{1}{2}m\langle v^2 \rangle \frac{N}{V}. \tag{1.6}$$

The pressure is proportional to the average translational kinetic energy and to the number density.

An empirical gas law

In the span from the seventeenth to the nineteenth centuries, experiments gave us an empirical gas law:

$$P = \frac{N}{V}kT. \tag{1.7}$$

The constant k is *Boltzmann's constant*,

$$k = 1.381 \times 10^{-23} \text{ joules/kelvin}, \tag{1.8}$$

and is entirely independent of the type of gas. The temperature T is the *absolute temperature*, whose unit is the *kelvin*, for which the abbreviation is merely K. Precisely how the absolute temperature is defined will be a major point in later chapters. You may know a good deal about that topic already. For now, however, we may regard T as simply what one gets by adding 273.15 to the reading on a mercury thermometer that is calibrated in degrees Celsius.

A brief history of this empirical gas law runs as follows. Around 1660, the Englishmen Robert Boyle, Henry Power, and Richard Towneley found that the product PV remains constant when air is compressed at constant temperature. In the years

1802–1805, the French chemist Joseph Louis Gay-Lussac showed that, at fixed pressure, the volume is a linear function of the temperature. Gay-Lussac used literally mercury thermometers and the Celsius scale. The accuracy of his measurements— good for those days—led him to infer that absolute zero was approximately 267 °C below the melting point of ice, close to the modern value of -273.15 °C.

By 1809, Gay-Lussac had found that reacting gases combine in a numerically simple fashion. For example, one volume of oxygen requires two volumes of hydrogen for complete combustion and yields two volumes of water vapor (all volumes being measured at a fixed pressure and temperature). This information led Amadeo Avogadro to suggest (in 1811) that equal volumes of different gases contain the same number of molecules (again at given pressure and temperature). In modern language, the number density N/V is the same for all dilute gases (at given pressure and temperature). The empirical gas law, as displayed in (1.7), uniquely incorporates the experimental insights of Boyle, Gay-Lussac, and Avogadro. All the functional dependences were known and well-established before the first quarter of the nineteenth century was over.

Our version of the empirical law is microscopic in the sense that the number N of individual molecules appears. Although data for a microscopic evaluation of the proportionality constant were available before the end of the nineteenth century, we owe to Max Planck the notation k and the first evaluation. In his study of blackbody radiation in 1900, Planck introduced two new constants, h and k, calling them "Naturconstanten:" constants of Nature. To determine their numerical values, he compared his theory with existing data on radiation (as will be described in section 6.3). Then, incorporating some work on gases by Ludwig Boltzmann, Planck showed that his radiation constant k was also the proportionality constant in the microscopic version of the empirical gas law.

Equation (1.7) is sometimes called the *ideal gas law* or the *perfect gas law*. Thus far, I have chosen to use the phrase "empirical gas law" to emphasize that equation (1.7) arose from experiments that actually measured P, V, and T. Although the relationship is accurate only for dilute gases, it is thoroughly grounded in experiment. As it enters our development, there is nothing hypothetical about it. So long as the gas is dilute, we can rely on the empirical gas law and can build on it.

Nevertheless, from here on I will conform to common usage and will refer to equation (1.7) as the "classical ideal gas law" or, for short, the "ideal gas law."

Average translational kinetic energy

Both kinetic theory and the ideal gas law provide expressions for the pressure. Those expressions must be numerically equal, and so comparison implies

$$\tfrac{1}{2}m\langle v^2\rangle = \tfrac{3}{2}kT. \tag{1.9}$$

We deduce that the average translational kinetic energy of a gas molecule is $\tfrac{3}{2}kT$, independent of the kind of gas.

To be sure, we must note the assumptions that went into this derivation. In working out the kinetic theory's expression for pressure, we assumed that the gas may be treated by classical Newtonian physics, that is, that neither quantum theory nor relativity theory is required. Moreover, the ideal gas law fails at low temperatures and high densities. Equation (1.9) is valid provided that the temperature is sufficiently high, but not too high, and that the particle number density is sufficiently low. Some of these criteria will be made more specific later, in chapters 5 and 8.

1.3   The First Law of Thermodynamics

The microscopic view of matter (where the "matter" might be a gas of diatomic oxygen) sees matter as a collection of atomic nuclei and electrons. The gas can possess energy in many forms:

- translational kinetic energy of the oxygen molecules,
- kinetic energy of nuclei vibrating and rotating relative to the molecular center of mass,
- kinetic energy of the electrons relative to the nuclei,
- electrical potential energy of the nuclei and electrons within a molecule,
- intermolecular potential energy (primarily electrical in origin),
- magnetic potential energy if an external magnetic field is present (because a diatomic oxygen molecule has a permanent magnetic dipole moment).

There can also be energies associated with the motion and location of the center of mass (CM) of the entire gas: translational kinetic energy of the CM, kinetic energy of bulk rotation about the CM, and gravitational potential energy of the CM, for example. Usually the energies associated with the center of mass do not change in the processes we consider; so we may omit them from the discussion. Rather, we focus on the items in the displayed list (and others like them), which—collectively—constitute the *internal energy* of the system.

The internal energy, denoted by E, can change in fundamentally two ways:

1. by our heating (or cooling) the system;
2. by the system's doing work on its surroundings as one or more of its external parameters change.

(If particles are permitted to enter and leave what one calls "the system," then their passage may also change the system's energy. In the first five chapters, all systems have a fixed number of particles, and so—for now—no change in energy with particle passage need be included.) An infinitesimal change ΔE in the internal energy is connected to items 1 and 2 by conservation of energy:

$$\begin{pmatrix} \text{energy input} \\ \text{by heating} \end{pmatrix} = \Delta E + \begin{pmatrix} \text{work done by system} \\ \text{on surroundings} \end{pmatrix}. \tag{1.10}$$

The energy that is transferred into the system by heating either increases the internal energy or provides energy for the work done by the system on its surroundings, or both. For sound historical reasons, equation (1.10) is called the *First Law of Thermodynamics*, although from a thoroughly modern point of view it is merely a statement of energy conservation. The word *thermodynamics* itself comes from "therme," the Greek word for "heat," and from "dynamics," the Greek word for "powerful" or "forceful." In the nineteenth century, the discipline of "thermodynamics" arose from the question, how can one best use heating processes to exert forces and to do work? Indeed, a 24-year-old William Thomson, later to become Lord Kelvin, coined the adjective "thermodynamic" in 1849 in a paper on the efficiency of steam engines.

Figure 1.3 and its caption contrast the nineteenth-century origin of thermodynamics with a modern view of the same subject matter.

We need a way to write the word equation (1.10) in succinct symbolic form. The lower case letter q will denote a small (or infinitesimal) amount of energy transferred by heating; the capital letter Q will denote a large (or finite) amount of energy so transferred. Analogously, the lower case letter w will denote a small (or infinitesimal) amount of work done by the system; capital W will denote a large (or finite) amount of work. Thus the First Law becomes

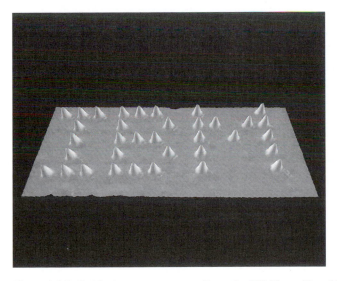

Figure 1.3 Individual xenon atoms spell out the IBM logo. Donald M. Eigler and Erhard K. Schweizer moved the 35 atoms into position (on a nickel surface) with a scanning tunneling microscope and then "took the picture" with that instrument. The work was reported in *Nature* in April 1990. When thermodynamics was developed in the nineteenth century, the very existence of atoms was uncertain, and so thermodynamics was constructed as a macroscopic, phenomenological theory. Today we can safely build a theory of thermal physics on the basis of atoms, electrons, nuclei, and photons.

By the way, xenon atoms are not shaped like chocolate kisses; the conical appearance is an artifact of the technique. [*Source*: D. M. Eigler and E. K. Schweizer, "Positioning single atoms with a scanning tunnelling microscope," *Nature* **344**, 524–6 (5 April 1990). Also, D. M. Eigler, private communication.]

$$q = \Delta E + w. \tag{1.11}$$

A further remark about notation is in order. The symbol Δ always denotes "change in the quantity whose symbol follows it." That change may be small (or infinitesimal), as in ΔE here. But, at other times, Δ may denote a large or finite change in some quantity. One needs to check each context.

The detailed expression for work done by the system depends (1) on which external parameter changes and (2) on whether the system remains close to thermal equilibrium during the change. Let us consider volume V and the pressure exerted by a gas in a hollow cylinder with a movable piston, as sketched in figure 1.4. For the small increment ΔV in volume, accomplished by slow expansion, the work done by the gas is this:

$$(\text{work done by gas}) = \text{force} \times \text{distance}$$

$$= \text{pressure} \times \text{area} \times \text{distance}$$

$$= P\Delta V. \tag{1.12}$$

The last line follows because the volume change ΔV equals the cross-sectional area times the distance through which the piston moves. The cylindrical shape helps us to derive the $P\Delta V$ form, but the expression is more general and holds for any infinitesimal (and slow) change in volume.

Thus, for a slow expansion, the First Law of Thermodynamics now takes the form

$$q = \Delta E + P\Delta V. \tag{1.13}$$

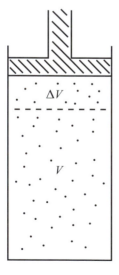

Figure 1.4 The gas expands from volume V to volume $V + \Delta V$, pushing on the piston as it goes.

1.4 Heat capacity

If we heat a dilute gas at constant volume, the molecules experience an increase in their average translational kinetic energy. Equation (1.9) tells us that the temperature increases also, and that agrees with common sense. The correlation of heating and temperature change is usefully captured in a ratio:

$$\frac{\left(\begin{array}{c}\text{energy input by heating}\\ \text{under specified conditions}\end{array}\right)}{(\text{ensuing change in temperature})}.$$

The ratio is called the system's *heat capacity* and is usually denoted by capital C, sometimes with a subscript to identify the specified conditions. Thus the generic expression is this:

$$C_X \equiv \left(\begin{array}{c}\text{heat capacity}\\ \text{under conditions } X\end{array}\right) \equiv \frac{\left(\begin{array}{c}\text{energy input by heating}\\ \text{under conditions } X\end{array}\right)}{(\text{ensuing change in temperature})}, \qquad (1.14)$$

where X may denote constant volume or constant pressure (or, for a magnetic system, constant external magnetic field).

Using equations (1.14) and (1.13), we can write the heat capacity at constant volume (and at constant values for any other external parameters) as

$$C_V \equiv \frac{\left(\begin{array}{c}\text{energy input by heating}\\ \text{at constant volume}\end{array}\right)}{\Delta T}$$

$$= \frac{q}{\Delta T} = \frac{\Delta E}{\Delta T} = \left(\frac{\partial E}{\partial T}\right)_V. \qquad (1.15)$$

Because all external parameters are held constant, no work is done, and so (1.13) implies that the energy input by heating, q, manifests itself entirely as a change ΔE in the internal energy. When a limit of infinitesimal transfer is taken, a partial derivative is required because we stipulate that the variation of internal energy with temperature is to be computed at constant external parameters; that is made explicit with parentheses and a subscript V. Succinctly, think of E as a function of T and V: $E = E(T, V)$; then differentiate with respect to T while holding V fixed.

If the gas is monatomic and if equation (1.9) holds, then

$$C_V = \frac{\partial(\frac{3}{2}kT\,N)}{\partial T} = \frac{3}{2}Nk, \qquad (1.16)$$

where N denotes the number of atoms.

As defined above, the heat capacity refers to the entire system. Frequently one finds the expression (1.14) divided by the system's mass or by the number of constituent particles. Such expressions give a heat capacity per unit mass or per particle. The expressions are called *specific heats*. There is nothing intrinsically new in them. Those quantities are merely more useful for tabulating physical properties and for comparing different materials.

Heat capacity at constant pressure

If a gas may expand while being heated but is kept at a constant pressure, the associated heat capacity—the heat capacity at constant pressure—is denoted C_P. Again using equations (1.14) and (1.13), we have

$$
C_P \equiv \frac{\left(\begin{array}{c} \text{energy input by heating} \\ \text{at constant pressure} \end{array} \right)}{\Delta T}
$$
$$
= \frac{\Delta E + P \Delta V}{\Delta T}. \tag{1.17}
$$

The quotients are easy to evaluate for an ideal gas (which may have polyatomic molecules) under conditions of temperature and number density such that the ideal gas law holds. We will call such a gas a *classical ideal gas*. Then the ideal gas law, $P = (N/V)kT$, implies

$$
V = \frac{N}{P} kT,
$$

whence

$$
\Delta V = \frac{N}{P} k \Delta T
$$

for small changes at constant pressure. The second term in the numerator of (1.17) will produce a quotient that is merely Nk. In the absence of intermolecular forces and when temperature and number density are such that the ideal gas law holds, the internal energy E depends on T, N, and the molecular species, but not on the volume. Thus the ratio $\Delta E / \Delta T$ in (1.17) is numerically the same ratio that appeared in (1.15) and gave C_V, the heat capacity at constant volume. In short, equation (1.17) may be written as

$$
C_P = C_V + Nk \tag{1.18}
$$

for a classical ideal gas.

The heat capacity is larger now because some energy goes into doing work as the gas is heated (and expands).

The mole

Tucked in here are a few paragraphs about the mole. To an accuracy of 1 percent, a *mole* of any isotope is an amount whose mass, measured in grams, is equal to the sum of the number of protons and neutrons in the isotope's nucleus. Rigorously, a mole of any isotope or naturally occurring chemical element is an amount that contains the same number of atoms as there are in 12 grams of the isotope ^{12}C, that is, the carbon atom that has six protons and six neutrons. The number of basic units itself is called *Avogadro's number* (or *constant*) and has the (approximate) value $N_A = 6.022 \times 10^{23}$ items per mole. A mole of a molecular species, such as water (H_2O) or carbon dioxide (CO_2), is an amount that contains N_A molecules of the species.

In the physics literature and in handbooks, heat capacities are often given in units of J/(K · mol), where "mol" is an abbreviation for "mole." Division by Avogadro's number N_A will convert to units of J/(K · molecule) or whatever the individual entities are: molecules, atoms, ions, etc.

The etymology of the word "mole" is curious and may help you to understand the use in physics and chemistry. According to the *Oxford English Dictionary*, the Latin word *moles* means "mass" in the loose sense of a large piece or lump of stuff. In the seventeenth century, the Latin diminutive *molecula* spawned the English word "molecule," meaning a small or basic piece. In 1900, the German chemist Wilhelm Ostwald lopped the "cule" off "molecule" and introduced the *mole* or *mol* in the sense defined two paragraphs back.

The ideal gas law can be expressed in terms of the number of moles of gas, and that version is common in chemistry. To see the connection, multiply numerator and denominator of equation (1.7) by Avogadro's number N_A and then factor as follows:

$$P = \frac{(N/N_A)}{V}(N_A k)T = \frac{\text{(number of moles)}}{V}RT. \tag{1.19}$$

The quotient N/N_A cites the amount of gas in moles. The symbol R denotes the *gas constant*: $R \equiv N_A k$.

1.5   An adiabatic process

The form of the First Law, as expressed in equation (1.10), suggests that there are two extremes for ways to change a system's internal energy E. In the preceding section we saw one of those extremes: hold all external parameters fixed, so that no work is done, and change E solely through energy input by heating. The opposite extreme consists of no heating but some work done. Any process in which no heating (or cooling) occurs is called *adiabatic*, from the Greek words *a* (not) + *dia* (through) + *bainein* (to go). That is, in an adiabatic process, no energy goes through an interface by heating or cooling. (If the adiabatic process occurs slowly, some work is sure to be done. If the process occurs rapidly, for example, as an expansion into a vacuum, it may be that no work is done on the environment.)

Another contrast is often made: between an adiabatic process and a process at constant temperature. The latter is called an *isothermal* process, for the adjective "isothermal" means "at constant temperature" or "for equal temperatures."

The adiabatic relation for a classical ideal gas

Consider a classical ideal gas. Figure 1.5 shows the effect of slow expansion under two different circumstances: isothermal conditions and an adiabatic process. When the temperature is fixed, the ideal gas law, $P = (N/V)kT$, asserts that the pressure drops because the volume increases. The gas's internal energy, however, does not change. In an expansion under adiabatic conditions, the internal energy drops because the gas does work on its surroundings but no energy input by heating is available to compensate. Thus the temperature drops. Consequently, in an adiabatic process, two factors cause the pressure to drop—an increase in volume and a decrease in temperature—and so the pressure drops faster.

To calculate the isothermal curve in figure 1.5, one supplements the ideal gas law with the relation $T = T_{initial}$, which is to hold throughout the entire expansion. For the adiabatic process, there must be an analogous supplementary relation. To derive it, we return to the First Law of Thermodynamics, equation (1.13), note that "adiabatic" implies $q = 0$, and write down energy conservation in the form

$$0 = \Delta E + P\Delta V. \tag{1.20}$$

The ideal gas law enables us to express the pressure P in terms of temperature and volume. As we noted in the preceding section, for a classical ideal gas the equation $\Delta E = C_V \Delta T$ holds, and so (1.20) takes the form

$$0 = C_V \Delta T + \frac{N}{V} kT \Delta V. \tag{1.21}$$

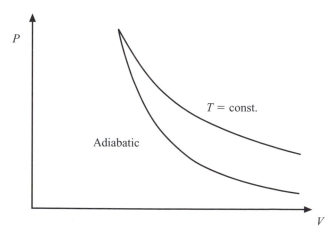

Figure 1.5 Comparing the run of pressure versus volume under two conditions: isothermal and adiabatic.

Now divide by $C_V T$ and prepare to integrate:

$$\int \frac{dT}{T} + \frac{Nk}{C_V} \int \frac{dV}{V} = 0. \tag{1.22}$$

Each integration will produce a logarithm. Both tradition and convenience suggest expressing Nk as the difference of two heat capacities; by (1.18),

$$\frac{Nk}{C_V} = \frac{C_P - C_V}{C_V} = \gamma - 1,$$

where γ is the traditional symbol for the ratio of heat capacities:

$$\gamma \equiv \frac{C_P}{C_V}. \tag{1.23}$$

Indefinite integration implies

$$\ln T + (\gamma - 1)\ln V = \text{const.}$$

Combining the terms on the left yields

$$\ln(TV^{\gamma-1}) = \text{const.}$$

The constancy of the right-hand side implies that the argument of the logarithm remains constant:

$$TV^{\gamma-1} = \text{new constant.} \tag{1.24}$$

This equation relates final and initial values of T and V during an adiabatic change of volume:

$$T_f V_f^{\gamma-1} = T_i V_i^{\gamma-1}. \tag{1.25}$$

Equation (1.24) *augments* the ideal gas law; *both* of them apply during the adiabatic process. The stipulation of no energy transfer by heating constrains the triplet T, P, and V more than the ideal gas law alone would and hence generates the additional relationship (1.24).

Precisely because the ideal gas law continues to hold, we may use it to eliminate T in (1.24) in terms of the product PV/Nk. The new form of what is really the same relationship becomes

$$\boxed{PV^\gamma = \text{another constant.} \qquad\qquad (1.26)}$$

This expression seems to be the easiest to remember. Starting from here, we recover equation (1.24) by eliminating P, the reverse of the recent step. Elimination of V from (1.26) with the aid of the ideal gas law demonstrates the constancy of $P^{1-\gamma}T^\gamma$. It suffices to remember one form and to know that you can get the others by elimination.

The expansion of a warm, low density cloud as it rises in the sky is (approximately) an adiabatic process. As the cloud rises into regions of lower pressure, it expands, does work on its surroundings, and thereby loses internal energy. Consequently, its temperature drops, and more moisture condenses.

How general are the relationships (1.24) and (1.26)? We specified a classical ideal gas and a slow process, so that the system remains close to thermal equilibrium. Those are two restrictions. Beyond that, the step from (1.21) to (1.22) assumed implicitly that C_V is independent of temperature. Often that assumption is a good approximation. For example, it is fine for diatomic oxygen and nitrogen under typical room conditions. If the temperature varies greatly, however, C_V will change; sections 11.3 and 13.4 explore this facet of diatomic molecules.

1.6   The meaning of words

If you glance back through the chapter, from its opening paragraph up to equation (1.10), you will find that I spoke of heating things and of "energy input by heating." Basically, I used verb-like forms of the word "heat." The focus was on energy transfer by heating (or cooling), a transfer produced by conduction or radiation, and that *process* is quite well-defined.

Historically, the word "heat" has been used as a noun as well, but such use—although common—is often technically incorrect. The reason is this: there is no way to identify a definite amount or kind of energy *in* a gas, say, as "heat." An example may clarify this point.

Figure 1.6 shows two sequences of events. In sequence (a), hot steam doubles in volume as it expands adiabatically into a vacuum. Because the water molecules hit only fixed walls, they always rebound elastically. No energy is gained or lost in the expansion, and so $E_{final} = E_{initial}$.

In sequence (b), the steam expands slowly and does work on a piston (which may be connected to other machinery, so that the work does something genuinely useful, such as turn the shaft of an electrical generator). The steam loses energy and also drops in temperature. To compensate for the energy loss, in the last stage a burner heats the water vapor, transferring energy by heating until the steam's energy returns to its initial value. Thus $E_{final} = E_{initial}$ in this sequence also.

If a definite amount or kind of energy in the water vapor could be identified as "heat," then there would have to be more of it at the end of sequence (b) than at the end of sequence (a), for only in sequence (b) has any heating occurred. But in fact, there is no difference, either macroscopic or microscopic, between the two final states.

An equivalent way to pose the problem is the following. The challenge is to identify a definite amount or kind of energy *in* a gas that (1) increases when the gas is heated by conduction or radiation and (2) remains constant during all adiabatic processes. No one has met that challenge successfully.

"Heat" as a noun flourished during the days of the caloric theory, but by 1865 at the latest, physicists knew that they should not speak of the "amount of heat" *in* a gas. Such a concept is untenable, as the analysis with figure 1.6 showed. Usage that is technically improper lingers nonetheless, and the phrase "heat capacity" is an

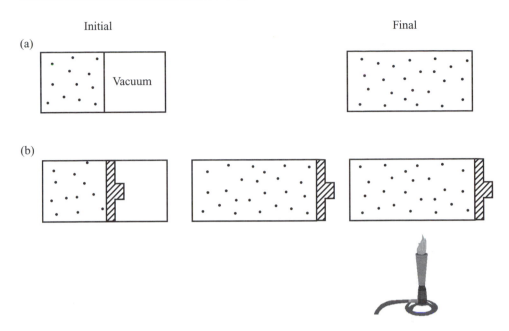

Figure 1.6 Two sequences from identical initial states of steam: (a) adiabatic expansion into a vacuum (after a barrier is removed); (b) slow adiabatic expansion, with work being done, followed by gentle heating. In both sequences, $E_{\text{final}} = E_{\text{initial}}$. Some time may be required for thermal equilibrium to be reached at the end of each sequence. Such time is incorporated into the sequences.

egregious example. The literal meaning of the phrase would be "capacity for holding heat," but that is not meaningful. The ratio in expression (1.14), however, is physically meaningful because it refers only to the amount of energy that is *transferred by heating*, and the *process* of heating is well-defined. From time to time, we will calculate a "heat capacity." Interpret the phrase not literally but rather in the sense of expression (1.14).

Even the words "heating" and "cooling" are used in more than one sense. Thus far, I have used them exclusively to describe the transfer of energy by conduction or radiation. The words are used also to describe any process in which the temperature rises or drops. Thus, if we return to figure 1.6, the first stage of sequence (b) could be described as "cooling by adiabatic expansion" because the temperature does drop. If one were to push the piston back in and return the gas to its initial state, one could describe that process as "heating by adiabatic compression" because the temperature would rise to its initial value. In both cases the adjective "adiabatic" means "no cooling or heating by conduction or radiation," but the temperature does change, and that alone can be the meaning of the words "cooling" and "heating." Such usage is not incorrect; it is just different, and one needs to be alert to such differences.

For a moment, imagine that you are baby-sitting your niece and nephew, Heather and Walter, as they play at the beach. The 4-year-olds are carrying water from the lake and pouring it into an old rowboat. Heather uses a green beach bucket; Walter improvises with the rectangular Tupperware container that formerly held the celery and carrot sticks. When you look in the rowboat, you see clear water filling it to a depth of several centimeters.

While Heather is carrying water, you can distinguish her lake water from Walter's—because it is in a green bucket—but once Heather has poured the water into the rowboat, there is no way to distinguish the water that she carried from that which Walter poured in or from the rainwater that was in the boat to start with.

The same possibilities and impossibilities hold for energy transferred by heating, energy transferred by work done (by or on the system), and internal energy that was present to start with. Energy that is being transferred by conduction or radiation may be called "heat." That is a technically correct use of the word and, indeed, a correct use as a noun. Once such energy has gotten into the physical system, however, it is just an *indistinguishable* contribution to the internal energy. Only energy *in transit* may correctly be called "heat."

Thermodynamics notes that certain attributes, such as temperature, pressure, total mass, density, and internal energy, serve adequately to define the macroscopic properties of a macroscopic system (when the system is in thermal equilibrium). Beyond that, thermodynamics notes that some of those attributes can be calculated from others (for example, via the ideal gas law). Such attributes are called *state functions* because, collectively, they define the macroscopic state and are defined by that state. "Heat," construed as a noun, is *not* a state function. The reason is this: the noun "heat" is defined only during the *process* of energy transfer by conduction or radiation. As we reasoned near the beginning of this section, one may *not* speak of the "amount of heat" *in* a physical system. These are subtle, but vital, points if one chooses to use the word "heat" as a noun. In this book, I will avoid the confusion that such usage invariably engenders and will continue to emphasize the process explicitly; in short, I will stick with the verb-like forms and will speak of "energy input by heating."

The language of thermodynamics is permeated with names constructed as though "heat" were a substance or a state function. It is neither. Historical usage, however, cannot be avoided, especially when you read other books or consult collections of tabulated physical properties. Stay alert for misnomers.

While we are on the subject of meanings, you may wonder, what is "thermal physics"? Broadly speaking, one can define *thermal physics* as encompassing every part of physics in which the ideas of heating, temperature, or entropy play an essential role. If there is any central organizing principle for thermal physics, then it is the Second Law of Thermodynamics, which we develop in the next chapter.

1.7   Essentials

This section collects essential ideas and results from the entire chapter. It is neither a summary of everything nor a substitute for careful study of the chapter. Its purpose is to emphasize the absolutely essential items, so that—as it were—you can distinguish the main characters from the supporting actors.

1. Think of *heating* as a process of energy transfer, a process accomplished by conduction or radiation. (No change in external parameters is required.)

2. Whenever two objects can exchange energy by heating (or cooling), one says that they are in *thermal contact*.

3. Any macroscopic environmental parameter that appears in the microscopic mechanical expression for the energy of an atom or electron is an *external parameter*. If you are familiar with quantum mechanics, then a more precise definition of an external parameter is this: any macroscopic environmental parameter that appears in the Schrödinger equation for an atom, electron, or entire physical system is an *external parameter*. Volume and external magnetic field are examples of external parameters. Pressure, however, is not an external parameter.

4. Energy transfer produced by a change in external parameters is called *work*.

5. We can order objects in a sequence that tells us which will gain energy (and which will lose energy) by heating when we place them in thermal contact. Of two objects, the object that loses energy is the "hotter" one; the object that gains energy is the "colder" one. *Temperature* is hotness measured on some definite scale. That is, the goal of the "temperature" notion is to order objects in a sequence according to their "hotness" and to assign to each object a number—its temperature—that will facilitate comparisons of "hotness." A *thermometer* is any instrument that measures the degree of hotness in a calibrated way.

6. The phrase *thermal equilibrium* means that a system has settled down to the point where its *macroscopic* properties are constant in time. Surely the microscopic motion of individual atoms remains, and tiny fluctuations persist, but no macroscopic change with time is discernible.

7. A classical non-relativistic analysis of a dilute gas implies

$$P = \tfrac{1}{3}\frac{N}{V}m\langle v^2 \rangle.$$

8. The *ideal gas law*,

$$P = \frac{N}{V}kT,$$

is an empirical gas law, valid for any dilute gas (of atoms or molecules). When the conditions of temperature and number density are such that a gas satisfies the ideal gas law, we will call the gas a *classical ideal gas*.

The shorter phrase, *ideal gas*, means merely that intermolecular forces are negligible. The questions of whether classical physics suffices or whether quantum theory is needed remain open.

9. Comparison of the items 7 and 8 implies

$$\tfrac{1}{2}m\langle v^2 \rangle = \tfrac{3}{2}kT.$$

The average translational kinetic energy of a gas molecule is $\tfrac{3}{2}kT$, independent of the kind of gas (provided that the motion is non-relativistic, that classical physics suffices, and that the gas is dilute).

10. The First Law of Thermodynamics is basically conservation of energy:

$$q = \Delta E + w.$$

11. If w, the small (or infinitesimal) amount of work done by the system, is performed by a volume expansion ΔV while the gas exerts a pressure P on a moving boundary, then the First Law becomes

$$q = \Delta E + P\Delta V.$$

12. A general definition of heat capacity is the following:

$$C_X \equiv \begin{pmatrix} \text{heat capacity} \\ \text{under conditions } X \end{pmatrix} \equiv \frac{\begin{pmatrix} \text{energy input by heating} \\ \text{under conditions } X \end{pmatrix}}{(\text{ensuing change in temperature})},$$

where X may denote constant volume or constant pressure (or, for a magnetic system, constant external magnetic field).

13. The heat capacity at constant volume may be expressed as

$$C_V = \left(\frac{\partial E}{\partial T} \right)_V \to \frac{3}{2} Nk \text{ for a monatomic classical ideal gas,}$$

where the arrow and the last expression refer to a monatomic classical ideal gas and non-relativistic motion. (Thus the last expression is *not* general.)

14. Similarly, the heat capacity at constant pressure may be expressed as

$$C_P = \frac{\Delta E + P\Delta V}{\Delta T} \to C_V + Nk \text{ for a classical ideal gas.}$$

15. A process in which *no heating* occurs is called *adiabatic*. For a classical ideal gas, the relationship

$$PV^{\gamma} = \text{constant}$$

holds during an adiabatic process (provided the process is performed slowly, so that the system remains close to thermal equilibrium). Here $\gamma \equiv C_P/C_V$ is the ratio of heat capacities, presumed to be independent of temperature. The relationship, $PV^{\gamma} = $ constant, holds in *addition* to the ideal gas law. Consequently, the latter may be used to eliminate pressure or volume from the former and hence to construct new versions of the adiabatic relationship.

16. Attributes that, collectively, define the macroscopic state and which are defined by that state are called *state functions*. Examples are internal energy E, temperature T, volume V, and pressure P.

17. "Heat," construed as a noun, is *not* a state function. The reason is this: the noun "heat" is defined only during the *process* of energy transfer by conduction or radiation. One may *not* speak of the "amount of heat" *in* a physical system.

Further reading

A marvelous resource for the historical development of thermal physics is provided by Martin Bailyn in his book, *A Survey of Thermodynamics* (AIP Press, Woodbury, New York, 1994).

Max Planck introduced the symbols h and k in his seminal papers on blackbody radiation: *Ann. Phys.* (Leipzig) **4**, 553–63 and 564–6 (1901). Thomas S. Kuhn presents surprising historical aspects of the early quantum theory in his book, *Blackbody Theory and the Quantum Discontinuity, 1894–1912* (Oxford University Press, New York, 1978).

"The use and misuse of the word 'heat' in physics teaching" is the title of a provocative article by Mark W. Zemansky in *The Physics Teacher* **8**, 295–300 (1970).

Problems

Note. Appendix A provides physical and mathematical data that you may find useful when you do the problems.

1. A fixed number of oxygen molecules are in a cylinder of variable size. Someone compresses the gas to one-third of its original volume. Simultaneously, energy is added to the gas (by both compression and heating) so that the temperature increases five fold: $T_{\text{final}} = 5\,T_{\text{initial}}$. The gas remains a dilute classical gas.

By what numerical factor does each of the following change:

(a) pressure,
(b) typical speed of a molecule,
(c) the number of impacts per second by molecules on 1 square centimeter of wall area?

Be sure to show your line of reasoning for each of the three questions.

2. *Radiation pressure.* Adapt the kinetic theory analysis of section 1.2 to compute the

pressure exerted by a gas of photons. There are N photons, each of energy $h\nu$, where h is Planck's constant and ν is a fixed frequency. The volume V has perfectly reflecting walls. Recall that a photon's momentum is $h\nu/c$ in magnitude. Express the pressure in terms of N, V, and the product $h\nu$. [For a photon gas in thermal equilibrium at temperature T, one would need to sum (or average) over a spectrum of frequencies. Section 6.2 will do that but from a different point of view.]

3. *Relativistic molecules.* Suppose the molecules of section 1.2 move with speed comparable to the speed of light c. (Suppose also that the molecules survive collision with the wall!)

(a) Adapt the kinetic theory analysis to this context; express the pressure in terms of the rest mass m and the relativistic energy $\varepsilon_{rel} = mc^2/\sqrt{1 - v^2/c^2}$. Eliminate the speed v entirely.
(b) After giving the general expression, examine the domain in which the strong inequality $\langle \varepsilon_{rel} \rangle \gg mc^2$ holds. If you know the pressure exerted by a photon gas, compare the limit here with the photon gas's pressure.

4. *Adiabatic compression.* A diesel engine requires no spark plug. Rather, the air in the cylinder is compressed so highly that the fuel ignites spontaneously when sprayed into the cylinder.

(a) If the air is initially at room temperature (taken as 20 °C) and is then compressed adiabatically by a factor of 15, what final temperature is attained (before fuel injection)? For air, the ratio of heat capacities is $\gamma = 1.4$ in the relevant range of temperatures.
(b) By what factor does the pressure increase?

5. *Adiabatic versus isothermal expansion.* In figure 1.5, the adiabatic curve drops away from the common initial point in the $P-V$ plane faster than the isothermal curve.

(a) Do the two curves ever have the same slope (at some larger, common value of V)?
(b) Do the two curves ever cross again (at larger volume)?

6. *Rüchardt's experiment: equilibrium.* Figure 1.7 shows a large vessel (of volume V_0) to which is attached a tube of precision bore. The inside radius of the tube is r_0, and the tube's length is ℓ_0. You take a stainless steel sphere of radius r_0 and lower it—slowly—down the tube until the increased air pressure supports the sphere. Assume that no air leaks past the sphere (an assumption that is valid over a reasonable interval of time) and that no energy passes through any walls.

Determine the distance below the tube's top at which the sphere is supported. Provide both algebraic and numerical answers. You have determined an equilibrium position for the sphere (while in the tube).

Numerical data: $V_0 = 10.15$ liters; $r_0 = 0.8$ cm; and $\ell_0 = 60$ cm. Mass density of

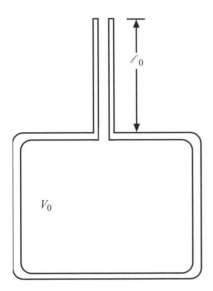

Figure 1.7 Apparatus for Rüchardt's experiment.

stainless steel $= 7.8$ times that of water. Take the ratio of heat capacities to lie in the range $1.3 \leqslant \gamma \leqslant 1.7$.

7. *Rüchardt's experiment: oscillation.* This question carries on from "Rüchardt's experiment: equilibrium." The physical context is the same, but now you release the steel ball (from rest) at the top of the tube.

(a) Determine the subsequent motion of the ball. In your quantitative calculations, ignore friction with the walls of the tightly fitting tube, but describe how the predicted evolution would change if you were to include friction. Note that the equilibrium location (for the "lowered" ball) is about half-way down the tube. After you have worked things out algebraically, insert numerical values.

(b) Describe how one could use the apparatus and analysis to determine the ratio of heat capacities for air under room conditions. Present an expression for γ in terms of the oscillation frequency together with known or readily measurable quantities.

8. A monatomic classical ideal gas of N atoms is initially at temperature T_0 in a volume V_0. The gas is allowed to expand slowly to a final volume $7V_0$ in one of three different ways: (a) at constant temperature, (b) at constant pressure, and (c) adiabatically. For each of these contexts, calculate the work done by the gas, the amount of energy transferred to the gas by heating, and the final temperature. Express all answers in terms of N, T_0, V_0, and k.

2 The Second Law of Thermodynamics

2.1 Multiplicity

2.2 The Second Law of Thermodynamics

2.3 The power of the Second Law

2.4 Connecting multiplicity and energy transfer by heating

2.5 Some examples

2.6 Generalization

2.7 Entropy and disorder

2.8 Essentials

Chapter 2 examines the evolution in time of macroscopic physical systems. This study leads to the Second Law of Thermodynamics, the deepest principle in thermal physics. To describe the evolution quantitatively, the chapter introduces (and defines) the ideas of multiplicity and entropy. Their connection with temperature and energy input by heating provides the chapter's major practical equation.

2.1 Multiplicity

Simple things can pose subtle questions. A bouncing ball quickly and surely comes to rest. Why doesn't a ball at rest start to bounce? There is nothing in Newton's laws of motion that could prevent this; yet we have never seen it occur. Why? (If you are skeptical, recall that a person can jump off the floor. Similarly, a ball could—in principle—spontaneously rise from the ground, especially a ball that had just been dropped and had come to rest.)

Or let us look at a simple experiment, something that seems more "scientific." Figure 2.1 shows the context. When the clamp is opened, the bromine diffuses almost instantly into the evacuated flask. The gas fills the two flasks about equally. The molecules seem never to rush back and all congregate in the first flask.

You may say, "That's not surprising." True, in the sense that our everyday experience tells us that the observed behavior is reasonable. But let's consider this "not surprising" phenomenon more deeply.

Could the molecules all go back? Certainly. There is nothing in Newton's laws, the irregular molecular motion, and the frequent collisions to prevent the molecules from all returning—and, indeed, from then staying in the original container. The collisions

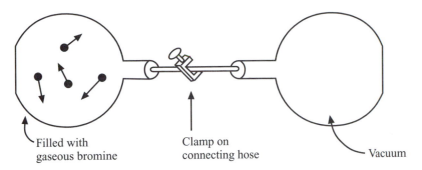

Filled with
gaseous bromine

Clamp on
connecting hose

Vacuum

Figure 2.1 The bromine experiment. Each flask has a volume of approximately one liter. The bromine is visible as a rusty brown gas. (In practice, the "bromine-filled" flask contains air also, but we focus exclusively on the bromine because it is visible.)

would have to be "just right," you say, for that to happen, but they could—in principle—be just right. And yet, although the collisions *could* be just right, such a situation is not probable.

If there were just three bromine molecules, we would expect a simultaneous return to the original container to occur from time to time, perhaps in a matter of minutes. But if we consider 6, then 60, next 600, and finally 10^{20} bromine molecules, the event of simultaneous return shrinks dramatically in probability. Table 2.1 provides a graphic analogy. We expect a more-or-less uniform spatial distribution of bromine molecules, and that is what we actually find.

We can say that a more-or-less uniform distribution of the molecules is much more probable than any other distribution. The simple phrase, "a more-or-less uniform distribution," is an instance of a large-scale, *macroscopic* characterization, as distinguished from a *microscopic* characterization (which would focus on individual molecules). Boldly, we generalize from the bromine experiment to macroscopic physics in general:

> **Macroscopic regularity**. When a physical system is allowed to evolve in isolation, some single macroscopic outcome is overwhelmingly more probable than any other.

This property, of course, is what makes our macroscopic physical world reproducible and predictable (to the large extent that it actually has those characteristics). The inference is another way of saying that a liter of water or a roomful of air has reproducible physical properties.

Multiplicity

Let us sharpen the inference. We distinguish between the "state of affairs" on the microscopic and on the macroscopic level. Here are two definitions:

Microscopic state of affairs, abbreviated *microstate*: the state defined by specifying in great detail the location and momentum of each molecule and atom.

Table 2.1 *An analogy of a tossed coin illustrates the probabilities of "all bromine molecules in one of two containers" and "more-or-less uniform distribution of bromine molecules."*

Number of tosses	Probability of all tosses being heads
1	$\frac{1}{2}$
2	$(\frac{1}{2})^2 = \frac{1}{4} = 0.25$
3	$(\frac{1}{2})^3 = \frac{1}{8} = 0.125$
6	$(\frac{1}{2})^6 = \frac{1}{64} = 1.6 \times 10^{-2}$
60	$(\frac{1}{2})^{60} = 10^{-18}$
600	$(\frac{1}{2})^{600} = 10^{-181}$
10^{20}	$(\frac{1}{2})^{10^{20}} = 10^{-3 \times 10^{19}}$

If you toss a penny 60 times, you expect about 30 heads and 30 tails. For all 60 tosses to yield heads is possible but quite improbable. The situation with 10^{20} tosses would be extreme.

A related and even more relevant issue is this: if a penny is tossed many times, can we be pretty sure that about half the tosses will come out heads and about half will come out tails? In short, can we be pretty sure that we will find a more-or-less even distribution of heads and tails? The table below shows some probabilities for a more-or-less even distribution.

Number of tosses N	Probability that the number of heads is within 1 percent of the value $N/2$
10	0.246
100	0.080
1,000	0.248
10^4	0.683
10^5	0.998
10^6	$1 - 1.5 \times 10^{-23} \cong 1$
10^8	$1 - 2.7 \times 10^{-2174} \cong 1$

As soon as the number of tosses is 100,000 or so, the probability of a more-or-less even distribution of heads and tails is nearly unity, meaning that such a distribution is a pretty sure outcome. By a million tosses, the more-or-less even distribution is overwhelmingly more probable than any and all significant deviations from that distribution, and the numbers quickly become unimaginable. If we were to consider a number of tosses comparable to the number of bromine atoms—10^{20} tosses—then the probability of a more-or-less even distribution would be so close to a certainty as to make any deviation not worth accounting for.

[You may wonder, how were these probabilities computed? Once the total number of tosses N has been specified, each conceivable sequence of heads and tails—such as HHTHT ... —has a probability of $(1/2)^N$. To compute the probability that the number of heads will be within 1 percent of $N/2$, one just counts up the number of different sequences that meet the 1 percent criterion and then multiplies by the probability of each sequence, $(1/2)^N$. Fortunately, there are some tricks for doing that arithmetic efficiently, but to go into them would take us too far afield.]

Macroscopic state of affairs, abbreviated *macrostate*: the state defined by specifying a few gross, large-scale properties, such as pressure P, volume V, temperature T, or total mass.

Here is an analogy. Given four balls labeled A, B, C, D and two bowls, what are the different ways in which we can apportion the balls to the two bowls? Table 2.2 sets out the possibilities.

Some macrostates have many microstates that correspond to them; others, just a few. This is a vital quantitative point. It warrants a term:

$$\begin{pmatrix} \text{the } \textit{multiplicity} \\ \text{of a macrostate} \end{pmatrix} \equiv \begin{pmatrix} \text{the number of microstates} \\ \text{that correspond} \\ \text{to the macrostate} \end{pmatrix}. \qquad (2.1)$$

The even distribution of balls has the largest multiplicity, but, because the number of items is merely four here, the multiplicity for the even distribution is not yet overwhelmingly greater than the multiplicity for the quite uneven distributions.

Table 2.2 *Microstates, macrostates, and multiplicity.*

Apportionment in detail[a]		Specification merely of the number of balls in each bowl		Number of detailed arrangements corresponding to each gross specification
Balls in left-hand bowl	Balls in right-hand bowl	Left	Right	
ABCD	None	4	0	1
ABC	D			
ABD	C	3	1	4
ACD	B			
BCD	A			
AB	CD			
AC	BD			
AD	CB	2	2	6
BC	AD			
BD	AC			
CD	AB			
A	BCD			
B	ACD	1	3	4
C	ABD			
D	ABC			
None	ABCD	0	4	1
Any single such apportionment is a microstate.		Any single such gross specification is a macrostate.		The multiplicity

[a]Position *within* a bowl does not matter.

2.2   The Second Law of Thermodynamics

The next step in our reasoning is easiest to follow if I draw the picture first and make the connections with physics later. Imagine a vast desert with a few oases. A "mindless" person starts at an oasis and wanders in irregular, thoughtless motion. With overwhelming probability, the person wanders into the desert (because there is so much of it around) and remains there (for the same reason). Figure 2.2 illustrates this scene.

If we make some correspondences, the desert picture provides an analogy for the behavior of a molecular system. Table 2.3 shows the correspondences.

With the aid of the analogy, we can understand the diffusion of bromine. Common sense and the example with the four balls tell us that the macrostate with a more-or-less uniform distribution of molecules has the largest multiplicity, indeed, overwhelmingly so. When the clamp was opened, the bromine—in its diffusion—evolved through many microstates and the corresponding macrostates. By molecular collisions, the bromine was (almost) certain to get to some microstate corresponding to the macrostate of largest multiplicity. Why? Simply because there are so many such

Table 2.3 *Correspondences in the desert analogy.*

A point on the map	A specific microstate
The desert	The macrostate of largest multiplicity
An oasis	A macrostate of small multiplicity
The "mindless" person	A system of many molecules
The person's path	The sequence of microstates in the evolution of the system
The initial oasis	The initial macrostate

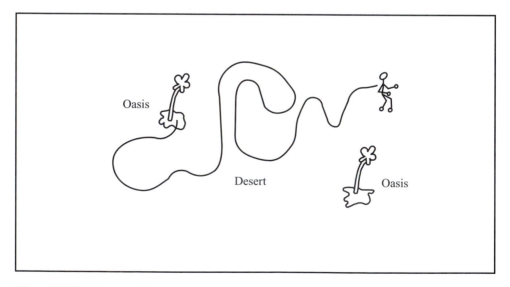

Figure 2.2 The desert analogy.

microstates. Thereafter, changes of microstate will certainly occur, but further change of macrostate is extremely unlikely. The continual changes of microstate will almost certainly take the bromine from one to another of the many microstates that correspond to the more-or-less uniform distribution.

The desert analogy suggests a refinement of our tentative inference about macroscopic regularity, a refinement presented here along with its formal name:

The Second Law of Thermodynamics. If a system with many molecules is permitted to change, then—with overwhelming probability—the system will evolve to the macrostate of largest multiplicity and will subsequently remain in that macrostate. Stipulation: allow the system to evolve in isolation. (The stipulation includes the injunction, do not transfer energy to or from the system.)

As for the stipulation, let us note that we left the bromine alone, permitting it to evolve by itself once the clamp was opened, and so prudence suggests that we append the stipulation, which contains implicitly the injunction about no energy transfer.

2.3 The power of the Second Law

Our statement of the Second Law may need a little more attention (to make it more quantitative), but already we have a powerful law, and we can even use it to save money for the government.

It costs a great deal of money to launch rockets for astronomical observations. We have to pay for a lot of energy (in the form of combustible liquids and solids) in order to put a satellite into orbit, and we have to pay even more to enable a satellite to escape the Earth's gravitational pull.

An economy-minded astronomer has an idea: let's use some of the vast amount of energy stored in the warmth of the sea. Figure 2.3 illustrates the proposal. The

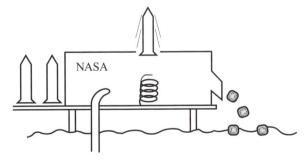

Figure 2.3 The astronomer's proposal. The mechanism inside the building is *not* permitted to use up anything, such as batteries. Rather, after each satellite-carrying projectile has been launched, the mechanism must return fully to its original state. *Only outside* the building is net change permitted. (Moreover, only change to the water and to the state of motion of the projectiles is permitted.)

astronomer wants to pump in sea water (which is well above freezing), extract some energy, and then throw the water back as ice. The extracted energy will be used to launch the satellite-carrying projectiles, perhaps by compressing a gigantic spring.

Now I will admit that there are some engineering difficulties here, but people have solved engineering problems before. Will the scheme work, even in principle? Should we give the astronomer a grant? No. We can tell the engineers not to exert themselves: no matter what fancy mechanism they put into the building, the proposal will not work.

By the First Law of Thermodynamics, the scheme is all right. Energy would be conserved. The kinetic energy of the projectiles would be provided by the energy extracted from the warm ocean water.

The Second Law, however, says no, the proposal will not work. Here is the logic: (1) We ask, *if* the proposal did work, what would result? (2) The Second Law says that such a result will not occur. (3) Therefore the proposal will not work. Now we go through the logic in detail, as follows.

The "system" consists of the projectiles, the building, and the water, in both liquid and solid form. An engineer can push one button to set in motion the launching of ten projectiles; thus there is no need for continuous human intervention, and we can meet the conditions under which the Second Law may be applied.

To bring the Second Law to bear on this issue, we need to compare multiplicities. Figures 2.4 and 2.5 show us what we need to see. How many ways can one arrange water molecules to get something that looks like sea water in liquid form? How many ways for an ice crystal? The multiplicity for the liquid greatly exceeds that for the crystal. The projectile multiplicities, however, are unchanged, for the following reason. During the launch, each atom of each projectile acquires a velocity increment of 10^4 meters/second (say) in the vertical direction. To each old microstate of a projectile, there corresponds one and only one new microstate, which differs from the old microstate in only one respect: each atom has acquired an increment of velocity in the vertical direction. The number of microstates does not change, and so the multiplicity for each projectile remains the same.

If the proposal worked, the system would go from a macrostate with high multiplicity to a macrostate with low multiplicity. The Second Law, however, says that (with overwhelming probability) a system will evolve to the macrostate of largest multiplicity (and then remain there). Therefore the proposal will not work.

We started with a bouncing ball and some bromine. By now we have a law, based on

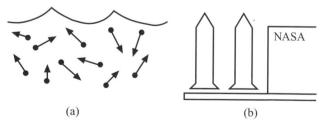

(a)   (b)

Figure 2.4 This is the situation we start with. (a) Sea water: molecules in helter-skelter disarray. (b) Projectiles all lined up.

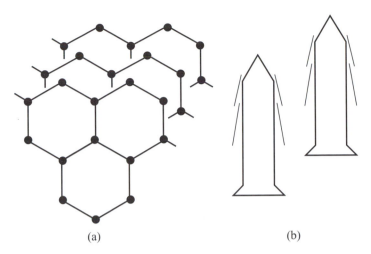

Figure 2.5 *If* the proposal worked, we would end up with this. (a) Ice: water molecules all neatly arranged in a crystalline pattern. (b) Projectiles possessing the transferred energy in the orderly motion of their upward flight.

an amalgam of experience, microscopic description, and statistical outlook, that can make powerful statements about macroscopic phenomena, the things we directly perceive.

2.4 Connecting multiplicity and energy transfer by heating

Now we go on to establish the connection between the Second Law and energy transfer by heating. We consider a confined classical ideal gas and study how the multiplicity changes when we allow the gas to expand by a small amount. Figure 2.6 sets the scene. To avoid a change in how the various molecular momenta (or velocities) contribute to the multiplicity (a calculation that would be difficult to handle now), let us keep the gas at constant temperature T while it expands. How can one do this? By heating the gas judiciously (perhaps with a large warm brick). As we slowly put in energy by heating, we let the gas (slowly) expand, do work, and thereby neither drop nor rise in temperature. Thus we have the verbal equation

$$\begin{pmatrix} \text{energy in} \\ \text{by heating} \end{pmatrix} = \begin{pmatrix} \text{energy out} \\ \text{as gas does work} \\ \text{on the external world} \end{pmatrix} \tag{2.2}$$

in this particular context. We will return to this equation, but first we need to learn how it can help us to make a connection with multiplicity.

Succinctly, more space available to the gas implies more ways to arrange the molecules, and that, in turn, implies a larger number of microstates and hence a larger multiplicity. Moreover, we reason that the number of spatial locations for a single molecule is proportional to the volume (V or $V + \Delta V$) in which the molecule can be.

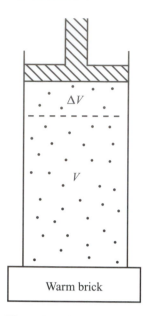

Figure 2.6 The gas expands from volume V to volume $V + \Delta V$ at a constant temperature as a small amount of energy q is put into the gas by heating. (Because we arrange to keep the temperature constant while the volume increases, the gas pressure will drop. Some device is used to control the piston's motion, and that device must take into account the change in gas pressure, but we need not concern ourselves with the device.)

For N molecules, the multiplicity will then be proportional to (volume)^N. [Note. If there were, say, 10 spatial locations for a single molecule, then there would be 10×10 different arrangements for two molecules, that is, 10^2. For three molecules, there would be $10 \times 10 \times 10$ different arrangements, yielding 10^3, and so on. The gas—as an ideal gas with molecules of infinitesimal size—is so dilute that we need not worry about molecules getting in each other's way or using up "spatial locations." Moreover, all that we will need is a *proportionality* to (volume)^N.]

Although we have reasoned that multiplicity is proportional to (volume)^N, we do not know the proportionality factor. If we form the ratio of multiplicities, however, the unknown proportionality factor cancels out, and we have

$$\frac{\text{final multiplicity}}{\text{initial multiplicity}} = \frac{(V + \Delta V)^N}{V^N} = \left(\frac{V + \Delta V}{V}\right)^N$$

$$= \left(1 + \frac{\Delta V}{V}\right)^N. \tag{2.3}$$

(The proportionality factor included whatever contribution the momenta make to the multiplicity. Because we kept the temperature constant, that contribution remained constant and canceled out.) Equation (2.3) gives us the ratio of multiplicities in terms of the original volume V and the volume change ΔV. That is a noteworthy result in itself, but our ultimate goal in this section is to establish a connection between change in multiplicity and energy transfer by heating. We can achieve that goal by relating ΔV to the energy input by heating, as follows.

Section 1.3 showed that, for the small increment ΔV in volume, the work done by the gas is $P \Delta V$. Thus the energy balance equation, which was displayed as equation (2.2), becomes

$$q = P \Delta V, \tag{2.4}$$

where q denotes the small energy input by heating. We need to eliminate the pressure P, and the ideal gas law, $P = (N/V)kT$, enables us to do that. Substitution for P in equation (2.4) first gives

$$q = \frac{N}{V} kT \Delta V;$$

then, after we divide on both sides by NkT, the equation becomes

$$\frac{q}{NkT} = \frac{\Delta V}{V}. \tag{2.5}$$

Now we use this result to substitute for $\Delta V / V$ in equation (2.3):

$$\frac{\text{final multiplicity}}{\text{initial multiplicity}} = \left(1 + \frac{q}{NkT}\right)^N. \tag{2.6}$$

The quotient q/NkT is small because $\Delta V/V$ is small—recall equation (2.5) here—but the exponent N is enormous—perhaps 10^{20} or even larger—and so we need to be circumspect in assessing numerical values.

Whenever one is confronted with a large exponent, taking logarithms may make the expression easier to work with. So we take the natural logarithm of both sides of equation (2.6):

$$\ln\left(\frac{\text{final multiplicity}}{\text{initial multiplicity}}\right) = \ln\left(1 + \frac{q}{NkT}\right)^N = N \ln\left(1 + \frac{q}{NkT}\right)$$

$$= N \frac{q}{NkT} = \frac{q}{kT}. \tag{2.7}$$

The step to the second line is permitted by an excellent approximation: $\ln(1 + a) \cong a$, provided a is small (in magnitude) relative to 1. (If you are not familiar with this approximation, you can find it derived in appendix A.) Because our analysis has in mind an infinitesimal value for $\Delta V/V$ and hence for q/NkT, the approximation is entirely justified.

If we multiply equation (2.7) through by k and write the logarithm of a ratio as a difference of the individual logarithms, we get

$$k \ln(\text{final multiplicity}) - k \ln(\text{initial multiplicity}) = \frac{q}{T}$$

$$= \frac{\left(\begin{array}{c} \text{energy input} \\ \text{by heating} \end{array} \right)}{T}. \tag{2.8}$$

We see that q is connected with the logarithms of multiplicities. The logarithm, we will find, is so useful that it merits its own symbol S,

$$S \equiv k \ln(\text{multiplicity}), \tag{2.9}$$

and its own name: the *entropy*. We can write the consequence of our slow expansion at constant temperature T as the equation

$$S_{\text{final}} - S_{\text{initial}} = \frac{q}{T}. \tag{2.10}$$

The German physicist Rudolf Clausius coined the word "entropy" in 1865. Looking for a word similar to the word "energy," Clausius chose the Greek word "entropy," which means (in Greek) "the turning" or "the transformation." The coinage is indeed apt: if we ask, "which transformation of a physical system will occur spontaneously?", then the multiplicity of each of the various possible macrostates is crucial, for the system will evolve to the macrostate of largest multiplicity. (This conclusion is the essential content of our purely verbal statement of the Second Law, presented in section 2.2.) The same is true, of course, if we describe the situation with the logarithm of the multiplicity. If we permit an otherwise-isolated system to change, it will evolve to the macrostate for which the logarithm of the multiplicity is largest, that is, to the macrostate of largest entropy.

A pause to consolidate is in order. Our detailed calculations in this subsection culminate in equation (2.8); that equation connects change in multiplicity with the energy input by heating—in the specific context of a classical ideal gas that expands slowly and at constant temperature. Equation (2.9) merely introduces a new symbol, and equation (2.10) just expresses the content of equation (2.8) in the new language. The notion of multiplicity remains primary, but the language of "entropy" will be increasingly useful.

Some generalization

While we have the context of this gaseous system, let us consider what would result if we increased the volume by ΔV *suddenly* (by removing a partition and letting the gas rush into a vacuum) and if we did *not* put in any energy by heating. In the expansion into a vacuum, the molecules would strike only stationary walls and hence no energy would be lost from the gas.

Altogether, the energy of the gas would not change—just as before—and the final situation, after the gas settled down, would be just what it was at the end of the slow expansion. Continuing the comparison with the slow expansion, we can say that we would start with the same S_{initial} and end up with the same S_{final}. Hence the change in S would be the same as before and would be some positive quantity. The major differences would be these: (1) the expansion would be rapid, not slow, and (2) no energy would be transferred by heating. Succinctly, the sudden expansion would yield

$$S_{\text{final}} - S_{\text{initial}} = \Delta S \neq 0, \qquad \text{indeed, } \Delta S > 0,$$

but no energy transfer by heating: $q = 0$ here.

We set out to find a connection between change in multiplicity and energy input by heating. So far we have two specific instances, both for a classical ideal gas:

$$\Delta S = \frac{q}{T}$$

when the expansion is slow, and

$$\Delta S > 0 \text{ but } q = 0$$

when the expansion is fast (and the gas departs—temporarily—from equilibrium). The two instances can be combined into a single mathematical form:

$$\Delta S \geqslant \frac{\left(\begin{array}{c}\text{energy input} \\ \text{by heating}\end{array}\right)}{T} \tag{2.11}$$

with equality if the change occurs slowly (and hence the gas remains close to equilibrium). Could this relationship be valid in general? It is neat enough and simple enough for that to be plausible. Indeed, the generality of this relationship can be established by reasoning from our verbal statement of the Second Law of Thermodynamics; we will do that in section 2.6. Right now, however, it is better to interrupt the sequence of derivations and to study some examples. So, *provisionally*, we take equation (2.11) to be valid for any physical system—gas, liquid, solid, or mixture of these—which at least starts from a state of thermal equilibrium. Moreover, the temperature (of both system and environment) may change during the system's evolution. Such a broad domain, we will find, is the equation's scope of applicability.

2.5 Some examples

Working out some examples will help you to grasp the ideas, and so this section is devoted to that project.

Example 1. Melting ice

The multiplicity of liquid water, in comparison with that of ice, was crucial in our analysis of the astronomer's proposal. Suppose we slowly melt an ice cube at 0 °C. By what factor does the multiplicity change?

Recall that "entropy" is just the logarithm of a multiplicity. If we can calculate the change in entropy, then we can readily determine the change in multiplicity. So, to start with, let us look at the entropy change,

$$S_{\text{liquid}} - S_{\text{ice}},$$

that is, look at ΔS. Working from the definition of entropy, we rewrite the difference as follows:

$$\Delta S = k \ln(\text{multiplicity}_{\text{liquid}}) - k \ln(\text{multiplicity}_{\text{ice}})$$

$$= k \ln \left(\frac{\text{multiplicity}_{\text{liquid}}}{\text{multiplicity}_{\text{ice}}} \right). \tag{2.12}$$

Thus, once we have calculated ΔS, we can determine the ratio of multiplicities.

Moreover, we can relate ΔS to the energy input by heating that occurs during the melting; our basic equation, equation (2.11), enables us to write:

$$\Delta S = \frac{Q}{T} \text{ if we melt the ice slowly at } T = 273 \text{ K.}$$

[Because the melting is specified to occur slowly, an equality sign applies in the step to Q/T, and because a finite amount of energy is transferred (all at the same temperature T), the symbol Q appears, rather than q. Recall the convention adopted in section 1.3: the lower case letter q denotes a small (or infinitesimal) amount of energy transferred by heating; the capital letter Q denotes a large (or finite) amount of energy so transferred.] The value of Q depends on the size of the ice cube. A cube from a typical refrigerator has a volume of approximately 18 cm³. At a density of about 1 gram/cm³, that means a mass of 18 grams or 18×10^{-3} kg. So we can go on:

$$\Delta S = \frac{Q}{T} = \frac{(3.34 \times 10^5 \text{ J/kg to melt ice})(18 \times 10^{-3} \text{ kg})}{273 \text{ K}}$$

$$= 22 \text{ J/K.}$$

Combine this result with equation (2.12) and divide by k:

$$\ln \left(\frac{\text{multiplicity}_{\text{liquid}}}{\text{multiplicity}_{\text{ice}}} \right) = \frac{1}{k} \frac{Q}{T}$$

$$= \frac{22 \text{ J/K}}{1.4 \times 10^{-23} \text{ J/K}} = 1.6 \times 10^{24}. \tag{2.13}$$

Next, to extract the ratio of multiplicities, recall the identity $x = e^{\ln x}$. If we let x stand for the ratio of multiplicities, then the identity implies

$$(\text{ratio of multiplicities}) = e^{\ln(\text{ratio of multiplicities})}.$$

Equation (2.13) gives us a numerical value for the natural logarithm of the ratio, and so the identity implies

$$\frac{\text{multiplicity}_{\text{liquid}}}{\text{multiplicity}_{\text{ice}}} = e^{1.6 \times 10^{24}}.$$

You may find the number on the right-hand side easier to appreciate if we express it as 10 raised to some power. We start with

$$e = 10^{\log_{10} e} = 10^{0.434}$$

and go on to

$$e^{1.6 \times 10^{24}} = (10^{0.434})^{1.6 \times 10^{24}} = 10^{0.434 \times 1.6 \times 10^{24}} = 10^{6.9 \times 10^{23}}.$$

Thus

$$\frac{\text{multiplicity}_{\text{liquid}}}{\text{multiplicity}_{\text{ice}}} = 10^{6.9 \times 10^{23}} = 10^{690,000,000,000,000,000,000,000}.$$

This is a staggering increase in multiplicity. There are vastly many more ways in which one can arrange the water molecules to look like a liquid than to look like an ice crystal. (You may find it worthwhile to glance back at figures 2.4 and 2.5 for visual confirmation.)

Before we leave this subsection, let us glance back over the logic of the calculation. Given that we study a slow process, the essential relationships are these:

$$k \ln \left(\frac{\text{multiplicity}_{\text{final}}}{\text{multiplicity}_{\text{initial}}} \right) = S_{\text{final}} - S_{\text{initial}} = \Delta S = \frac{Q}{T}. \tag{2.14}$$

If we read this set of equations from right to left, we commence with the relationship between a change in entropy and energy transfer by heating. The next equality just spells out what ΔS symbolizes. The left-most equality comes from the definition of entropy as $S = k \ln(\text{multiplicity})$ and from a property of logarithms: the difference of two logarithms is equal to the logarithm of the ratio of the arguments.

Example 2. Slow adiabatic expansion

Let us return to the classical ideal gas of section 2.4. Again we allow the gas to expand slowly, but now we specify *no heating* by an external source of energy; rather, the gas is thermally isolated. The expansion is both slow and adiabatic.

As the gas does work against the piston, the energy of the gas decreases and the temperature of the gas drops. Does the entropy change? Section 2.6 will prove that equation (2.11) is generally valid, even for processes in which the temperature changes; so we can employ that equation to calculate the change in entropy, using the equal sign because the process is slow and using $q = 0$ because there is no heating. Thus, for every small increment in volume, the entropy change is zero. [That the temperature decreases throughout the expansion is numerically irrelevant because the numerator in equation (2.11) is always zero.] We deduce that the total entropy change (for a finite change in volume) is zero.

Yet we know that an increase in volume implies an increase in the number of possible spatial locations for molecules and hence an increase in the spatial part of the multiplicity. How, then, can the entropy *not* change?

We must remember the momentum part of the multiplicity, namely, the idea that the molecules may have different directions and magnitudes for their momenta. During the adiabatic expansion, the total kinetic energy of the molecules decreases. The less energy, the fewer the ways to share it (as kinetic energy of individual molecules). Therefore the momentum part of the multiplicity decreases. Figure 2.7 illustrates this.

The preceding two paragraphs provide qualitative reasons for why—in a slow, adiabatic expansion—the spatial part of the multiplicity increases and the momentum part decreases. The effects operate in opposing directions. Only our use of equation (2.11), however, can tell us that the changes cancel each other exactly, leaving the over-all multiplicity (and hence the entropy) at its original value.

Figure 2.8 summarizes the three different expansions that we have studied.

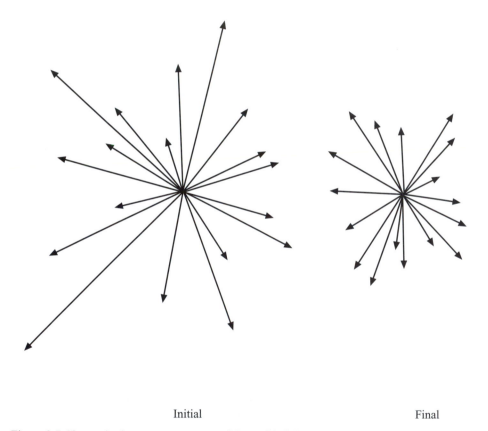

Initial   Final

Figure 2.7 Change in the momentum part of the multiplicity. In the ususal graphics, one draws the momentum vector of each molecule as emanating from the molecule's current spatial location. For the present purposes, move the tail of each momentum vector to a single, fixed location. Then, collectively, the momenta form a bristly object, like a sea urchin or a porcupine. The slow, adiabatic expansion cools the gas and shrinks the bristly object, as illustrated here. In a significant sense, the momentum vectors occupy a smaller spherical volume in a "momentum space," and so the multiplicity associated with different arrangements of the momentum arrows decreases.

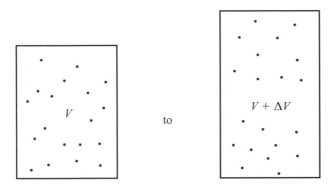

Figure 2.8 The three expansions.

Sequence 1. Slow isothermal expansion. Heating: $q > 0$. $T = $ constant. $S_{final} > S_{initial}$ because the spatial part of the multiplicity increases. Detailed derivation produced the equation

$$\Delta S = \frac{q}{T}.$$

Sequence 2. Extremely fast adiabatic expansion (into a vacuum). No heating: $q = 0$. $T_{final} = T_{initial}$. $S_{final} > S_{initial}$ because the initial and final macrostates are the same as those in Sequence 1, and that sequence shows an increase in entropy.

$$\Delta S > \frac{q}{T}$$

because $\Delta S > 0$ but $q = 0$. Sequences 1 and 2 suggest the generalization

$$\Delta S \geqslant \frac{\left(\begin{array}{c} \text{energy input} \\ \text{by heating} \end{array}\right)}{T}$$

with equality when the process occurs slowly.

Sequence 3. Slow adiabatic expansion. No heating: $q = 0$. Temperature drops: $T_{final} < T_{initial}$. Our generalization implies

$$\Delta S = \frac{q}{T} = \frac{\text{zero}}{T} = 0,$$

and so $S_{final} = S_{initial}$. Can this be consistent with what we know? Yes. Lower energy and temperature imply smaller average molecular speed and hence a smaller value for the momentum part of the multiplicity. That *decrease* compensates exactly for the *increase* in the spatial part of the multiplicity.

2.6 Generalization

Our route to full generalization proceeds in two stages. First we learn how the entropy of a classical ideal gas changes when its temperature changes. Then we go on to entropy changes in much more general physical systems, not merely an ideal gas.

Variation with temperature

In equilibrium, the macrostate of a monatomic classical ideal gas is determined by the volume V and the temperature T, provided we know independently the number N of atoms. After all, we would then have the information necessary to compute the pressure P (via the ideal gas law) and to evaluate the energy E (via $E = \frac{3}{2}NkT$). From our analysis in section 2.4, we know that the multiplicity is proportional to V^N. How does the multiplicity depend on the temperature? We can reason as follows.

The average kinetic energy of a gas atom is

$$\langle \tfrac{1}{2}mv^2 \rangle = \tfrac{3}{2}kT,$$

and so a specific component of linear momentum, such as p_x, satisfies the equation

$$\left\langle \frac{p_x^2}{2m} \right\rangle = \tfrac{1}{3} \times \tfrac{3}{2}kT$$

because each of the three Cartesian components of momentum must make the same *average* contribution to the kinetic energy. Thus

$$\langle p_x^2 \rangle = mkT.$$

The typical size of the momentum component p_x is thus $(mkT)^{1/2}$ and hence is proportional to $T^{1/2}$.

In a loose but correct fashion, we may say that a momentum component p_x is pretty sure to fall in the range

$$-\text{few} \times (mkT)^{1/2} \leqslant p_x \leqslant \text{few} \times (mkT)^{1/2}, \tag{2.15}$$

where "few" denotes a number like 3 or so. The important point is that the size of the range grows with temperature as $T^{1/2}$.

Now we reason by analogy with the spatial part of the multiplicity. For a cubical box of edge length L, the volume V is L^3, and the spatial part of the multiplicity is proportional to $(L^3)^N$. Analogously, if a momentum component is pretty sure to fall in a momentum range proportional to $(mkT)^{1/2}$, then—taking that range to be analogous to L—we reason that the momentum part of the multiplicity is proportional to $\{[(mkT)^{1/2}]^3\}^N$, that is, proportional to $T^{3N/2}$. (The sea urchin or porcupine diagrams with momentum arrows in figure 2.7 may help you to visualize both the "volume" in *momentum space* and also the volume's dependence on temperature.)

Each arrangement of momenta in the momentum space may be paired with each spatial arrangement of the atoms in the literal volume V. Thus the spatial and momentum parts of the multiplicity combine as a product of two factors to give the full multiplicity. Using what we know about the spatial and momentum parts of the multiplicity, we find that the entropy of a monatomic classical ideal gas takes the form

$$S = k \ln(\text{constant} \times T^{3N/2} \times V^N). \tag{2.16}$$

The "constant" is a constant of proportionality for the actual multiplicity. Its numerical value remains unknown to us, but we shall not need the value here; later, in section 5.6, the value will emerge from a complete calculation.

Next, suppose we heat the gas slowly and only a little bit but do so at constant volume; thereby we change the temperature by an amount ΔT. How much does the entropy change? In preparation, we expand the logarithm in (2.16) as

$$S = k \ln(\text{constant} \times V^N) + k\tfrac{3}{2} N \ln T. \tag{2.17}$$

The change in $\ln T$ is

$$\Delta \ln T = \frac{d \ln T}{dT} \Delta T = \frac{1}{T} \Delta T. \tag{2.18}$$

Then equations (2.17) and (2.18) imply

$$\Delta S = k \frac{3}{2} N \frac{1}{T} \Delta T = \frac{1}{T} \times \frac{3}{2} N k \Delta T.$$

The combination $\tfrac{3}{2} N k \Delta T$ is precisely the change in the energy of the gas, and that change arises from the energy input by heating, q. Thus we find

$$\Delta S = \frac{q}{T} \tag{2.19}$$

when a monatomic classical ideal gas is slowly heated at constant volume. A glance back at equation (2.11) shows that our present result is consistent with that earlier, tentative relationship and thus supports it. But our ultimate goal is to establish firmly that equation (2.11) is valid in general. Equation (2.19) and its context, stated right after the equation, provide the crucial ingredient that we need in the next stage of generalization, to which we now turn.

Further generalization

Our statement of the Second Law was restricted by the stipulation, "allow the system to evolve in isolation." We added the comment that "the stipulation includes the injunction, do not transfer energy to or from the system." Equation (2.11), however, explicitly describes a situation where the system may receive energy by heating or lose it by cooling. To derive from our statement of the Second Law a consequence like that in equation (2.11), we need to enlarge our view to include a source (or receiver) of energy transfer by heating, for example, in the way shown in figure 2.9. Imagine removing some barrier that initially prevents change—such as permitting a small amount of the gases in the chemically reactive system to mix and to react. Simultaneously, permit energy exchange (by heating) between the chemical system and the helium, a monatomic gas. (Note that there is no chemical reaction *with* the helium.) Then wait until things settle down.

If there was real change in that finite time interval, our verbal form of the Second Law implies that the entropy of the combined system—the chemical system plus the helium—increases because the combined system evolves to a macrostate of larger multiplicity. Symbolically,

$$\Delta S_{\text{chem system plus helium}} > 0, \tag{2.20}$$

and we may split this relationship as

$$\Delta S_{\text{chem system}} + \Delta S_{\text{helium}} > 0. \tag{2.21}$$

Our strategy now is to evaluate ΔS_{helium} and thus learn about $\Delta S_{\text{chem system}}$.

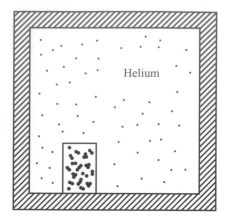

Figure 2.9 The context for generalization. The chemically reactive system is in the small container (which holds molecules of odd shapes), and it is the system of interest. It is surrounded by a great deal of gaseous helium at, say, $T = 293$ K. The dashed outer wall (filled with thermal insulation) prevents any transfer of energy to the external world or from it.

[Note. You may wonder about the step from equation (2.20) to (2.21). The multiplicity of the composite system is the product of the individual multiplicities,

$$\text{multiplicity}_{\text{chem system plus helium}} = \text{multiplicity}_{\text{chem system}} \times \text{multiplicity}_{\text{helium}},$$

because every microstate of the chemical system can be paired with every microstate of the helium. The logarithm of a product equals the sum of the logarithms of the factors. So, when one takes a logarithm to form the entropy, one finds

$$S_{\text{chem system plus helium}} = S_{\text{chem system}} + S_{\text{helium}}.$$

Thus equation (2.21) follows directly from equation (2.20).

The property that the entropy of a composite system is the sum of the entropies of macroscopic component subsystems is called the *additivity of entropy*.]

The large amount of helium can exchange some energy by heating without appreciable change in its temperature T. Moreover, the helium remains close to thermal equilibrium. Thus we may say

$$\Delta S_{\text{helium}} = \frac{(\text{energy into helium by heating})}{T}, \tag{2.22}$$

based on equation (2.19) and the "slow change" context.

How is that energy related to "energy *into* the chemical system by heating," which we will denote by q? Energy conservation implies

$$\left(\begin{matrix} \text{change in energy} \\ \text{of the helium} \end{matrix} \right) + \left(\begin{matrix} \text{change in energy} \\ \text{of chemical system} \end{matrix} \right) = 0,$$

that is, one change is the negative of the other. Thus

$$(\text{energy into helium}) = -q. \tag{2.23}$$

Because I chose to describe each change in energy with the words "energy into ...", one of the "energy into ..." expressions is negative, meaning that energy actually leaves. The minus sign is a nuisance to deal with, but thermodynamics is ultimately easier to comprehend if one sticks with "energy into ..." expressions, and so I have chosen that *modus operandi* here.

We use equation (2.23) in (2.22) and then the latter in equation (2.21), finding

$$\Delta S_{\text{chem system}} - \frac{q}{T} > 0,$$

which implies

$$\Delta S_{\text{chem system}} > \frac{q}{T}. \tag{2.24}$$

Whenever there is a real spontaneous change, we can count on a finite increase in the total entropy, $\Delta S_{\text{total}} > 0$, and that increase leads to the strict inequality in equation (2.24). Slow change, however, when taken to the limit of a vanishing *rate* of change, can yield $\Delta S_{\text{total}} = 0$. The line of reasoning for this conclusion is the following.

If change occurs at an infinitesimal rate, allowing the total system always to be no more than infinitesimally away from equilibrium, then the total system is not changing to a macrostate of finitely larger multiplicity and so, in the limit, $\Delta S_{\text{total}} = 0$. Here is the evidence: under the prescribed conditions, one can—by an infinitesimal change in the circumstances—reverse the changes in the system, but that would not be possible if the system had gone to a macrostate of finitely larger multiplicity. (Note that finite changes in a volume or in the number of molecules that have reacted chemically are permitted even in the limit of vanishing *rate* of change. One just has to wait a long time—in principle, infinitely long in the limit.)

The limit of slow change is, of course, an idealization, but it is extremely fruitful. We incorporate the limit by generalizing equation (2.24) to

$$\Delta S_{\text{chem system}} \geq \frac{q}{T}, \tag{2.25}$$

where the equality sign holds when the change occurs slowly and the inequality holds otherwise.

Note that we used no special properties of the "chemical system;" it could be a dilute gaseous system or a dense one or a liquid or have some solids present. The change need not be a chemical reaction, but could be melting, say. Hence one may generalize yet again, turning equation (2.25) into

$$\Delta S_{\text{any system}} \geq \frac{\left(\begin{array}{c} \text{energy into the} \\ \text{system by heating} \end{array} \right)}{T}, \tag{2.26}$$

with equality if the process is slow and inequality otherwise. (Section 3.6 will elaborate on the conditions required for the equality sign to hold.)

A few words should be said about the temperature T that appears in the denominator on the right-hand side of equation (2.26). When the change is slow, the system of interest and the source (or receiver) of energy transfer by heating have the same temperature (or virtually so), and hence T refers to their common temperature. When the process is fast, however, the system of interest may be so far from equilibrium that we cannot ascribe a temperature to it. Nonetheless, the heating (or cooling) source has a well-defined temperature (a supposition implicit in the derivation), and the symbol T refers to the source's temperature. Indeed, in the derivation in this section, the symbol T appears first in equation (2.22), where it refers explicitly to the source's temperature.

2.7   Entropy and disorder

Our analysis led from the notion of multiplicity to a special role for the logarithm of a multiplicity (more specifically, the logarithm times Boltzmann's constant). That logarithmic quantity we called "the entropy," and it has a perfectly clear meaning in terms of the fundamental notion of "the multiplicity of a macrostate." You may, however, sometimes hear entropy characterized as "a measure of disorder." Let us see what the connection might be.

The words "order" and "disorder" are colloquial and qualitative; nonetheless, they describe a distinction that we are likely to recognize in concrete situations, such as the state of someone's room. The connection with multiplicity becomes clear if we use the notion of "correlation" as a conceptual intermediary, as indicated in figure 2.10.

Imagine a bedroom with the usual complement of shoes, socks, and T-shirts. Suppose, further, that the room is one that we intuitively characterize as "orderly." Then, if we see one black dress shoe of a pair, we know—without looking—that the other shoe is right next to it. If we see one clean T-shirt, then the others are in a stack just below it. There are strong spatial correlations between the shoes in a pair or the T-shirts on the dresser. Those correlations limit severely the ways in which shoes and T-shirts can be distributed in the room, and so the objects exhibit a small multiplicity and a low entropy.

Now take the other extreme, a bedroom that we immediately recognize as "disorderly." If we see one jogger, we have no idea where the other jogger is. Under the

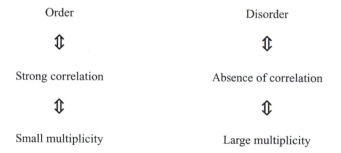

Figure 2.10 "Correlations" enable us to establish a correspondence between the notions of entropy and "disorder". The double-headed arrow signifies that each of the notions implies the other.

dresser? Behind the bed? Lost in the pile of dirty T-shirts? And, for that matter, what a about the T-shirts? If we see a clean one on the dresser, the next clean one may be on the desk or in the easy chair. Correlations are absent, and the objects enjoy a large multiplicity of ways in which they may find themselves distributed around the room. It is indeed a situation of high entropy.

There is usually nothing wrong with referring to entropy as "a measure of disorder." The phrase, however, doesn't take one very far. To gain precision and something quantitative, one needs to connect "disorder" with "absence of correlations" and then with multiplicity. It is multiplicity that has sufficient precision to be calculated and to serve as the basis for a physical theory.

[To be sure, P. G. Wright issues strictures on the use of "disorder" to characterize entropy in his paper, "Entropy and disorder," *Contemporary Physics*, **11**, 581–8 (1970). Dr. Wright provides examples where an interpretation of entropy as disorder is difficult at best; most notable among the examples is crystallization from a thermally isolated solution when the crystallization is accompanied by a *decrease* in temperature. In a private communication, Dr. Wright cites a supersaturated solution of calcium butanoate in water and also sodium sulfate (provided that the solid crystallizing out is anhydrous sodium sulfate).]

But enough of order and disorder. When a process occurs rapidly, how does one calculate a definite numerical value for the entropy change? Our central equation, equation (2.26), would appear to provide only an inequality for such a process. Section 3.4 addresses the issue of rapid change. Indeed, chapter 3 fills in some gaps, provides practice with entropy, and introduces several vital new ideas. We turn to those items now.

2.8   Essentials

1. The primary concept is the *multiplicity* of a macrostate:

$$\left(\begin{array}{c} \text{the } \textit{multiplicity} \\ \text{of a macrostate} \end{array} \right) \equiv \left(\begin{array}{c} \text{the number of microstates} \\ \text{that correspond} \\ \text{to the macrostate} \end{array} \right).$$

2. *Entropy* is basically the logarithm of the multiplicity:

$S \equiv k \ln(\text{multiplicity}).$

3. The major dynamical statement is the *Second Law of Thermodynamics*. It can be formulated in terms of multiplicity or, equivalently, in terms of entropy. The latter formulation is the following:

> **The Second Law of Thermodynamics**. If a system with many molecules is permitted to change, then—with overwhelming probability—the system will evolve to the macrostate of largest entropy and will subsequently remain in that

macrostate. Stipulation: allow the system to evolve in isolation. (The stipulation includes the injunction, do not transfer energy to or from the system.)

For all practical purposes, the one-line version of the Second Law is this: An isolated macroscopic system will evolve to the macrostate of largest entropy and will then remain there.

4. The chapter's major equation connects a change in entropy with the energy input by heating and with the temperature:

$$\Delta S_{\text{any system}} \geq \frac{\left(\begin{array}{c}\text{energy into the}\\ \text{system by heating}\end{array}\right)}{T},$$

with equality if the process is slow and inequality otherwise.

5. The temperature T on the right-hand side in item 4 refers to the temperature of the source of energy input by heating. When the equality sign holds, the system and the source have the same temperature (or virtually so), and hence one need not draw a distinction about "whose temperature?".

6. Section 2.6 established the *additivity of entropy*: if two macroscopic systems are in thermal equilibrium and in thermal contact, then the entropy of the composite system equals the sum of the two individual entropies.

Further reading

The Second Law of Thermodynamics has prompted several equivalent formulations and a vast amount of controversy. A comprehensive discussion of the formulations can be found in Martin Bailyn, *A Survey of Thermodynamics* (AIP Press, Woodbury, New York, 1994). A fine introduction to the debates in the late nineteenth century is provided by Stephen G. Brush, *Kinetic Theory, Vol. 2: Irreversible Processes* (Pergamon Press, New York, 1966). There one will find papers by Boltzmann, Poincaré, and Zermelo, ably introduced and set into context by Brush.

Rudolf Clausius followed a route to entropy quite different from our "atomistic" route based on the idea of multiplicity. That was, of course, almost an historical necessity. In a clear fashion, William H. Cropper describes Clausius's analysis in "Rudolf Clausius and the road to entropy," *Am. J. Phys.* **54**, 1068–74 (1986). More about the etymology of the word "entropy" can be found in the note, "How entropy got its name," Ralph Baierlein, *Am. J. Phys.* **60**, 1151 (1992).

Problems

Note. Equation (2.7) was essential to the logical development in this chapter, but it is rarely a good way to approach a homework problem. The reason is that the equation is accurate only when q/NkT is extremely small (which that ratio was when we used the equation). Just as the scaffolding around a marble sculpture is removed when its function has been served, so it is best to put equation (2.7) aside and to use other equations from the chapter, especially equation (2.26).

1. *Computer simulation of macroscopic regularity.* Imagine a volume V partitioned into 10 bins of equal volume. Use a computer's random number generator to toss molecules into the bins randomly. Plot a bar chart of the number of molecules in each bin after the computer has tossed in $N = 100$ molecules. Next, increase to $N = 1,000$ and then to $N = 10^4$. Try $N = 10^4$ several times. Also, compare results with those gotten by your classmates. Do you find macroscopic regularity emerging from these simulations of an ideal gas?

2. Work out the analog of table 2.2 but with $N = 6$ labeled balls. Draw a bar graph of multiplicity versus macrostate, the latter being specified by the number of balls in the right-hand bowl (together with the total number of balls, N). Using symmetry will expedite your work. Using the known total number of microstates (which you can reason to be 2^6) provides either a check on your arithmetic or a way to skip one multiplicity computation. Do you see the more-or-less even distribution growing in numerical significance?

3. Figure 2.11 shows a long, hollow cylinder mentally divided into three regions of equal length. Let each region correspond to "a spatial location," in the sense used in section 2.4. (A region may contain more than one molecule.)

(a) If you put only one molecule (labeled A) into the cylinder, how many different "spatial arrangements" are there and, consequently, how many different microstates are there? Show the different spatial arrangements with sketches.
(b) Repeat the analysis with two molecules (labeled A and B and hence regarded as distinguishable).
(c) Repeat with three molecules (labeled A, B, and C). Now you may skip the sketches—except as an aid to your thinking.

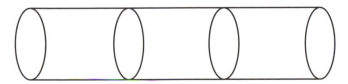

Figure 2.11 The volume, split up into three "spatial locations."

(d) Generalize to N labeled molecules.
(e) If the cylinder were expanded lengthwise, so that there were now four regions (each of the original size), how would your answers to parts (a) through (d) change?
(f) We can reason that the number of regions is proportional to the cylinder's volume. How, then, is the number of microstates for N molecules related to the volume—on the basis of your analysis here?

4. A penny, made of crystalline copper, is heated at 1,083 °C and melts. The density of copper is 8.5 grams per cubic centimeter. The energy needed to melt one gram of copper is 200 joules. You will need to estimate the volume of a penny.
  By what factor does the multiplicity of the copper change when the penny is melted? Express your answer as $10^{(\text{some power})}$.

5. In a volume $V = 0.3$ cubic meters is a number $N = 10^{24}$ of helium atoms initially at $T = 400$ K. While the volume is kept fixed, the gas is gently heated to 403 K.

(a) By what numerical factor does the multiplicity change?
(b) Starting from the initial situation, we could produce the same temperature increase *without* transferring any energy by heating: just gently compress the gas adiabatically. In this process, the entropy of the gas would not change. How can we understand this—that is, the "no change in entropy"—in terms of the various contributions to the total multiplicity?

6. On a cold winter day, a snowflake is placed in a large sealed jar in the sunshine (the jar otherwise being full of dry air). The sunlight turns the snow directly to water vapor, all this occurring at $T = 260$ K, well below freezing, so that no liquid water is ever formed. (There is no need—here—to concern yourself with the temperature of the sun's surface.)
  It takes 3,000 joules per gram to vaporize snow in this fashion. The snowflake has a mass of 10^{-3} grams.

(a) By what numerical factor does the multiplicity of the water change? (The physical system here is the water, existing first as ice and then as vapor.) Express your answer as $10^{(\text{some power})}$.
(b) Why, in simple microscopic terms, does the multiplicity change as you found it to in part (a)? That is, give one or more *qualitative* reasons why a change in multiplicity should occur.

7. Gaseous helium at $T = 300$ K and atmospheric pressure fills a volume V, as shown in figure 2.12. Separated from the gas by a wall with a hole—initially closed—is a region of total vacuum. Here are some additional facts:

$V = 1.4$ liters $= 1.4 \times 10^{-3}$ m^3,

$V_{\text{vacuum}} = $ volume of vacuum region $= 0.2\ V$.

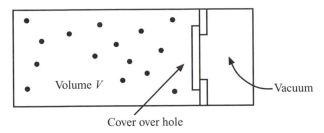

Figure 2.12 The context.

(a) Compute the total number of helium atoms.
(b) The cover is now slid off the hole (which is about the size of a dime). When everything has settled down again, by what factor has the multiplicity changed? Briefly, why? Also, what is the change in entropy? (Take the walls to be thermally insulating.)
(c) If we did *not* uncover the hole (and hence kept the gas volume at its initial value) but wanted to produce an entropy change of the same numerical amount, what should we do? Describe the process verbally and be quantitatively precise wherever you can be.

8. A cylindrical container of initial volume V_0 contains N atoms of a classical ideal gas at room temperature: $T = 300$ kelvin, to use a round figure. One end of the container is movable, and so we can compress the gas slowly, reducing the volume by 2 percent while keeping the temperature the same (because the container's walls are in contact with the air in the room). Specifically, $V_0 = 10^{-3}$ m^3 and $N = 3 \times 10^{22}$ atoms.

(a) What is the change in entropy of the confined gas? (Provide first an algebraic expression and then a complete numerical evaluation.)
(b) How much work do we do in compressing the gas?
(c) How much energy was absorbed by the environment (through heating)?
(d) What was the entropy change of the environment?

[Note. You do *not* have to do the four parts in the sequence in which the questions are posed, but the sequence (a) through (d) is a convenient route. For each part, be sure to provide a numerical answer.]

9. Section 2.4 described both a slow expansion and a rapid expansion of a classical ideal gas. What happens for an intermediate process, specifically, for the following process?

(a) Expand from the initial volume V to volume $V + \frac{1}{2}\Delta V$, slowly and at constant temperature T.
(b) Then expand from $V + \frac{1}{2}\Delta V$ to volume $V + \Delta V$, rapidly, indeed, by letting the gas expand into an evacuated region (of additional volume $\frac{1}{2}\Delta V$) and while preventing any energy transfer by heating.

After the gaseous system has settled down, what is the size of ΔS relative to the entropy change in the entirely slow expansion discussed in section 2.4? What is the energy input by heating relative to that for the slow expansion? Do you find ΔS equal to "the energy input by heating divided by T"? Should you expect to? Or what is the relationship?

10. In section 2.5, we calculated the change in multiplicity when an ice cube melts. (That was for 18 grams of ice.)

(a) Now calculate the subsequent change in multiplicity when the water at 0 °C is heated to 1 °C. It takes 4.2 joules of energy to raise the temperature of one gram of water by one degree Celsius.

(b) Which change in multiplicity is larger, the change associated with melting or the change associated with a temperature increase of one degree Celsius? Can you think of a reason why?

(c) Calculate the entropy change when water is heated from 0 °C to 70 °C. How does this change compare numerically with the entropy change in part (a)?

11. If 20 grams of water are heated slowly from 10 °C to 40 °C, what is the numerical change in the water's entropy? Take the specific heat of water to have the constant value 4.2 J/gram·K. Provide a qualitative explanation for the change that you calculate.

3 Entropy and Efficiency

3.1 The most important thermodynamic cycle: the Carnot cycle

3.2 Maximum efficiency

3.3 A practical consequence

3.4 Rapid change

3.5 The simplified Otto cycle

3.6 More about reversibility

3.7 Essentials

The chapter begins with a classic topic: the efficiency of heat engines. The topic remains relevant today—for environmental reasons, among others—and it also provides the foundation for William Thomson's definition of absolute temperature, an item discussed later (in chapter 4). Next, the chapter develops a method for computing the entropy change when a process occurs rapidly. An extended example—the Otto cycle—and a discussion of "reversibility" conclude the chapter.

3.1 The most important thermodynamic cycle: the Carnot cycle

Engineers and environmentalists are interested in cyclic processes, because one can do them again and again. Figure 3.1 shows the essence of the cycle for water and steam in a typical power plant that generates electricity. A simpler cycle, however, is more instructive theoretically, and so let us consider the cycle shown in figure 3.2, performed *slowly throughout*. We may take the substance to be some real gas, such as nitrogen (as distinguished from an ideal gas). There are four stages, characterized as follows.

Stage 1 to 2: isothermal expansion. From state 1 to state 2, the gas is allowed to expand while being maintained at constant, high temperature T_{hot}. The pressure drops because the volume increases. [The ideal gas law, $P = (N/V)kT$, valid for a dilute gas, suggests such a pressure drop—qualitatively—for a real, not-necessarily-dilute gas, too.] The gas does work as it expands and would drop in temperature if energy were not supplied by heating. We denote by Q_{hot} the total amount of energy that is supplied by heating at temperature T_{hot}.

Stage 2 to 3: adiabatic expansion. During the expansion from state 2 to state 3, there is no transfer of energy by heating; thus $q = 0$ for each small step in volume and pressure. The pressure drops faster now as the volume increases; this is so because the temperature drops as the gas loses energy by doing work. (The gas is no longer heated

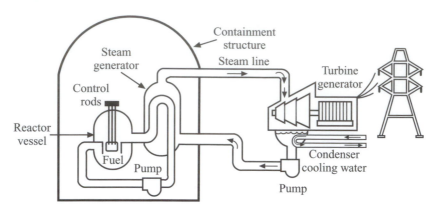

Figure 3.1 The cycle for water and steam in a typical nuclear power plant. Nuclear fission heats water in the reactor vessel; that hot water, at temperature T_{hot}, transfers energy to the liquid water in the central loop and produces steam. The steam, at high pressure, turns the blades of the turbine, doing work; that work is used to run the electrical generator and give electrical potential energy to electrons. To ensure a difference of pressure across the turbine, the steam—after passing through the turbine—must be cooled and condensed back into liquid water. The partial loop labeled "Condenser Cooling Water," at the cold temperature T_{cold}, cools the steam, and the ensuing liquid water is pumped back to the steam generator, ready to start the cycle again. From a theoretical point of view, it is the water (in both liquid and vapor form) in the central loop that goes through the essential cyclic process. (The condenser cooling water often comes from a river or the ocean and then returns to that place of origin; sometimes the condenser cooling water is run through separate cooling towers and dumps some or all of the energy it acquired from the steam into the atmosphere by evaporation.)

to compensate for the work done.) Adiabatic expansion is carried out until the gas temperature drops enough to equal the temperature of the cold reservoir (at state 3).

Stage 3 to 4: isothermal compression. We now compress the gas at constant temperature T_{cold}. Energy now *leaves* the gas and goes into the cold reservoir; the symbol Q_{cold} denotes the total amount of energy that enters the cold reservoir.

Stage 4 to 1: adiabatic compression. The final, adiabatic compression takes the gas back to its original values of pressure, volume, and temperature.

The cycle is called a *Carnot cycle*, after the nineteenth-century French engineer Sadi Carnot, who made it the central piece in his study of heat engines, published in 1824. (The pronunciation is "Car-no," for the "t" is silent.) By the way, the phrases, "a reservoir at temperature T" or "a heat reservoir," mean any object or system that can transfer substantial amounts of energy by heating (or by cooling) and can do so without significant change in its own temperature. A large lake provides a heat reservoir for a nuclear power plant; a one-liter can of cold ethylene glycol (which is an anti-freeze) provides a heat reservoir for a picnic lunch of tuna fish sandwiches and soda pop in mid-summer. To be sure, the phrase, "a heat reservoir," is another confusing vestige of an earlier era. The reservoir possesses internal energy, not "heat," but it is capable of transferring some of that internal energy by conduction or radiation, and so the reservoir is capable of heating another object.

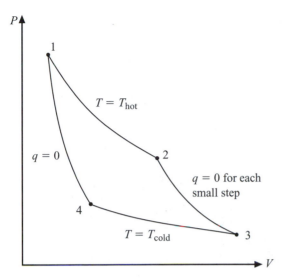

Figure 3.2 The Carnot cycle. During the expansion from state 1 to state 2, the gas is in thermal contact with a large reservoir of hot water at temperature T_{hot}; the gas absorbs energy Q_{hot}. During the compression from state 3 to state 4, the gas is in contact with a large reservoir of cold water at temperature T_{cold} and loses energy Q_{cold}. The expansion from state 2 to state 3 occurs adiabatically, that is, under conditions of no energy transfer by heating. Similarly, the compression from state 4 to state 1 occurs adiabatically. All processes are specified to occur slowly. (For a real gas like diatomic nitrogen, the diagram is qualitatively faithful, but the steepness of the adiabatic portions has been exaggerated for the sake of clarity.)

But back to the Carnot cycle itself. The gas does work on the external world when it expands. To compress the gas, the external world must do work on the gas. What is the net effect? The expansion is at higher pressure (for each value of the volume) than is the compression, and so the net work done by the gas is a positive quantity. We denote the net work by W and write our statement symbolically as

net work done by gas $\equiv W > 0$.

What about the net change in entropy of the gas? The gas returns to its initial macrostate (state 1) and hence returns to whatever multiplicity it had originally. Because entropy is $k \ln(\text{multiplicity})$, the entropy of the gas returns to its original value.

What about the over-all entropy change, gas plus hot reservoir and cold reservoir? Table 3.1 helps us to reason out the answer. During stage 1 to 2, which is the isothermal expansion, the gas acquires energy Q_{hot} by heating at temperature T_{hot}; so, by equation (2.26), the entropy of the gas increases by Q_{hot}/T_{hot}. Simultaneously, the hot reservoir loses an equal amount of energy (by cooling), and so its entropy drops by $-Q_{hot}/T_{hot}$. During stage 2 to 3, the gas is thermally isolated and expands slowly (per specification); hence, by equation (2.26), its entropy does not change. During stage 3 to 4, energy Q_{cold} enters the cold reservoir, and so the reservoir's entropy increases by

Table 3.1 *A tabulation of entropy changes. Because the gas returns to its initial macrostate and the corresponding multiplicity, the sum of the entropy changes for the gas must equal zero. The two reservoir columns,* when combined, *describe an entropy change that is precisely the negative of the sum for the gas column; because the gas column yields zero, so must the combined reservoir columns.*

Stage	ΔS_{gas}	$\Delta S_{\text{hot reservoir}}$	$\Delta S_{\text{cold reservoir}}$
1 to 2	$\dfrac{Q_{\text{hot}}}{T_{\text{hot}}}$	$-\dfrac{Q_{\text{hot}}}{T_{\text{hot}}}$	
2 to 3	zero		
3 to 4	$-\dfrac{Q_{\text{cold}}}{T_{\text{cold}}}$		$\dfrac{Q_{\text{cold}}}{T_{\text{cold}}}$
4 to 1	zero		
Net value	zero	Reservoir columns *together* sum to zero	

$Q_{\text{cold}}/T_{\text{cold}}$. Simultaneously, the gas loses an equal amount of energy (by cooling), and so its entropy drops by $-Q_{\text{cold}}/T_{\text{cold}}$. Stage 4 to 1 is a slow process with no energy transfer by heating, and so no entropy changes occur (just as was true for stage 2 to 3).

To tot up the column of ΔS_{gas} entries is easy. Two paragraphs back we reasoned that the net entropy change of the gas is zero because the gas returns to its initial macrostate and the corresponding multiplicity. Each reservoir experiences a lasting change in entropy. The sum of those changes, however, is zero. Why? Because the sum for the two reservoirs (taken together) is equal in magnitude to the sum of the ΔS_{gas} entries, and the latter sum we showed to be zero.

Indeed, it is worthwhile to display here what is implied by the second column of table 3.1. The net value of zero for the sum of the ΔS_{gas} entries implies

$$\frac{Q_{\text{cold}}}{T_{\text{cold}}} = \frac{Q_{\text{hot}}}{T_{\text{hot}}}. \tag{3.1}$$

We will use this relationship when we work out the efficiency of a Carnot cycle, to which we now turn.

The efficiency

By the *efficiency* of a Carnot cycle, one means the ratio of *net work output* to *energy input* at the hot reservoir by heating. Thus

$$\text{efficiency} \equiv \frac{\text{net work output}}{\text{energy in at } T_{\text{hot}}}$$

$$= \frac{W}{Q_{\text{hot}}}. \tag{3.2}$$

With a little algebraic effort, we can express the efficiency in terms of the two temperatures, T_{hot} and T_{cold}, and that is worth doing.

First, we use energy conservation to express the net work W in terms of Q_{hot} and Q_{cold}. Because the total energy output must equal the total energy input in a cyclic process, we may write

$$W + Q_{cold} = Q_{hot}, \tag{3.3}$$

whence $W = Q_{hot} - Q_{cold}$. We use this relation to substitute for W in equation (3.2) and then use equation (3.1) to get to the second line:

$$\begin{pmatrix} \text{efficiency of} \\ \text{Carnot cycle} \end{pmatrix} = \frac{Q_{hot} - Q_{cold}}{Q_{hot}} = 1 - \frac{Q_{cold}}{Q_{hot}}$$

$$= 1 - \frac{T_{cold}}{T_{hot}} = \frac{T_{hot} - T_{cold}}{T_{hot}}. \tag{3.4}$$

The efficiency of a Carnot cycle is equal to the difference in reservoir temperatures divided by the hotter temperature.

A pause to reflect is in order. The crucial elements in a Carnot cycle are these: (a) two stages at constant temperature, separated by adiabatic stages; (b) all stages are performed slowly, so that both the system of interest (for us, the gas) and the reservoirs remain close to equilibrium; (c) a return, of course, by the system of interest to its initial macrostate. The material that is taken through the cycle (for us, a real gas such as nitrogen) is called the *working substance*. The crucial elements, (a) through (c), require only the existence of the working substance (called there the "system of interest"), not any specific properties. In short, the nature of the working substance plays no role. Therefore equation (3.4) for the efficiency holds for every working substance in a Carnot cycle: pure diatomic nitrogen, air, water vapor, liquid water plus water vapor, helium, photons, fluorocarbons, etc. (It isn't even necessary that the working substance be a gas and that work be done by expansion and compression. One can design Carnot cycles in which electric and magnetic forces replace gas pressure—but that would take us too far afield.)

3.2   Maximum efficiency

The logic of the next steps is worth laying out here. The objective is to answer the question, "Given heat reservoirs at the temperatures T_{hot} and T_{cold}, what is the maximum efficiency—among all possible cycles—that thermodynamics allows?" [As before, efficiency is defined as in equation (3.2).] We make our way to the objective in three steps, described succinctly as follows.

1. Illustrate the idea of a *reversible* heat engine.
2. Show that a reversible heat engine has the maximum efficiency.
3. Gain our objective.

But first, what is a "heat engine"? A Carnot cycle provides the prime example of a heat engine, and from that cycle we can extract the basic definition. A *heat engine* is any device that operates in a cyclic fashion, that absorbs energy by heating during one portion of the cycle, that loses energy by cooling in another portion of the cycle, and that performs a positive net amount of mechanical work, for instance, lifts a weight or turns the shaft of an electrical generator.

Reversibility

What happens if we try to run the Carnot cycle of figure 3.2 in reverse? That is, the gas is to go through the states in the sequence 1, 4, 3, 2, 1.

We do not receive any useful work; indeed, we have to supply net work, precisely as much as we got when the cycle was performed in the "forward" direction. The cold reservoir gives up the energy it got, and the hot reservoir recovers the energy that it gave. We restore the world to its original configuration. We can achieve this restoration because, when the cycle is run in the forward direction, the net change in entropy of the gas is zero *and also* so is the net change in entropy of the two reservoirs, taken together. (Table 3.1 provides a reminder of these facts.) Thus the "global" change in entropy—the change for gas, hot reservoir, and cold reservoir—is zero for the forward cycle, not some positive quantity. The backward cycle can undo all the changes and leave behind no residual effect because the global change in entropy for the forward cycle is only zero. Recall here our statement of the Second Law of Thermodynamics, namely, that evolution proceeds in the direction of increasing entropy (provided "the system" is so encompassing that it is effectively isolated). The forward cycle can be undone by a backward cycle because we are never asked to orchestrate a global *decrease* in entropy, which would be a practical impossibility.

A cycle for which this "return of the world" to its original state is possible is called a *reversible cycle*.

If we were to acknowledge friction between a piston and a cylinder wall, say, the cycle would not be reversible. The work needed for the backward cycle would be more than what we got from the forward cycle. We may, however, focus on an ideal, frictionless system and see what consequences we can draw from it.

Note that the adjective "reversible," as used here, does not mean that the total system returns to its original state spontaneously. "Reversible" means merely that we can arrange such a return, at least as an idealization, and that no residual change is required in parts of the universe outside the system.

A Carnot cycle is reversible, but reversibility is *not* typical of cycles. Figure 3.3 illustrates a cycle—a simplified Otto cycle, having only two heat reservoirs and using the same gas again and again—that is not reversible. Section 3.5 analyzes the Otto cycle in detail, but here we continue the quest for maximum efficiency.

A reversible engine has maximum efficiency

Now we will show that a reversible heat engine provides the maximum possible efficiency for given values of T_{hot} and T_{cold}. As sketched in figure 3.4, we hitch up a

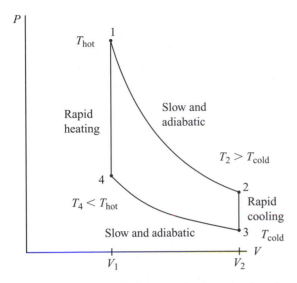

Figure 3.3 The simplified Otto cycle. A gas is taken through a cycle. The four stages are the following: from state 1 to 2, slow adiabatic expansion; 2 to 3, cooling at constant volume; 3 to 4, slow adiabatic compression; 4 to 1, heating at constant volume. To preserve as much of a parallel with the Carnot cycle as possible, specify that the heating occurs by putting the gas in thermal contact with a single large reservoir of hot water at temperature T_{hot}; the cooling, with a similar single reservoir of cold water at T_{cold}. Then the heating during stage 4 to 1 occurs over a range of gas temperatures, from well below T_{hot} up to T_{hot}, and will not be a slow process. During stage 2 to 3, the cooling will occur over a range of gas temperatures, from well above T_{cold} down to T_{cold}, and will not be a slow process. The rapid processes make this simplified Otto cycle irreversible.

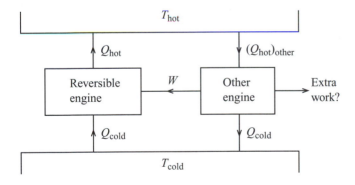

Figure 3.4 A reversible heat engine (running "backward") and some "other" heat engine (running "forward") are connected to reservoirs at temperatures T_{hot} and T_{cold}. The "extra work," if there is any, could be used to launch satellite-carrying projectiles.

reversible heat engine and any "other" heat engine; we run the other engine forward, but run the reversible engine backward. Indeed, we run the other engine so that it dumps precisely energy Q_{cold} into the cold reservoir in the time that the reversible engine extracts energy Q_{cold}. Moreover, the other engine supplies the work W needed to run the reversible engine backward; there may be some "extra work" left over, to be used for some other purpose. The reversible engine dumps energy Q_{hot} into the hot reservoir while the other engine absorbs an amount $(Q_{hot})_{other}$ from that reservoir. (In this context, the symbols W, Q_{hot}, and Q_{cold} are numerically equal to the values of work, energy input by heating, and energy given off through cooling that the reversible engine would have if running forward.) These specifications set the scene.

Just to be sure that we understand the context and the symbols, let us suppose that the other engine is precisely as efficient as the reversible engine. Then there would be no extra work, and $(Q_{hot})_{other}$ would equal Q_{hot}.

Next, we suppose the contrary of what we intend to prove and show that the contrary situation is impossible. That is, we suppose that the other engine is more efficient than the reversible engine. Then, for the same value Q_{cold} of energy dumped into the cold reservoir, the other engine would produce a total work output greater than W [and its energy input $(Q_{hot})_{other}$ would be greater than Q_{hot}]. The numerical values must all be consistent with conservation of energy; nonetheless, there would be some extra work available. Such a process, however, would convert irregular motion (of molecules in the hot reservoir) to regular motion (of launched satellite-carrying projectiles, for example) with no other change. It would achieve the astronomer's dream of section 2.3, which we showed to be impossible. The process would lead to a global *decrease* in entropy, but that cannot happen (as a practical matter). Thus the other engine cannot be more efficient than the reversible engine.

[If the other engine is precisely as efficient as the reversible engine, then there is no extra work, and also $(Q_{hot})_{other} = Q_{hot}$. Consequently, the other engine is perfectly equivalent to the reversible engine (when the latter is run forward), and so the other engine is reversible, also.]

Putting it all together

A Carnot cycle provides a reversible heat engine, and so—by the analysis of the last subsection—its efficiency is equal to the maximum efficiency. Recalling the efficiency of a Carnot cycle from equation (3.4), we may write

$$\left(\begin{array}{c} \text{maximum efficiency} \\ \text{of heat engine} \end{array} \right) = \frac{T_{hot} - T_{cold}}{T_{hot}}. \tag{3.5}$$

We have gained our objective, but a verbal summary may be welcome. When operating between reservoirs at temperatures T_{hot} and T_{cold}, no heat engine can have an efficiency greater than the efficiency of a reversible heat engine. Therefore, for given T_{hot} and T_{cold}, all reversible heat engines have the same efficiency. Moreover, because a Carnot cycle is a reversible cycle, the maximum efficiency is given by the Carnot efficiency, displayed in equation (3.5).

Conventional heat engines operate with temperatures that satisfy the inequalities

$$T_{hot} > T_{cold} > 0.$$

Thus the maximum efficiency is necessarily less than 1 (or, equivalently, less than 100 percent). This rigorous limit holds even in the optimum situation in which the environment's over-all change in entropy is zero. In that case, the hot reservoir experiences a decrease in entropy (as energy is removed by cooling), but the cold reservoir compensates by its increase in entropy (as it absorbs energy by heating), as displayed in table 3.1.

If the environment's total change in entropy is nonzero, then the Second Law requires the total change to be positive. Compare individual changes with the reversible cycle of the preceding paragraph. For the same entropy decrease of the hot reservoir, the cold reservoir must now experience a greater increase in entropy. That means more energy must be dumped into the cold reservoir, and so the engine's efficiency will be less. This analysis provides a more algebraic reason why any "other" engine will be less efficient than a reversible engine.

The next section looks further into the practical limits of efficiency.

3.3   A practical consequence

Connecticut Yankee was a nuclear power plant 10 miles down the Connecticut River from Middletown, Connecticut. It began commercial operation in 1968 and was decommissioned in 1997. In its early years, Connecticut Yankee had an enviable record for reliability of operation; twice it held the world's record among commercial nuclear plants for the longest continuous period of operation. Yet the efficiency with which Connecticut Yankee converted the energy in a uranium nucleus into electrical energy flowing out of the plant was only 33 percent. Should we fault the engineers and the management for egregiously inefficient use of energy?

Rather than jump to a conclusion, we should ask, how efficient could Connecticut Yankee have been—according to the Second Law of Thermodynamics?

The kinetic energy arising from nuclear fission was used to heat water and to form steam. The temperature of the "hot reservoir" was approximately 500 degrees Fahrenheit or, better, approximately 533 K. The power plant used water from the Connecticut River to condense the steam after it had passed through the turbine. The temperature of the river water varied with the seasons, colder in winter than in summer. A good average figure was 55 degrees Fahrenheit or, better, 286 K. Thus, from our analysis of the Carnot cycle, the maximum efficiency that Connecticut Yankee could have had is this:

$$\frac{T_{hot} - T_{cold}}{T_{hot}} = \frac{533 - 286}{533} = 0.46,$$

that is, a mere 46 percent. Given unintended thermal conduction here and there and especially given the need to operate more rapidly than "slowly," a realized efficiency of 33 percent was a quite reasonable achievement for the engineers and the staff.

Here is a bit more detail about realistic goals for efficiency. Energy must flow from the hot reservoir to the working substance (which is water here) through a strong metal partition. To produce that flow by conduction and at an industrially significant rate, the working substance must have a temperature lower than T_{hot}, the temperature of the hot reservoir. In the other isothermal stage, energy must flow through another metal partition to the cold reservoir, and so the working substance must have a temperature higher than T_{cold} (the temperature of the cold reservoir) during that portion of the cycle. From the point of view of the working substance, its effective "T_{cold}/T_{hot}" ratio is higher than that of the actual reservoirs. Efficiency falls as the temperature ratio approaches 1 (for at the value 1, no work is done), and so the actual industrial efficiency will be less than the Carnot efficiency.

The oil-fired power plant on the river within sight of Middletown had an efficiency similar to that of the nuclear plant, perhaps a few percent higher. In general, fossil fuel power plants have an actual efficiency in the range 35–40 percent. As a safety precaution, nuclear plants run at lower temperatures (and pressures) for their hot reservoirs and hence have slightly smaller efficiencies. Overall, if we consider any power plant that generates electrical energy either by burning fossil fuel to heat water or by using nuclear fission to make the steam, then the plant has an efficiency of roughly 1/3—and hence must dump 2/3 of the energy into the environment, be it a river, the ocean, or the atmosphere. This is a sobering consequence of the Second Law.

Indeed, for nuclear power plants, the consequence is ironic as well. The kinetic energy of the fission fragments and extra neutrons comes from electrical potential energy originally stored within the uranium nucleus. The objective of the power plant is to give electrical potential energy to the electrons in the wires that emanate from the plant. The process of converting the "nuclear electrical potential energy" into the electrons' electrical potential energy tosses away 2/3 of the energy. According to the Second Law, most of that dissipation is inevitable (if steam is an intermediary), but there is irony nonetheless.

3.4   Rapid change

This section addresses the general question, how can we compute the entropy change, $S_{final} - S_{initial}$, if the actual evolution of the physical system is *not* slow? The word "slow" implies that the system and its surroundings remain close to equilibrium throughout the process. Thus, in the present section, the system is allowed to depart substantially from thermal equilibrium. We do require, however, that the initial and final states be ones of thermal equilibrium.

Equation (2.26) is not *directly* useful here because the equality sign applies only if the process is slow. To learn how to proceed, we return to the very definition of entropy as the logarithm of a multiplicity. The multiplicity is characteristic of the macrostate itself and does not depend on the route by which the physical system evolved to the macrostate. The same, then, is true for the entropy. This insight enables us to answer the question, as follows.

Procedure for coping with rapid change

To calculate $S_{\text{final}} - S_{\text{initial}}$ for a rapid process, find (or invent) a *slow* process that starts at the specified initial situation and ends at the desired final situation; then use the equation

$$\Delta S = \frac{q}{T} \tag{3.6}$$

in step-wise fashion to compute the change $S_{\text{final}} - S_{\text{initial}}$ for the *slow* process. The change in the system's entropy is the same for the rapid process as for the slow process because multiplicity (and hence entropy) is characteristic of each macrostate itself and does not depend on the route by which the physical system evolved to the macrostate. Thus we cope with rapid change by using the equality

$$\left(\begin{array}{c} S_{\text{final}} - S_{\text{initial}} \\ \text{for a rapid process} \end{array} \right) = \left(\begin{array}{c} S_{\text{final}} - S_{\text{initial}} \\ \text{for any slow process whose} \\ \text{initial and final macrostates} \\ \text{are the same as those for} \\ \text{the rapid process} \end{array} \right). \tag{3.7}$$

Example

Suppose N atoms of helium are in a container of fixed volume V and cold initial temperature T_{i}. Now heat the gas by placing the container in contact with a huge hot brick of temperature T_{brick}, where $T_{\text{brick}} \gg T_{\text{i}}$. There will be rapid transfer of energy (by heating) until the gas settles down to a new equilibrium at a final temperature T_{f}, which is equal to T_{brick}.

To invent a slow process that connects the specified initial and final situations, we *imagine* a sequence of many bricks, each brick a little hotter than the preceding one and ranging in temperature from T_{i} to T_{f}. We imagine placing the container in contact with the bricks sequentially, so that the temperature difference between container and current brick is always small (infinitesimal, in principle) and the process of heating is a slow one.

If the difference in temperature between one brick and the next is ΔT, we can readily calculate how much energy is transferred by heating at each step. The total energy of the gas is $E = \frac{3}{2}NkT$. Raising the temperature by ΔT increases the energy by $\frac{3}{2}Nk\Delta T$, and so that much energy must be provided by the hotter brick:

$$q = \tfrac{3}{2}Nk\Delta T.$$

Now we can turn to equation (2.26), use the equality sign, and compute the total change in entropy as follows:

$$S_{final} - S_{initial} = \text{sum of } \frac{q}{T} \text{ for each brick}$$

$$= \text{sum of } \frac{\frac{3}{2}Nk\Delta T}{T_{current\ value}} \text{ for each brick}$$

$$= \int_{T_i}^{T_f} \frac{3}{2}Nk \frac{dT}{T} = \frac{3}{2}Nk(\ln T_f - \ln T_i). \tag{3.8}$$

In the limit as the number of intermediary bricks goes to infinity, the sum becomes an integral. Equation (3.8) gives the total change in entropy as calculated for the slow process. Equation (3.7) asserts that we may use the algebraic result here for the total entropy change in the rapid process, too.

For a system more complicated than a monatomic classical ideal gas, one would use the system's heat capacity C. To increase the system's temperature by dT, one must supply energy $C\,dT$ by heating. The ensuing entropy change is $(C\,dT)/T$. Thus the total change in entropy is

$$S_{final} - S_{initial} = \int_{T_i}^{T_f} \frac{C(T)}{T} dT. \tag{3.9}$$

The heat capacity may vary with temperature, and so it is written as $C(T)$. Equation (3.9) is useful for computing the entropy as a function of temperature whenever the heat capacity is known (and also the entropy is known at some specific temperature, such as at absolute zero).

Before we go on, let me address a question that may have occurred to you. The procedure codified in equation (3.7) says that, as far as the entropy change of the system is concerned, there is no difference between the actual rapid process and the imagined slow one. If that is so, then where *is* there a difference? The answer is this: the difference occurs in the environment. If one literally replaced the rapid process with a slow process, the effect of the process on the environment would be different, and the change in the environment's entropy would be different. The example in the next section may make this distinction clearer, and so we turn to it.

3.5   The simplified Otto cycle

Figure 3.3 displayed the simplified Otto cycle, and the legend asserted that rapid processes make the cycle irreversible. Accordingly, the entropy change of gas plus environment should be positive when the cycle is run through once. Also, the efficiency should be less than that of a Carnot cycle operating between the same two temperatures. This section offers detailed calculations that confirm those assertions.

The given data consist of the temperature and volume at state 1, T_{hot} and V_1, and the temperature and volume at state 3, T_{cold} and V_2. The gas is specified to behave like a classical ideal gas and to have a constant heat capacity C_V.

The temperature T_2 at state 2 follows from the adiabatic relation (1.24) for the expansion from state 1 to state 2:

$$T_2 V_2^{\gamma-1} = T_{\text{hot}} V_1^{\gamma-1},$$

whence

$$T_2 = T_{\text{hot}} \left(\frac{V_1}{V_2}\right)^{\gamma-1}. \tag{3.10}$$

Similar reasoning applied to the adiabatic compression from state 3 to state 4 yields

$$T_4 = T_{\text{cold}} \left(\frac{V_2}{V_1}\right)^{\gamma-1}. \tag{3.11}$$

We will need two energies, as follows:

$$Q_{\text{hot}} = \left(\begin{array}{c} \text{energy input to gas by} \\ \text{heating, state 4 to state 1} \end{array}\right) = \int_{T_4}^{T_{\text{hot}}} C_V \, dT$$

$$= C_V(T_{\text{hot}} - T_4); \tag{3.12}$$

$$Q_{\text{cold}} = \left(\begin{array}{c} \text{energy input to cold reservoir by} \\ \text{cooling of gas, state 2 to state 3} \end{array}\right)$$

$$= C_V(T_2 - T_{\text{cold}}). \tag{3.13}$$

For a classical ideal gas, the change in internal energy is simply the heat capacity C_V times the temperature change. The definition of Q_{cold} here is analogous to the definition used with the Carnot cycle and ensures that Q_{cold} is a positive quantity. That completes the preliminaries.

Total entropy change

To assess the total change in entropy, we split the cycle into its four natural stages and examine them separately.

Stage 4 to 1: rapid heating of the gas. Precisely because the gas is heated rapidly, we must use the indirect procedure of section 3.4 to calculate $\Delta S_{\text{gas},4\to1}$. We *imagine* slow heating by placing the gas in thermal contact with a succession of warm reservoirs whose temperatures range from T_4 to T_{hot}. While the gas is in contact with a reservoir at temperature T, the energy transferred by heating is the infinitesimal amount $C_V \, dT$, and the entropy change is $(C_V \, dT)/T$. Adding up all those contributions by integration gives

$$\Delta S_{\text{gas},4\to1} = \int_{T_4}^{T_{\text{hot}}} \frac{C_V \, dT}{T} = C_V \ln(T_{\text{hot}}/T_4). \tag{3.14}$$

Although the literal heating of the gas between states 4 and 1 is a rapid process for the gas, the huge reservoir placidly supplies energy without any change in its temperature. Thus we may evaluate the entropy change for the hot reservoir as

$$\Delta S_{\text{hot reservoir}} = \frac{-Q_{\text{hot}}}{T_{\text{hot}}} = -C_V \frac{(T_{\text{hot}} - T_4)}{T_{\text{hot}}}. \tag{3.15}$$

How do these two entropy changes compare? If the denominator of the integrand in (3.14) were constant, the integral of the numerator would give $C_V(T_{\text{hot}} - T_4)$, which is Q_{hot}. But the denominator T varies and is generally smaller than T_{hot}, which is the denominator in (3.15). Therefore $\Delta S_{\text{gas},4\to 1}$ exceeds the magnitude of $\Delta S_{\text{hot reservoir}}$, and we may write

$$\Delta S_{\text{gas},4\to 1} + \Delta S_{\text{hot reservoir}} > 0. \tag{3.16}$$

Stage 1 to 2: slow adiabatic expansion. Provided the adiabatic expansion is slow, which has been specified, the gas experiences no change in entropy.

Stage 2 to 3: rapid cooling of the gas. The cold reservoir acquires energy Q_{cold} by heating, but it maintains its temperature T_{cold}. Thus the reservoir's entropy change is

$$\Delta S_{\text{cold reservoir}} = \frac{Q_{\text{cold}}}{T_{\text{cold}}} = C_V \frac{(T_2 - T_{\text{cold}})}{T_{\text{cold}}}. \tag{3.17}$$

For the gas, the rapid cooling requires that we *imagine* a process of slow cooling. Thinking of a sequence of cool reservoirs, we follow the pattern set by equations (3.7) and (3.14):

$$\Delta S_{\text{gas},2\to 3} = \int_{T_2}^{T_{\text{cold}}} \frac{C_V \, dT}{T}$$

$$= C_V \ln(T_{\text{cold}}/T_2) = -C_V \ln(T_2/T_{\text{cold}}). \tag{3.18}$$

The ratio T_2/T_{cold} is greater than 1, and so the entropy change is negative. Because the gas cools, a negative change was to be expected.

To compare the magnitudes of the entropy changes here, we proceed as before. The denominator T_{cold} in (3.17) is smaller than every value of the integrand's denominator in (3.18), except at the T_{cold} end of the integration range. Thus $\Delta S_{\text{cold reservoir}}$ is greater than the magnitude of $\Delta S_{\text{gas},2\to 3}$, and so

$$\Delta S_{\text{gas},2\to 3} + \Delta S_{\text{cold reservoir}} > 0. \tag{3.19}$$

Stage 3 to 4: slow adiabatic compression. As in the other adiabatic stage, there are no entropy changes.

The sum of the left-hand sides of inequalities (3.16) and (3.19) gives us the total change in entropy. Because both of those left-hand sides are greater than zero, we may add them and may conclude that

$$\Delta S_{\text{total}} = \Delta S_{\text{gas}} + \Delta S_{\text{hot reservoir}} + \Delta S_{\text{cold reservoir}} > 0. \tag{3.20}$$

The positive result is what we expected.

A check on the computation is readily made, and a check is always worthwhile. When the gas makes a full cycle, it returns to its initial macrostate and hence to its initial entropy. So our calculation should imply $\Delta S_{\text{gas}} = 0$. The contributions to the net entropy change for the gas come from equations (3.14) and (3.18). To see whether those two contributions cancel each other, we need to compare two temperature ratios. Eliminating the volume ratio between equations (3.10) and (3.11) implies

$$\frac{T_2}{T_{\text{cold}}} = \frac{T_{\text{hot}}}{T_4}, \tag{3.21}$$

and so the two contributions do cancel, yielding zero for the net entropy change of the gas.

Efficiency

To compute the efficiency, we proceed as we did with the Carnot cycle. Energy conservation enables us to express the net work output as $Q_{\text{hot}} - Q_{\text{cold}}$. Then the efficiency follows as

$$\left(\begin{array}{c} \text{efficiency of} \\ \text{simplified Otto cycle} \end{array} \right) = \frac{\text{net work output}}{\text{energy input by heating}}$$

$$= \frac{Q_{\text{hot}} - Q_{\text{cold}}}{Q_{\text{hot}}}$$

$$= 1 - \frac{T_{\text{cold}}}{T_{\text{hot}}} \times \frac{T_2}{T_{\text{cold}}}. \tag{3.22}$$

To obtain the last line, first use equations (3.12) and (3.13) and then equation (3.21). The expression for efficiency here differs from the Carnot efficiency by the factor T_2/T_{cold}. Because that factor is greater than unity and multiplies a negative term, the Otto cycle's efficiency is less than the Carnot cycle's.

The connection with automobiles

The simplified Otto cycle is a useful approximation to the cycle in a typical automobile engine. Figure 3.5 displays the latter cycle, which consists of the following stages.

(a) During the *intake* stroke, the piston is pulled down by the crankshaft and connecting rod. A mixture of gasoline and fresh air rushes into the cylinder through the intake valve.

(b) In the *compression stroke*, both intake and exhaust values are closed. The piston is pushed up and compresses the mixture.

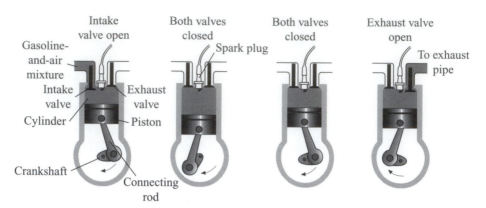

(a) Intake stroke   (b) Compression stroke   (c) Power stroke   (d) Exhaust stroke

Figure 3.5 The internal combustion engine of a typical automobile. Alphonse Beau de Rochas proposed the four-stroke cycle in 1862 but never built an engine. In 1876, the German engineer Nikolaus Otto developed the idea into a commercial model; since then, hundreds of millions of such engines have been manufactured. After its developer, this four-stroke cycle is called the *Otto cycle*.

(c) Near the top of the compression stroke, the spark plug fires. The spark ignites the mixture, which burns quickly and then expands, pushing the piston vigorously downward in the *power stroke*.

(d) To get rid of the combustion products, the crankshaft again pushes the piston upward, this time while the exhaust valve is open. The *exhaust stroke* cleans out the cylinder and prepares it for the next cycle.

To see the connection between this cycle, which is the literal Otto cycle, and the simplified Otto cycle, we return to figure 3.3 and start at state 3. The adiabatic compression to state 4 is the analog of the compression stroke. The rapid heating that takes the gas from state 4 to state 1 corresponds to the burning of the gasoline-and-air mixture. The burning occurs so swiftly that the piston hardly moves during that period, which can be approximated as heating at constant volume. The adiabatic expansion from state 1 to state 2 is a fine analog of the power stroke. The simplified cycle has cooling at constant volume as the means to get from state 2 to state 3. This process is intended to approximate the net effect of the exhaust and intake strokes. The new mixture of gasoline and air is indeed cooler than the combustion products were. The exhaust stroke starts at maximum volume, and the intake stroke ends at maximum volume; so maximum volume is appropriate for representing the net effect of those two strokes. Of course, in the actual automobile cycle, the old air is replaced by fresh air, and new gasoline is introduced, whereas in the simplified version the same molecules remain in the cylinder, cycle after cycle. Nonetheless, when tested against measured engine efficiencies, the simplified cycle provides a surprisingly good approximation to actual performance.

If you are an automobile buff, you may know that the ratio V_2/V_1, the ratio of maximum volume to minimum volume, is called the *compression ratio* and that it is a critical determinant of efficiency. We are only a step away from seeing that conclusion emerge from our derivation of the efficiency. Return to equation (3.22), cancel the two factors of T_{cold}, and use (3.10) to eliminate T_2/T_{hot} in favor of the compression ratio:

$$\left(\begin{array}{c} \text{efficiency of} \\ \text{simplified Otto cycle} \end{array} \right) = 1 - \left[\frac{1}{(\text{compression ratio})} \right]^{\gamma-1}. \tag{3.23}$$

Recall that γ, the ratio of heat capacities, is greater than 1, and so the exponent $\gamma - 1$ is positive. Thus, the larger the compression ratio, the higher the efficiency.

Typical passenger cars have a compression ratio in the range of 8 to 10. Sports cars run up to 11 or 12.

3.6 More about reversibility

Section 3.2 defined a reversible cycle: a cycle that can be entirely reversed, returning both system and environment to their original states. The entire "world" is returned to its original state. A sufficient condition for such a return is that the total change in entropy of system and environment be zero when the cycle is run forward.

The notion of reversibility can be extended to a process that stops short of being a cycle. For example, the slow isothermal expansion in the Carnot cycle can be totally reversed. The work that was done during the expansion is used to compress the gas isothermally from state 2 to state 1. Energy flows out of the gas and into the hot reservoir, decreasing the gas's entropy and increasing the reservoir's to their original values. Again, the property that $\Delta S_{\text{gas}} + \Delta S_{\text{environment}} = 0$ for the forward process is what enables one to reverse the process entirely.

The general situation can be expressed as follows.

(a) Definition: a *process* is called *reversible* if the system and environment can be restored to their original state, leaving no residual changes anywhere.

$$\tag{3.24}$$

(b) A sufficient condition for reversibility is that $\Delta S_{\text{system}} + \Delta S_{\text{environment}} = 0$ for the forward process. $\tag{3.25}$

Given the definition in (3.24), a reversible cycle becomes a special case: it is a reversible process in which the system itself returns to its original state, whence the equation $\Delta S_{\text{system}} = 0$ holds automatically. Thus a sufficient condition for reversibility of a cycle is merely $\Delta S_{\text{environment}} = 0$.

In the preceding statements, the conditions for reversibility are couched in terms of entropy changes. Can one be more direct and specify experimental conditions? Yes. A process will be reversible if the following two conditions are met.

(a) The process is carried out slowly, so that the system is always close to equilibrium. $\tag{3.26a}$

(b) No dissipative processes, such as frictional rubbing or viscous damping of fluid motion, accompany the process. (3.26b)

The limit of an infinitely slow process is usually required, and so a reversible process is an idealization.

Of the two conditions here, "slowness" is the more important, for slowness itself sometimes eliminates the irreversible effect of a dissipative process. For example, suppose one needs to transfer 5.7 coulombs of charge across an electrical resistance R. The dissipative process is resistive heating, produced at the rate $I^2 R$ joules per second, where I is the electric current. Transfer the charge slowly, letting the current be $I = 5.7/t_{long}$, where t_{long} is the long time during which a tiny current will flow and will transfer the desired amount of charge. That is,

$$(\text{charge transferred}) = \int_0^{t_{long}} I\, dt = 5.7,$$

independent of the duration t_{long}. The total dissipation of electrical energy will be given by another integral:

$$(\text{total dissipation}) = \int_0^{t_{long}} I^2 R\, dt = \frac{(5.7)^2 R}{t_{long}} \to 0;$$

the total dissipation, being quadratic in the current I, goes to zero as the transfer time goes to infinity.

The processes that I characterized as "slow" in sections 2.4 and 2.6 are more properly specified to be "slow and free of dissipation," conditions that ensure reversibility. Slow and free of dissipation are the conditions that entitle one to use the equality sign in the central equation,

$$\Delta S_{\text{any system}} \geq \frac{\left(\begin{array}{c} \text{energy into that} \\ \text{system by heating} \end{array} \right)}{T}. \tag{3.27}$$

For most purposes, slowness is sufficient, and I took that simpler view when we started.

Moreover, now we can say also that reversibility is sufficient for the equal sign in equation (3.27). Here is the logic. In section 2.6, we reasoned that a slow (and dissipation-free) process implies reversibility and that that property, in turn, implies $\Delta S_{\text{total}} = 0$. In the context of section 2.6, no change in total entropy means $\Delta S_{\text{chem system}} + \Delta S_{\text{helium}} = 0$. That equation leads to the equal sign in equation (2.25), which then carries over to equations (2.26) and (3.27).

Summary

For the topics studied in this book, the three statements, (1) the process is performed slowly and without dissipation, (2) the process is reversible, and (3) $\Delta S_{\text{system}} +$

$\Delta S_{\text{environment}} = 0$, are equivalent. Any one of them implies the other two. Moreover, any one of the statements implies that the equal sign holds in equation (3.27).

In subsequent chapters, I will often use the phrases "reversible" and "occurs slowly" interchangeably. The phrase "reversible" is the formal technical term, but "occurs slowly" has a wonderful immediacy and reminds us that we are talking about experimental conditions.

3.7   Essentials

1. The *Carnot cycle* consists of two isothermal stages (at temperatures T_{hot} and T_{cold}) separated by two adiabatic stages. All stages are executed slowly.

The adiabatic intervals ease the system's temperature to that of the next isothermal stage. Such adiabatic intervals are necessary in order to avoid fast, irreversible processes that would occur if the system and reservoir had finitely different temperatures.

2. A cycle is called *reversible* if, when the cycle is run in reverse, the *environment* can be restored to its original configuration. (The "working substance" of the cycle is restored to its original configuration when the cycle is run in either the forward or the reverse direction. One needs to focus on the environment.)

3. The efficiency of the Carnot cycle is

$$\begin{pmatrix} \text{efficiency of} \\ \text{Carnot cycle} \end{pmatrix} = \frac{T_{\text{hot}} - T_{\text{cold}}}{T_{\text{hot}}}.$$

4. When operating between reservoirs at temperatures T_{hot} and T_{cold}, no heat engine can have an efficiency greater than the efficiency of a reversible heat engine. Therefore, for given T_{hot} and T_{cold}, all reversible heat engines have the same efficiency. Moreover, because a Carnot cycle is a reversible cycle, the maximum efficiency is given by the Carnot efficiency.

5. To compute the entropy change for a rapid process, use the equation

$$\begin{pmatrix} S_{\text{final}} - S_{\text{initial}} \\ \text{for a rapid process} \end{pmatrix} = \begin{pmatrix} S_{\text{final}} - S_{\text{initial}} \\ \text{for any slow process whose} \\ \text{initial and final macrostates} \\ \text{are the same as those for} \\ \text{the rapid process} \end{pmatrix}.$$

6. When one knows the heat capacity as a function of temperature, one can calculate an entropy change as

$$S_{\text{final}} - S_{\text{initial}} = \int_{T_i}^{T_f} \frac{C(T)}{T}\, dT.$$

If one knows the entropy at the initial temperature T_i (for example, at absolute zero), then the equation serves to determine the entropy itself at the final temperature T_f.

7. The *simplified Otto cycle* consists of two constant-volume stages separated by two adiabatic stages. Energy transfer by heating (or cooling) occurs rapidly during the constant-volume stages [where the reservoirs are at temperatures T_{hot} and T_{cold}, but the working substance is not at either of those temperatures (except at the end points of the stages)]. The cycle is not reversible, and its efficiency is less than that of a Carnot cycle operating between the same temperatures.

8. A *process* is called *reversible* if the system and environment can be restored to their original state, leaving no residual changes anywhere. A sufficient condition for reversibility is that $\Delta S_{\text{system}} + \Delta S_{\text{environment}} = 0$ for the forward process.

In terms of experimental conditions, a process will be reversible if (a) the process is carried out slowly, so that the system is always close to equilibrium, and (b) no dissipative processes, such as frictional rubbing or viscous damping of fluid motion, accompany the process. Often, "slowness" is sufficient.

For the topics studied in this book, the three statements, (1) the process is performed slowly and without dissipation, (2) the process is reversible, and (3) $\Delta S_{\text{system}} + \Delta S_{\text{environment}} = 0$, are equivalent. Any one of them implies the other two. Moreover, any one of the statements implies that the equal sign holds in the equation connecting ΔS and q:

$$\Delta S = q/T.$$

For brevity's sake, I will use the phrases "reversible" and "occurs slowly" interchangeably.

Further reading

D. S. L. Cardwell, *From Watt to Clausius: The Rise of Thermodynamics in the Early Industrial Age* (Cornell University Press, Ithaca, New York, 1971). Cardwell tells an engrossing tale of success and error as thermodynamics grew out of a uniquely productive mix: practical engineering and laboratory science.

Carnot's paper is reprinted in *Reflections on the Motive Power of Fire, by Sadi Carnot, and other Papers on the Second Law of Thermodynamics by É. Clapeyron and R. Clausius*, edited by E. Mendoza (Peter Smith, Gloucester, Massachusetts, 1977). Carnot used little mathematics and wrote engagingly; his magnum opus—actually only 59 pages in this translation—is a pleasure to read.

Industrial cycles must produce substantial amounts of work in finite time intervals. Hence the cycles cannot proceed slowly, cannot be reversible, and cannot have maximum efficiency. What efficiency can they attain? "Efficiency of a Carnot engine at maximum power output" is the title of a specific study by F. L. Curzon and B. Ahlborn, *Am. J. Phys.* **43**, 22–4 (1975).

Problems

1. A Carnot engine operates between two heat reservoirs of temperature 550 °C and 30 °C, respectively.

(a) What is the efficiency of this engine?
(b) If the engine generates 1,500 joules of work, how much energy does it absorb from the hot reservoir? And how much does it reject into the cold reservoir?

2. The context is the Carnot engine of the preceding question. As an engineer, you are able to change the temperature of one (but only one reservoir) by 5 °C. To get the greatest increase in efficiency, which temperature should you change and in which sense (that is, hotter or colder)? Explain your reasoning.

3. In each cycle, an engine removes 150 joules from a reservoir at 100 °C and rejects 125 joules to a reservoir at 20 °C.

(a) What is the efficiency of this engine?
(b) Does the engine achieve the maximum efficiency possible (given the two temperatures)? If not, by what percentage does the engine fall short of ideal behavior?

4. A system (which you may think of as a gas or liquid) absorbs 200 J of energy from a reservoir at 400 K and also 300 J from a reservoir at 300 K. It interacts with a third reservoir whose temperature is T_3. When the system returns to its original state, it has done net work in the amount of 100 J.

(a) What is the entropy change of the system (i.e., the gas or liquid) for the complete cycle?

Now specify that the cycle is reversible.

(b) Sketch a possible path for the cycle in the pressure–volume plane. Label each qualitatively distinct portion of the cycle.
(c) What is the numerical value of the temperature T_3?

5. During a portion of the Carnot cycle, a dilute gas of diatomic nitrogen is compressed slowly at constant temperature. Here are some data:

$$V_{\text{initial}} = 0.3 \text{ m}^3, \qquad\qquad V_{\text{final}} = 0.1 \text{ m}^3,$$
$$N = \text{number of molecules} = 3 \times 10^{24}, \qquad T = 280 \text{ K}.$$

(a) What is the change in multiplicity of the gas? Express the factor by which the multiplicity changes as $10^{(\text{some power})}$.

(b) What is the change in entropy of the gas?

(c) How much energy was transferred to the environment through heating or cooling? Be sure to specify whether the energy of the environment decreased or increased.

(d) How much work was done on the gas while the gas was being compressed?

[Note. You do *not* have to do the four parts in sequence, but (a) through (d) is a convenient route. For each part, be sure to provide a numerical answer.]

6. *Total work from finite heat reservoirs.* Specify that the finite reservoirs of a Carnot cycle start at initial temperatures $T_{\text{hot, initial}}$ and $T_{\text{cold, initial}}$. Acknowledge the consequences of finiteness: the hot reservoir will drop in temperature, and the cold reservoir will grow in temperature. The two temperatures will converge to a final common temperature, T_{common}, and then the engine will cease to function.

Take the heat capacities of the two reservoirs to be equal and constant; each has the value $C_{\text{reservoir}}$. Assume negligible change in each reservoir's temperature during any one cycle of the engine.

(a) Determine T_{common}.

(b) Determine the total work done by the engine.

7. In figure 3.2, focus on two stages in the Carnot cycle: the isothermal expansion and the adiabatic expansion. The system (the "working substance" in the engine) consists of $N = 10^{25}$ atoms of helium (which you may regard as a classical ideal gas). As the gas "moves" slowly from state 1 to state 3, it absorbs 10^5 joules of energy from a heat reservoir at temperature 800 K. The Carnot engine lifts a small load of bricks. Provide numerical answers to all of the following questions.

(a) For the helium, what are the numerical values of the following ratios?

$$\frac{\text{multiplicity at state 2}}{\text{multiplicity at state 1}} \quad \text{and} \quad \frac{\text{multiplicity at state 3}}{\text{multiplicity at state 1}}.$$

(b) How much work did the gas do as it moved from state 1 to state 2? (Note: state 2, *not* state 3.)

(c) When the gas moves from state 1 all the way to state 3, what is the net entropy change of the environment (which consists of the hot reservoir, the cold reservoir, and the load of bricks)? Explain your reasoning.

8. *Refrigerator.* When run backward, a Carnot cycle provides an ideal refrigerator: the cycle extracts energy from the cold reservoir and dumps it into the hot reservoir. The left-hand side of figure 3.4 illustrates this cyclic process. Work from an external source is required to run the refrigerator, and the energy associated with that work is also dumped into the hot reservoir. A room air conditioner is fundamentally a refrigerator, and the questions below examine it.

(a) The temperature inside the room is $T_{inside} = 25\,°C$, and the temperature outside the house is $T_{outside} = 32\,°C$. The temperature difference causes energy to flow into the room (by conduction through the walls and window glass) at the rate $3,000\,J/s$. To return this energy to the outside by running an ideal refrigerator, how much electrical energy must be supplied to the refrigerator (to perform the external work)?

(b) If the outside temperature grows to $39\,°C$, so that $\Delta T \equiv T_{outside} - T_{inside}$ doubles, by what factor must the supply of electrical energy increase? Take the inflow of energy to be proportional to ΔT. Does this calculation help you to appreciate the problems facing an electrical power company on a hot summer afternoon?

9. *Overall change in entropy.* A copper penny, initially at temperature T_i, is placed in contact with a large block of ice that serves as a heat reservoir and has a constant temperature T_{res} (well below freezing). Take the penny's heat capacity to have the constant value C, and specify $T_i \neq T_{res}$ (by a finite amount). The following questions pertain after the joint system has come to thermal equilibrium.

(a) What are the entropy changes of the penny and of the ice block?
(b) What sign does the total change in entropy have [according to your calculations in part (a)]?
(c) Is the sign independent of whether the penny was hotter or colder than the ice block?

10. A large, thermally isolated container is initially partitioned into two volumes, V_0 and $2V_0$, as shown in figure 3.6. Gaseous helium and neon (which you may consider to be classical ideal gases) have pressures P_0 and $3P_0$, respectively, in the two regions. The temperature is uniform throughout the entire volume and has the initial value T_0. The thin aluminum partition is now allowed to slide to one side or the other, and complete equilibrium is ultimately established. Answer the following questions in terms of V_0, P_0, and T_0.

(a) What is the final temperature?
(b) What is the final pressure?
(c) What is the change in the total entropy?

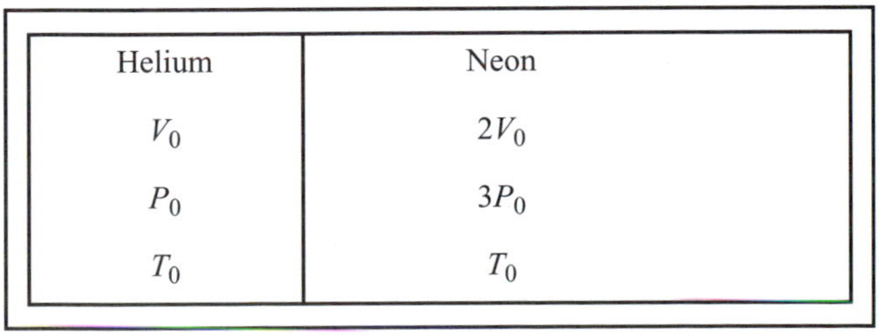

Helium	Neon
V_0	$2V_0$
P_0	$3P_0$
T_0	T_0

Figure 3.6 The context.

11. An inventor claims to have developed a heat engine that produces 2 joules of work for every joule of energy that is discarded. The engine is designed to use reservoirs at temperatures of 400 °C and 0 °C. Would you recommend investing in the stock of this high-tech company? Why?

12. A heat engine is run with a large block of hot metal as the hot reservoir and with the ocean as the cold reservoir. The metal has initial temperature T_i and a heat capacity C that is independent of temperature. The ocean remains at temperature T_0.
   Calculate the maximum amount of work that can be done by the engine.

(a) Express your answer in terms of T_i, T_0, and C (and no other parameters or physical constants).
(b) Check for reasonableness (for example, for correct sign!) by inserting the numerical values $T_i = 1,200$ K, $T_0 = 280$ K, and $C = 7 \times 10^6$ J/K.

13. *Substituting a slow process.* In part (a) of figure 1.6, a gas undergoes a rapid process but comes ultimately to thermal equilibrium. To determine the entropy change, substitute the slow process displayed in part (b). Take the gas to be a classical ideal gas with constant heat capacity C_V. What are the changes in entropy, energy, and temperature for the two stages of the slow process? Does the over-all entropy change agree with previous results? For the over-all entropy change, express your answer in terms of the initial volume V_i, the final volume V_f, and the number of gas molecules N.

14. *Efficiency and entropy.* Specify a heat engine that operates between two reservoirs at temperatures T_{hot} and T_{cold}, but do *not* assume reversibility.

(a) For a forward cycle, express the entropy change of the environment, ΔS_{env}, in terms of Q_{hot}, Q_{cold}, and the two reservoir temperatures.
(b) Express the engine's efficiency in terms of ΔS_{env}, Q_{hot}, and the two temperatures.
(c) What conclusions can you draw about maximum efficiency?

4 Entropy in Quantum Theory

4.1 The density of states
4.2 The quantum version of multiplicity
4.3 A general definition of temperature
4.4 Essentials

This chapter has two goals. The first is to develop the quantum version of multiplicity. That will show us how entropy is expressed in quantum theory. The second goal is to develop a general quantitative definition of temperature. Entropy plays a key role in that definition.

4.1 The density of states

When a physical system has reached thermal equilibrium, its macroscopic properties do not change with time. In quantum theory, the energy eigenstates of an isolated system provide predictions and estimates that are constant in time; therefore such states are appropriate for a quantum description of thermal equilibrium.

To be sure, the information at hand will not enable us to select a single state as uniquely the correct state to use. We will be driven to consider many states and to form sums over them. This section develops a mathematical technique for working with such sums.

When the system is both isolated and of finite size (as we shall specify here), the energy eigenstates form a discrete set, whose members we can arrange and label in order of increasing energy. (If any two distinct states happen to have the same energy, we just assign them consecutive labels.) Typically, the states will be densely spaced in energy, and so a sum over a range of states can often be approximated adequately by an integral with respect to energy, provided that we have constructed an appropriate *density of states*: a function that specifies the number of energy eigenstates per unit energy interval. To make these abstract words more meaningful, we construct such a density of states in detail.

The density of states for a single spinless particle

A single spinless particle is confined to a cubical box; the edge length of the box is L. In terms of the linear momentum $\mathbf{p}$, the particle's energy ε is

$$\varepsilon = \frac{p^2}{2m} = \frac{1}{2m}(p_x^2 + p_y^2 + p_z^2), \tag{4.1}$$

where m denotes the rest mass. We take the wave function that describes the particle to be a standing wave. Provisionally, imagine a standing wave with variation purely along the x-axis and with wavelength λ_x. To ensure zero values of the wave at both ends of the length L, an integral number of half wavelengths must fit along the length L. Figure 4.1 illustrates this constraint. Thus

$$\frac{L}{(\lambda_x/2)} = n_x, \tag{4.2}$$

where n_x is a positive integer. The associated momentum component is then

$$p_x = \frac{h}{\lambda_x}, \tag{4.3}$$

where h is Planck's constant. To get p_x, we solve (4.2) for $1/\lambda_x$ and substitute in (4.3):

$$p_x = n_x \frac{h}{2L}. \tag{4.4}$$

In a general standing wave pattern in three dimensions, similar results hold for p_y and p_z, but each expression is entitled to a distinct integer: n_y or n_z. Each single-particle state has its own set of three positive integers: $\{n_x, n_y, n_z\}$.

Substitute for $\{p_x, p_y, p_z\}$ into (4.1) and note that the cubical volume V equals L^3, so that $L^2 = V^{2/3}$. One finds

$$\varepsilon_\alpha = \frac{h^2}{8m} \frac{1}{V^{2/3}} (n_x^2 + n_y^2 + n_z^2)_\alpha; \tag{4.5}$$

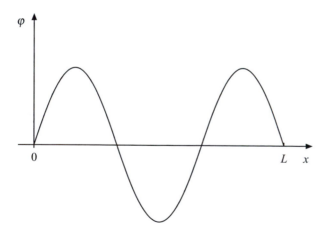

Figure 4.1 An integral number of half wavelengths must fit along the length L. The zero values of the wave function at the ends ($x = 0$ and $x = L$) enable the wave function to match smoothly to zero values in the region outside the box (where the probability of finding the particle is always zero).

the subscripts α indicate that the results pertain to the single-particle state φ_α labeled by the subscript α. For example, the state φ_3 has $n_x = 1$, $n_y = 2$, $n_z = 1$, and $\varepsilon_3 = 6h^2/(8mV^{2/3})$. Further detail is shown in table 4.1.

Each of the numbers n_x, n_y, and n_z may range separately over the positive integers. Specifying a set of three such integers specifies the single-particle state φ_α and its associated energy ε_α. We may also think of the three integers as specifying a single point in a three-dimensional space with axes labeled n_x, n_y, and n_z. This is illustrated in figure 4.2. Because of the restriction to positive integers, only one octant of the space is relevant. Except at the edges of the octant, there is one quantum state per unit volume in this purely mathematical space.

If we need to sum a function $A(\varepsilon_\alpha)$ over all single-particle states, we may often convert the sum to an integral as follows:

Table 4.1 *A few single-particle states.*

Wave function	n_x	n_y	n_z	Comments
φ_1	1	1	1	Single-particle ground state
φ_2	2	1	1	These three states have the same energy
φ_3	1	2	1	but different spatial behavior, and so they are
φ_4	1	1	2	distinct, different states.
φ_5	2	2	1	

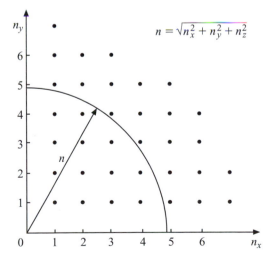

Figure 4.2 A slice in the plane $n_z =$(some positive integer) of the three-dimensional mathematical space with axes labeled n_x, n_y, and n_z. Each point indicates a triplet of values $\{n_x, n_y, n_z\}$ and thus gives a distinct single-particle state. The symbol n denotes the "radial distance" in the space.

$$\sum_{\text{states } \varphi_a} A(\varepsilon_a) = \sum_{\text{positive triplets } \{n_x, n_y, n_z\}} A(\varepsilon_a)$$

$$= \tfrac{1}{8} \int A(\varepsilon) 4\pi n^2 \, dn. \tag{4.6}$$

The factor $4\pi n^2 \, dn$ gives integration over spherical shells in the mathematical space of all triplets of integers; the factor $1/8$ cuts down the integral to the size appropriate for integration over the octant with positive integers. [If a concrete example for the function A would help you, take

$$A(\varepsilon_a) = \exp(-\varepsilon_a/kT) \quad \text{and} \quad A(\varepsilon) = \exp(-\varepsilon/kT).$$

This form for A will later play a prominent role.]

The next step is to convert to energy ε as the integration variable. In terms of the radial distance n, equation (4.5) tells us that the energy is

$$\varepsilon = \frac{h^2}{8m} \frac{1}{V^{2/3}} n^2, \tag{4.7}$$

whence

$$n = \left(\frac{8mV^{2/3}}{h^2} \right)^{1/2} \varepsilon^{1/2}. \tag{4.8}$$

Thus

$$dn = \frac{dn}{d\varepsilon} d\varepsilon = \left(\frac{8mV^{2/3}}{h^2} \right)^{1/2} \tfrac{1}{2} \varepsilon^{-1/2} \, d\varepsilon. \tag{4.9}$$

After using both (4.8) and (4.9) in equation (4.6), we emerge with

$$\sum_{\text{states } \varphi_a} A(\varepsilon_a) = \int A(\varepsilon) \frac{2\pi (2m)^{3/2}}{h^3} V \varepsilon^{1/2} d\varepsilon. \tag{4.10}$$

From the integrand, we can read off the density of states $D(\varepsilon)$ as

$$\left(\begin{array}{c} \text{number of single-particle states} \\ \text{per unit energy interval near energy } \varepsilon \end{array} \right) \equiv D(\varepsilon) = \frac{2\pi (2m)^{3/2}}{h^3} V \varepsilon^{1/2}. \tag{4.11}$$

This density holds for a single spinless particle, and its shape is sketched in figure 4.3. Often we will replace a sum over single-particle states by an integration with $D(\varepsilon)$, as follows:

$$\sum_{\text{states } \varphi_a} A(\varepsilon_a) = \int A(\varepsilon) D(\varepsilon) \, d\varepsilon. \tag{4.12}$$

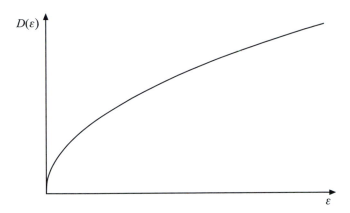

Figure 4.3 The density of states $D(\varepsilon)$ for a single spinless particle (when the only energy is kinetic energy).

By the way, the explicit expression (4.11) for the density of states is somewhat more general than the derivation might suggest. It holds for a macroscopic container of any reasonable shape and volume V, not just the cubical box used here. As soon as the de Broglie wavelength is substantially shorter than any "diameter" of the container, the number of states in any given small energy range is independent of the shape of the boundary, at least to good approximation.

Generalization to the entire system

An entire macroscopic system, consisting of many particles, will be described by one or another energy eigenstate Ψ_j having energy E_j, where j is an index. (Thus we use φ_α, ε_α, and ε to denote states and energies for a single particle, but Ψ_j, E_j, and E to denote the corresponding quantities for the entire system.) The notation is summarized in table 4.2. If we need to sum a function $B(E_j)$ over the states of the system, we may often convert the sum to an integral over the total energy E with the notation

$$\sum_{\text{states } \Psi_j} B(E_j) = \int B(E)D(E)\,dE. \tag{4.13}$$

The use of a capital E here distinguishes the density of states $D(E)$ for the entire

Table 4.2 *Some notation for states and energies.*

Quantity	Single particle	Entire system
Energy eigenstate (or wave function)	φ_α	Ψ_j
Energy of that state	ε_α	E_j
Energy in general	ε	E

system from $D(\varepsilon)$, the single-particle density of states. Just as $D(\varepsilon)$ depends on the volume, so may $D(E)$; moreover, the latter will certainly depend on the number N of particles in the system, but the notation will not show that explicitly unless the context fails to suffice.

Rarely will any details about $D(E)$ be needed, and so we can go on immediately.

4.2   The quantum version of multiplicity

To express the entropy S in quantum theoretical terms, we need the quantum version of multiplicity: the number of microstates that correspond to the particular macrostate. An entirely general expression is not needed, for we will study in detail only systems that are in thermal equilibrium or that depart only infinitesimally from such equilibrium. Because we know the macrostate, we may presume that we know the system's energy E, though only to some imprecision δE naturally associated with a macroscopic measurement. Each of the energy eigenstates in a range δE around the energy E is a microstate that we could associate with the given macrostate. The density of states $D(E)$ enables us to express this idea succinctly:

$$\begin{pmatrix} \text{the number of microstates} \\ \text{that correspond to} \\ \text{the particular macrostate} \end{pmatrix} = D(E)\delta E. \tag{4.14}$$

The system's entropy is then

$$S = k\ln[D(E)\delta E]. \tag{4.15}$$

Lest you worry unnecessarily, let me say that we will assign a precise value to the energy range δE later, in chapter 5. For now, just regard δE as a typical experimental uncertainty range.

The discussion following equation (4.13) noted that $D(E)$ will typically depend on the volume V and will certainly depend on the number N of particles in the system. At the moment, we do not know the details of any of these dependences; yet it is worthwhile to note their existence and to write symbolically

$$S = S(E, V, N), \tag{4.16}$$

a relationship that will be valuable in the next section and again in chapter 10.

4.3   A general definition of temperature

Suppose you run some cold water from the kitchen faucet into an insulated container, as sketched in figure 4.4. Next, you take a handful of ice cubes from the freezer, put them in a plastic bag, tie up the bag snugly, and drop the ice into the water. Finally, you cover the container with a lid and wait.

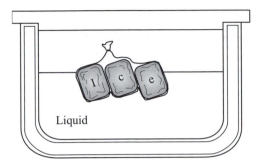

Figure 4.4 Coming to thermal equilibrium. Ice (in a plastic bag) and the surrounding liquid water are the center of our attention. (The air and water vapor above them are inconsequential, except in so far as they provide a pressure of 1 atmosphere.) The thick lid and double-glass-walled container provide isolation from the environment.

Initially, the ice and the liquid water are at different temperatures. As time goes on, the liquid will transfer energy to the ice (by conduction across the plastic bag). The ice warms up, and some of it melts; the water outside the bag cools. Ultimately, an equilibrium is reached in which the ice and melt water in the bag and the water outside the bag have come to the same temperature.

Let us now analyze this process from the perspective of entropy and the Second Law of Thermodynamics. The isolated, composite system will evolve to the macrostate of largest entropy. What does that imply—as a mathematical relationship between the two subsystems?

We will stay on familiar ground if we think of the bag as rigid (but ample in volume) so that the heating occurs at constant volume for the system inside the bag. The total entropy S_{total} is a sum of the entropy of the ice and melt water inside the bag plus the entropy of the liquid water outside the bag:

$$S_{\text{total}} = S_{\text{inside}}(E_{\text{inside}}) + S_{\text{outside}}(E_{\text{outside}}). \tag{4.17}$$

(Section 2.6 established the additivity of entropy for macroscopic systems.) Each entropy is a function of the corresponding energy, as we infer from the dependence on E that is displayed in equations (4.15) and (4.16). According to the Second Law, energy will be transferred across the plastic until the total system has reached the macrostate of largest entropy. To find the maximum of S_{total} subject to conservation of energy, we must incorporate the condition

$$E_{\text{total}} = E_{\text{inside}} + E_{\text{outside}} = \text{constant}. \tag{4.18}$$

Therefore we first write E_{outside} in terms of E_{total} and E_{inside}:

$$S_{\text{total}} = S_{\text{inside}}(E_{\text{inside}}) + S_{\text{outside}}(E_{\text{total}} - E_{\text{inside}}).$$

Then we set the derivative with respect to E_{inside} equal to zero:

$$\frac{\partial S_{\text{total}}}{\partial E_{\text{inside}}} = \frac{\partial S_{\text{inside}}}{\partial E_{\text{inside}}} + \frac{\partial S_{\text{outside}}}{\partial E_{\text{outside}}} \frac{\partial (E_{\text{total}} - E_{\text{inside}})}{\partial E_{\text{inside}}} = 0.$$

(Partial derivatives appear because the entropy depends on variables other than the energy, such as the number of water molecules, although those variables are not explicitly displayed here. The chain rule enables one to differentiate $S_{outside}$ with respect to its entire energy argument and then differentiate that argument with respect to E_{inside}.) The last derivative equals -1. Thus the maximum entropy is reached when the total energy has been distributed such that

$$\frac{\partial S_{inside}}{\partial E_{inside}} = \frac{\partial S_{outside}}{\partial E_{outside}}. \tag{4.19}$$

In short, this equation states the condition for thermal equilibrium, and it must imply equality of temperatures inside and outside the bag.

Indeed, if we look back to a key result from chapter 2, namely

$$\Delta S = \frac{\left(\begin{array}{c} \text{energy input} \\ \text{by heating} \end{array}\right)}{T} = \frac{1}{T} \times \left(\begin{array}{c} \text{energy input} \\ \text{by heating} \end{array}\right), \tag{4.20}$$

provided the process occurs slowly, we see that $1/T$ gives the rate at which entropy changes as energy is transferred by heating. We achieve both consistency and generality if we say that the absolute temperature T is *always* given by the equation

$$\frac{1}{T} = \left(\frac{\partial S}{\partial E}\right)_{\text{fixed external parameters}}. \tag{4.21}$$

The partial derivative is to be taken while external parameters (such as volume or magnetic field) are held fixed and while the number of particles is kept constant. To avoid a clumsy notation, we will sometimes abbreviate the statement as

$$\frac{1}{T} = \left(\frac{\partial S}{\partial E}\right)_V \tag{4.22}$$

displaying only the subscript V for "fixed volume." The relationship (4.21) is to hold for any physical system that is in thermal equilibrium. The benefits of this generalization are many, as we will see.

First, however, note the following: if we use the relation (4.21) in equation (4.19), then the latter becomes

$$\frac{1}{T_{inside}} = \frac{1}{T_{outside}}.$$

Maximum entropy for the composite system implies that the temperatures of the macroscopic component subsystems have become equal.

Defining temperature: the sequence of steps

Let me recapitulate the critical steps that led us to the generalized definition of temperature. Section 1.1 characterized temperature as "hotness measured on some definite scale. That is, the goal of the 'temperature' notion is to order objects in a sequence according to their 'hotness' and to assign to each object a number—its temperature—that will facilitate comparisons of 'hotness.'" In section 1.2, the absolute temperature was defined *provisionally* as "what one gets by adding 273.15 to the reading on a mercury thermometer that is calibrated in degrees Celsius." That definition suffices for only a modest range of temperatures, but it is enough to give operational meaning to all the factors in the ideal gas law, $P = (N/V)kT$, over that range. The ideal gas law played a central role in chapter 2, where we connected change in entropy (and change in multiplicity) with energy input by heating, as displayed in (4.20). Whenever the provisional definition of the absolute temperature is experimentally realistic, the partial derivative, $(\partial S/\partial E)_V$, will yield 1 over that temperature. Hence the derivative definition connects seamlessly with the provisional definition and with the original "measure of hotness" concept. The route that we took builds from the familiar to the abstract, and therein lies its merit.

In the next development, William Thomson plays a major role, and so an interlude about him is in order. In 1846, at age 22, William Thomson became Professor of Natural Philosophy at Glasgow College, the undergraduate college of the University of Glasgow, Scotland. He served in that capacity for 53 years, many times refusing offers from institutions as prestigious as the University of Cambridge. Thomson's work focused primarily on thermodynamics and electromagnetism, but his research papers, numbering more than 300, bore on virtually every topic that physicists studied in the latter half of the nineteenth century. Thomson so greatly improved the submarine cable and the signaling apparatus that transoceanic telegraphy became a practical commercial operation. This brought him public fame, wealth, and a knighthood. In 1892, he was raised to the peerage as Baron Kelvin of Largs, the first British scientist to enter the House of Lords (on merit alone), and so we know him today primarily as "Lord Kelvin." In those days, Largs was merely a village on the Firth of Clyde, some 40 kilometers from Glasgow; Thomson had his country home there. The River Kelvin ran through Glasgow and past the university. Thomson chose a title that would reflect his long and heartfelt association with the city of Glasgow and its university. A modest and kindly man all his long life, he was an encouraging mentor to his many students.

Now we go on and consider other ways to implement the basic "measure of hotness" notion. In section 3.1, we found that, given two heat reservoirs at different temperatures, the efficiency of a Carnot cycle run between those reservoirs is independent of the working substance, be it diatomic nitrogen, air, fluorocarbons, and so on. In 1848, William Thomson suggested that one could use the Carnot cycle as a thermometer, at least in principle. In the modern version of his proposal, one reservoir would be a mixture of ice, liquid water, and pure water vapor when all three coexist in thermal equilibrium. This state is called the *triple point of water*; experiments with a mercury thermometer indicate that the three constituents coexist at only one temperature. The temperature of the triple point would be chosen freely, and the value adopted is 273.16 K, purely a matter of definition. To define and determine the temperature of a hotter reservoir, one would first measure the efficiency of a Carnot cycle run between the two reservoirs. Then one would use the relationship

$$\begin{pmatrix} \text{efficiency of} \\ \text{Carnot cycle} \end{pmatrix} = 1 - \frac{T_{\text{triple point}}}{T_{\text{hot}}}, \qquad\qquad (4.23)$$

which is extracted from (3.4), to determine the numerical value of T_{hot}. The beauty of Thomson's idea lies in this: the definition will yield the same value for T_{hot} regardless of what the working substance in the Carnot engine is. In this sense, the scale is "absolute:" it is independent of substance or system.

Let me illuminate the last point: the independence. After graduating from the University of Cambridge, Thomson worked in Victor Regnault's lab in Paris. There he came to appreciate accurate thermometry, and he also confronted the practical problem of constructing fine thermometers and calibrating them. To illustrate the conceptual problem, consider two familiar thermometers: mercury in glass and alcohol in glass. Suppose you calibrate each in the old Celsius fashion: set zero at the melting point of ice and set 100 at the boiling point of water (both calibrations being at atmospheric pressure); then mark off the intervening length into 100 intervals of equal length. If the mercury thermometer reads 40 in some warm water, will the alcohol thermometer also read 40? For most practical purposes, yes, but to all decimals, no. Why? Because no two liquids expand with temperature in precisely the same way, not even after allowance for a proportionality constant. Then, in principle, which thermometer—if either—should one believe? Kelvin's use of the Carnot cycle adroitly circumvents the question.

You may wonder, of course, how useful is a definition such as Thomson's? Remarkably useful—not directly, to be sure, but indirectly. Whenever the behavior of a physical system depends on temperature—for example, the vapor pressure of liquid helium—one can construct a chain of reasoning that links the behavior to temperature as defined by equation (4.23). Some links in the chain are experimental; others are theoretical. Many examples appear throughout this text; a preview of one will suffice here.

The hot metals in both a kitchen oven and a smelting furnace emit electromagnetic waves with a continuous spectrum of frequencies. The intensity (as a function of frequency) has a maximum at the frequency $\nu_{\text{max}} = 2.82 kT/h$, a result derived in section 6.2. If the experimental maximum is found to occur at the frequency $\nu_{\text{max}} = 1.82 \times 10^{14}$ Hz, say, then the temperature T is $T = h\nu_{\text{max}}/(2.82 k) = 3{,}100$ K. Was a Carnot engine employed? Not directly. Rather, the theoretical expression for ν_{max} is calculated from a probability distribution (derived in section 5.2) that incorporates temperature via equation (4.21): $1/T = (\partial S/\partial E)_V$. In turn, that definition of temperature is entirely consistent with temperature as defined by the Carnot efficiency, equation (4.23).

The equations (4.21) and (4.23) give a broad, device-independent definition of the quantitative aspect of temperature. They provide the framework that ensures consistency among a myriad ways of measuring actual temperatures in the lab, in astrophysics, and in geophysics.

Incidentally, you may wonder why the number 273.16 appears in the modern version of Thomson's proposal, but 273.15 appeared in our provisional definition of absolute temperature. The difference is not just sloppiness or a typographical error. On the Celsius scale of the nineteenth century, the temperature of ice and liquid water, when coexisting at atmospheric pressure, was assigned the value zero. The triple point of water, however, is the state in which ice, liquid water, and pure water vapor coexist at whatever pressure pure water vapor produces at mutual coexistence (which happens to be a mere 0.006 atmosphere). On the old Celsius scale, the triple point of water was found to be approximately 0.01 °C. Thus, to adequate approximation for our previous purposes, adding 273.15 to the value 0.01 °C read from a mercury thermometer yields the 273.16 K that is, by definition, the absolute temperature of water's triple point.

There is yet another facet to these numbers. The Celsius scale of the nineteenth century assigned zero to the melting point of ice and 100 to the boiling point of water (when the system is at atmospheric pressure). In 1989, the International Committee on Weights and Measures adopted the "International Temperature Scale of 1990" and redefined the Celsius scale to be

temperature on Celsius scale $\equiv$ temperature on Kelvin scale $- 273.15$.   (4.24)

The definition makes temperature on the Celsius scale independent of the material or system that is used as a thermometer. On the new Celsius scale, however, the melting point of ice and the boiling point of water (at atmospheric pressure) differ from zero and 100 in distant decimal places; they are approximately 0.0002 °C and 99.974 °C.

Individual versus relational

Our original definition of temperature (in section 1.1) focused on comparison among objects. To repeat yet again, "the goal of the 'temperature' notion is to order objects in a sequence according to their 'hotness' and to assign to each object a number—its temperature—that will facilitate comparisons of 'hotness.'" The focus is on a relational aspect of nature.

It would be nice to have, also, a conception of temperature that focuses on the physical system itself. For this, the key is the rate at which the system's entropy changes when its energy is changed: $(\partial S/\partial E)_V$. A rate of change can be an extremely fundamental notion. Witness Newtonian mechanics: the rate of change of velocity, $d\mathbf{v}/dt$, is central. Once we give $d\mathbf{v}/dt$ its own name and symbol—the acceleration, $\mathbf{a}$—it seems to take on a life of its own. So it is with temperature. In the equation $1/T = (\partial S/\partial E)_V$, the left-hand side is familiar; the right-hand side, relatively foreign. Yet one should develop a sense that the rate of change on the right-hand side is the truly fundamental notion; via a reciprocal, the left-hand side gives the rate its own symbol and name.

The rate $(\partial S/\partial E)_V$ serves also a comparative function. Recall that the Second Law of Thermodynamics implies increasing multiplicity and hence increasing entropy (in the context of evolution in isolation). Thus, when two systems are placed in thermal contact (but are otherwise isolated), the system with the greater rate $(\partial S/\partial E)_V$ will gain energy (by heating) from the other (so that the total entropy will increase). The system with the greater value of $(\partial S/\partial E)_V$ will be the colder system, and so $(\partial S/\partial E)_V$ provides a way to order systems according to their hotness. Of course, this paragraph largely recapitulates the ice-and-water scene that introduced this section, but temperature is a surprisingly subtle notion, and so a little repetition is a good thing.

Hierarchy

You may wonder, do we have three different quantitative definitions of temperature? No. Where their ranges of applicability overlap, the definitions are equivalent and mutually consistent. The provisional definition of section 1.2, namely, "For now ... we may regard T as simply what one gets by adding 273.15 to the reading on a mercury thermometer that is calibrated in degrees Celsius," is the least general. When mercury freezes or boils, the mercury thermometer is worthless, and one would not want to use it even close to either extreme. The Kelvin definition, displayed in equation (4.23), is applicable to all positive absolute temperatures and sufficed for all the physics of the nineteenth century. The entropy derivative definition, presented in equation (4.21), is the most general. It agrees with the Kelvin definition at all positive temperatures, and—in chapter 14—it will give meaning to even a negative absolute temperature.

4.4   Essentials

1. A sum over single-particle states may (often) be replaced by an integral:

$$\sum_{\text{states } \varphi_\alpha} A(\varepsilon_\alpha) = \int A(\varepsilon)D(\varepsilon)\,d\varepsilon,$$

where the density of single-particle states $D(\varepsilon)$ is given by

$$\left(\begin{array}{c} \text{number of single-particle states} \\ \text{per unit energy interval near energy } \varepsilon \end{array} \right) \equiv D(\varepsilon) = \frac{2\pi(2m)^{3/2}}{h^3}V\varepsilon^{1/2}$$

for a spinless particle in non-relativistic motion.

2. The system's entropy may be expressed in terms of $D(E)$, the density of states for the entire system:

$$S = k\ln[D(E)\delta E].$$

3. The general quantitative definition of (absolute) temperature is

$$\frac{1}{T} = \left(\frac{\partial S}{\partial E} \right)_{\text{fixed external parameters}}.$$

4. William Thomson's definition of absolute temperature uses the efficiency of a Carnot cycle and the assigned temperature of the triple point of water. For example, the relationship

$$\left(\begin{array}{c}\text{efficiency of} \\ \text{Carnot cycle}\end{array}\right) = 1 - \frac{T_{\text{triple point}}}{T_{\text{hot}}}$$

determines the temperature T_{hot} of the hot reservoir when the cold reservoir is water at its triple point. The merit of Thomson's prescription is this: because the efficiency is independent of the working substance, the "thermometer" gives a reading that is independent of its material constituents.

Problems

1. Consider a helium atom in a volume of one liter: $V = 10^{-3}$ m^3.

(a) How many single-particle energy eigenstates are there in the energy range $0 \leqslant \varepsilon \leqslant 1/40$ electron volt? (You may need to convert the energy to SI units.)
(b) How many such states are in the range $0.025 \leqslant \varepsilon \leqslant 0.026$ electron volt?

2. A Carnot cycle operates with one reservoir at the triple point of water and the other reservoir at the boiling point of liquid oxygen (under atmospheric pressure). If the measured efficiency is 0.67, at what temperature does liquid oxygen boil?

3. *Thermal paradox*. Consider a Carnot cycle (or engine) in which the hot and cold reservoirs consist of large volumes of the same material (water or iron, say). The hot reservoir has a higher ratio of entropy to mass than does the cold reservoir. Consequently, one might think that the energy in the hot reservoir is less available for conversion to work than the energy in the cold reservoir. Nonetheless, energy in the hot reservoir can be turned into work (at least partially) although (in the present context) the energy in the cold reservoir cannot. How can one resolve this paradox?

4. *Estimating $D(E)$*.

(a) Equation (2.16) gave the entropy of a monatomic classical ideal gas—that is, the dependence on T and V. Eliminate the temperature in terms of the system's total energy E and then infer the dependence of the density of states $D(E)$ on E and V. It would be reasonable to assume that the energy range δE is proportional to E.
  The "constant" in equation (2.16) may depend on the number N of atoms, and so this route does *not* enable you to infer the *entire* dependence of $D(E)$ on N.
(b) With sketches, compare the behavior of $D(E)$ as a function of E for two cases: $N \gg 1$ and $N = 1$.

5. Provide a succinct and accurate definition of each of the following terms. Where equivalent definitions exist, give all of them (up to a maximum of three).

(a) Entropy
(b) Second Law of Thermodynamics (verbal form)
(c) Energy input by heating
(d) External parameters
(e) Temperature

5 The Canonical Probability Distribution

5.1 Probabilities

5.2 Probabilities when the temperature is fixed

5.3 An example: spin $\frac{1}{2}\hbar$ paramagnetism

5.4 The partition function technique

5.5 The energy range δE

5.6 The ideal gas, treated semi-classically

5.7 Theoretical threads

5.8 Essentials

If we know the temperature of a system and the values of its external parameters, how can we estimate its physical properties, such as energy, pressure, magnetic moment, and distribution of molecular velocities? The question is answered in this chapter: we derive the canonical probability distribution, learn some techniques for applying it efficiently, and work out two major examples.

5.1 Probabilities

Probabilities enter into thermal physics because the available data are insufficient to determine the individual properties of 10^{20} molecules. Moreover, even if such finely detailed data were available, no person or computer could cope with it. Out of necessity, one turns to a statistical analysis. A handful of data and some plausible reasoning lead to predictions whose success is nothing short of astonishing.

Before we look at what a "probability" means, we should note that probabilities always arise in a context. For example, the probability of a 4 appearing, *given that* I roll a die once, is $1/6$. The probability of a four appearing, if I were to count the number of letters that come in my daily mail, would be quite different. Sometimes the context is made explicit; at other times, it is left implicit; but always there is a context.

In a broad view of the subject, there are two distinct schools of thought on the question, what should the word "probability" mean?

1. Frequency meaning. According to the frequency school, a probability is a relative frequency in the long run, that is,

$$probability = \frac{number\ of\ successes}{number\ of\ tries}. \tag{5.1}$$

If one were rolling dice and looking for the appearance of 4s, one could use the ratio—after many rolls—to assess the probability. Similarly, a probability could be the relative frequency in a large collection of objects, that is,

$$probability = \frac{number\ of\ objects\ with\ property\ X}{total\ number\ of\ objects}. \tag{5.2}$$

The objects might be 10^{20} sodium atoms in a hot vapor, and property X might be the property of being in the first excited state. If 10^{16} atoms are indeed in that state, then the probability would be

$$probability = \frac{10^{16}}{10^{20}} = 10^{-4}.$$

But a quite different meaning can be attached to the word "probability," as follows.

  2. Degree of belief meaning. According to the degree of belief school, a probability is the rational degree of belief in the correctness of proposition A, given the context B. To construct an example, let us take proposition A to be the statement, "The first sodium atom I examine will be in the first excited state." Let the context B be the data in the preceding paragraph: 10^{20} sodium atoms in a hot vapor, of which 10^{16} atoms are in the first excited state. Then the rational degree of belief in proposition A, given context B, would be assigned the value 10^{-4}.

  Rational degrees of belief can be assigned numerical values. The scale for degree of belief goes from 0 (for zero degree of belief) to 1 (for complete conviction). As a general rule, whenever a situation permits one to construe a probability as a relative frequency, the degree of belief school will assign the same numerical value to what it calls the probability.

  Among the advocates of a degree of belief interpretation have been the economist John Maynard Keynes, the geologist Sir Harold Jeffreys, and the physicists Erwin Schrödinger, Richard T. Cox, and Edwin T. Jaynes. One can find other physicists who adopt the view at least occasionally and at least implicitly. For example, Hans Bethe, who was head of the Theoretical Division at Los Alamos during the Second World War, wrote the following in a letter to the *New York Times*, published on 28 February 1971:

> By February 1945 it appeared to me and to other fully informed scientists that there was a better than 90 percent probability that the atomic bomb would in fact explode, that it would be an extremely effective weapon, and that there would be enough material to build several bombs in the course of a few months.

  In Bethe's statement we can see an advantage of the degree of belief view, for surely there was only one first attempt to explode an atomic bomb, not 10^{20} attempts nor a long run of attempts. In thermal physics, one often wants to estimate the behavior of a *specific* system in the lab, for example, the lithium fluoride crystal that was grown last week and resides now in the core of the lab's only superconducting magnet. The degree of belief interpretation permits such "one of a kind" applications of probability theory. The frequency interpretation would require that one imagine a large number of replicas of last week's crystal and consider a relative frequency in that collection. Or it would require that one consider doing the experiment on the same crystal again and again. Some physicists find such constructions to be artificial. Nevertheless, the frequency view has been espoused for thermal physics by many eminent practitioners. Perhaps the most prominent of them was the American physicist J. Willard Gibbs, one of the founders of the entire subject, although it must be noted that the degree of belief school hardly existed as an alternative when Gibbs did his work (in the last decades of the nineteenth century).

Fortunately for us, almost all calculations in thermal physics come out the same, regardless of which interpretation of "probability" one adopts. Certainly the computations in this book all come out the same. Readers may make their own choices. Some annotated references to the two schools of thought appear at the chapter's end. Moreover, appendix C develops a framework for probability theory, primarily to provide some familiarity with a powerful system of inductive reasoning. Here it suffices to note that the rules for working symbolically or numerically with probabilities are independent of which meaning one has in mind. In short, there is only one set of rules.

5.2   Probabilities when the temperature is fixed

The goal in this section is simply stated: to describe, in probabilistic terms, a system whose temperature is fixed. An example from low temperature physics is illustrated in figure 5.1. The sample, perhaps a bit of cerium magnesium nitrate, is in good thermal contact with a relatively large copper disk, which we will call the "reservoir." The copper disk establishes and maintains the temperature of the sample. We regard the sample and reservoir as isolated from the environment and as sharing a total energy E_{tot}.

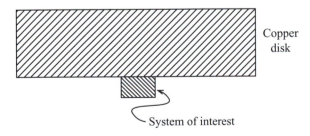

Figure 5.1 A side view of the sample, which is the system of interest, and the reservoir: a copper disk. The copper was cooled by liquid helium, which was pumped on (for additional evaporative cooling) until all the helium had evaporated. Both sample and disk are isolated from the environment (to adequate accuracy) in a cryostat: a double-walled stainless steel vessel (which is not shown).

The energy eigenstates of the sample (regarded as isolated from even the reservoir) are denoted by Ψ_j and have energy E_j. Although there is thermal contact with the reservoir and hence interaction, those states remain the best states that we can use to describe the thermal equilibrium of the sample. Let $P(\Psi_j)$ denote the probability that Ψ_j is the right state to use to describe the sample. We reason that $P(\Psi_j)$ is proportional to the multiplicity of the reservoir when it has energy $E_{tot} - E_j$:

$$P(\Psi_j) = \text{const} \times \left(\begin{array}{c} \text{multiplicity of reservoir when} \\ \text{it has energy } E_{tot} - E_j \end{array} \right). \tag{5.3}$$

The more ways to realize the state Ψ_j and *some* microstate of the reservoir, the larger the probability $P(\Psi_j)$ should be. In more detail, every joint state of sample and reservoir (that is allowed by energy conservation) must be assigned the same probability because there is no justification for any preferential treatment. For any chosen sample state Ψ_j and its energy E_j, the number of states of the reservoir is given by the reservoir's multiplicity, evaluated at energy $E_{tot} - E_j$. Thus the probability $P(\Psi_j)$ must be proportional to that multiplicity.

Note that, on the right-hand side of (5.3), the only dependence on Ψ_j or E_j is through the E_j that appears in $E_{tot} - E_j$. In particular, the factor "const" is independent of the index j.

The reservoir's entropy S_{res} is Boltzmann's constant k times the logarithm of its multiplicity, and we can productively express the multiplicity in terms of that entropy:

$$P(\Psi_j) = \text{const} \times \exp\left[\frac{1}{k} S_{res}(E_{tot} - E_j) \right]. \tag{5.4}$$

Because the reservoir is large relative to the sample, the argument $E_{tot} - E_j$ will remain close to E_{tot} for the sample states of primary interest. We can afford to make a Taylor series expansion about the value E_{tot}, retaining only the first two terms:

$$S_{res}(E_{tot} - E_j) = S_{res}(E_{tot}) + \left. \frac{\partial S_{res}}{\partial E} \right|_{E_{tot}} \times (-E_j)$$

$$= S_{res}(E_{tot}) - \frac{1}{T} E_j. \tag{5.5}$$

Section 4.3 showed that the derivative $\partial S_{res}/\partial E$ is $1/T$. (If you worry about our truncating the Taylor series so abruptly, problem 6 will help you to work out the justification.) When we insert the relationship (5.5) into equation (5.4), we can combine all factors that are independent of E_j into a new proportionality constant, "new constant," and write

$$P(\Psi_j) = (\text{new constant}) \times \exp(-E_j/kT). \tag{5.6}$$

Calculating with the reservoir's entropy has the great benefit that it introduces the temperature T in a simple, direct fashion. Moreover, the entropy, which is proportional to the logarithm of the multiplicity, varies much more slowly with energy than does the multiplicity itself. A truncated Taylor series for the entropy is justifiable, but such a step would not be acceptable for the multiplicity itself.

To determine the value of "new constant," we note that the probabilities, when summed over all energy eigenstates, must yield the value 1:

$$\sum_j P(\Psi_j) = 1. \tag{5.7}$$

(Note. The technical meaning of "all" energy eigenstates is that one should sum over a complete orthonormal set of energy eigenstates.) Now think of summing both sides of equation (5.6) over all states. The sum on the left-hand side should yield the value 1. Thus, on the right-hand side, "new constant" must be the reciprocal of a sum over all the exponentials. The upshot is the final form,

$$P(\Psi_j) = \frac{\exp(-E_j/kT)}{\sum_{j'} \exp(-E_{j'}/kT)}. \tag{5.8}$$

(The index j' in the denominator is distinct from the index j in the numerator and runs over all states.) This probability distribution, perhaps the most famous in all of thermal physics, is called the *canonical probability distribution*, a name introduced by J. Willard Gibbs in 1901. (The adjective "canonical" is used in the sense of "standard." In his *Elementary Principles in Statistical Mechanics*, Gibbs wrote, "This distribution, on account of its unique importance in the theory of statistical equilibrium, I have ventured to call *canonical*.") The numerator itself, $\exp(-E_j/kT)$, is called the *Boltzmann factor*. The denominator occurs often, is surprisingly useful, and has its own name and symbol. Max Planck called it the *Zustandsumme*, which means "sum over states", and the German name provides the symbolic abbreviation:

$$Z \equiv \sum_{j'} \exp(-E_{j'}/kT). \tag{5.9}$$

The common English name is the *partition function*, and that is what we will call the sum. Indeed, for future reference, the canonical probability distribution now takes the concise form

$$P(\Psi_j) = \frac{\exp(-E_j/kT)}{Z}. \tag{5.10}$$

Qualitatively stated, the probability of high-energy states is exponentially smaller than the probability of low-energy states. (The statement presumes that the temperature T is positive, which is usually the case, but chapter 14 will provide exceptions.)

Our derivation stipulated a context where the sample's temperature is *fixed* by a reservoir. What about a situation where a system of interest is isolated, but one *knows* its temperature? An example might be a can of lemonade, just removed from the refrigerator, which was set for 4 °C (equivalent to 277 K). In almost every way, nothing physical would change if the system were put in thermal contact with a reservoir whose temperature was the same as the sample's. To continue the analogy, one could put the lemonade can back in the fridge. Because the canonical probability distribution would be applicable if contact with a reservoir existed, it must be applicable even when the system is isolated, provided we know the system's temperature.

5.3   An example: spin $\frac{1}{2}\hbar$ paramagnetism

An example is in order. For simplicity's sake, we consider a single atom whose spin is $\frac{1}{2}\hbar$ (where $\hbar \equiv h/2\pi$) and whose location is absolutely fixed in some crystal lattice; motion relative to a lattice site is the subject of a later chapter. [The compound cesium titanium alum, $CsTi(SO_4)_2 \cdot 12H_2O$, provides a good example. The effective magnetic moment comes from the single titanium ion, which has a net angular momentum of $\frac{1}{2}\hbar$.] The atom's magnetic moment $\mathbf{m_B}$ arises from the net angular momentum of $\frac{1}{2}\hbar$. (Figure 5.2 reviews the idea of a magnetic moment.) The magnitude of $\mathbf{m_B}$ is

$$\mathbf{m_B} \equiv \frac{e\hbar}{2m_e} = 9.274 \times 10^{-24} \text{ joules per tesla,} \qquad (5.11)$$

where e is the magnitude of the electronic charge and m_e denotes the electron's rest mass. The quantity $\mathbf{m_B}$ is called the *Bohr magneton*. The atom is in an external

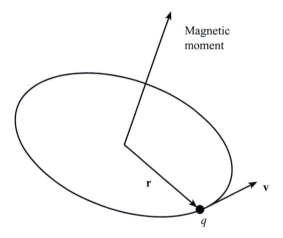

Figure 5.2 A classical orbital magnetic moment. A positive charge q moves in uniform circular motion with speed v around a circle of radius r. The number of round trips per second is $v/2\pi r$. An observer stationed on the circle would note an (average) electric current of $q \times (v/2\pi r)$. By definition, the associated magnetic moment has the following magnitude:

$$\text{(magnetic moment)} = \text{(current)(area enclosed)} = \frac{qv}{2\pi r}\pi r^2$$

$$= \frac{q}{2m}(mvr) = \frac{q}{2m}\text{(angular momentum)}.$$

The vectorial magnetic moment is $(q/2m)$ times the orbital angular momentum, where now q may be positive or negative.

magnetic field **B**. Our intention is to estimate the component of magnetic moment along **B** when the atom is part of a crystal in thermal equilibrium at temperature T.

Because we focus strictly on the atom's spin, the "system" has only two energy eigenstates. Table 5.1 displays the orientation of the magnetic moment, the component of magnetic moment along **B**, and the energy for each of the states. The associated probabilities are the following, where $\parallel$ denotes "parallel orientation:"

$$P_{\parallel} = \frac{e^{+m_B B/kT}}{Z},$$
(5.12a)

$$P_{\text{anti-}\parallel} = \frac{e^{-m_B B/kT}}{Z}.$$
(5.12b)

The state of lower energy is favored.

We estimate the magnetic moment by first weighting each possibility by the probability of occurrence, $P(\Psi_j)$, and then summing. Working from the table, we find

$$\left\langle \begin{array}{c} \text{magnetic moment} \\ \text{along } \mathbf{B} \end{array} \right\rangle = \sum_{j=1}^{2} \left(\begin{array}{c} \text{value of } \mathbf{m_B} \cdot \hat{\mathbf{B}} \\ \text{in state } \Psi_j \end{array} \right) P(\Psi_j)$$

$$= m_B \frac{e^{m_B B/kT}}{Z} + (-m_B) \frac{e^{-m_B B/kT}}{Z}$$
(5.13)

$$= m_B \tanh(m_B B/kT).$$

The step to the last line follows because the partition function is simply

$$Z = e^{m_B B/kT} + e^{-m_B B/kT} = 2\cosh(m_B B/kT).$$
(5.14)

Whenever we estimate a quantity by weighting each possible value by the probability of occurrence and then summing, angular brackets will denote the process and the outcome. The outcome is called an *expectation value* or an *expectation value estimate*, and the underlying idea is discussed further in appendix C. For now, it suffices to note that the present use of angular brackets is analogous to forming an average, a use that was introduced in section 1.2. Context will suffice to distinguish the two meanings.

Figure 5.3 displays the estimate both as a function of temperature at fixed field and also as a function of field at fixed temperature.

Table 5.1 *For the energy of a magnetic moment $\mathbf{m_B}$ in a field $\mathbf{B}$, the classical vectorial expression is $-\mathbf{m_B} \cdot \mathbf{B}$. The vector $\hat{\mathbf{B}}$ is a unit vector in the direction of $\mathbf{B}$.*

State index j	Orientation of moment relative to **B**	Component of moment along **B**: $\mathbf{m_B} \cdot \hat{\mathbf{B}}$	System energy E_j
1	Parallel	m_B	$-m_B B$
2	Anti-parallel	$-m_B$	$+m_B B$

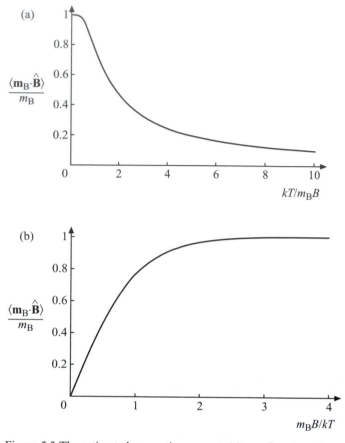

Figure 5.3 The estimated magnetic moment: (a) as a function of temperature at fixed field; (b) as a function of field at fixed temperature. When the ratio $m_B B/kT$ is large, parallel alignment is strongly favored.

Later we will learn that the same curves would emerge if we studied simultaneously all N paramagnets in a macroscopic sample. The curves would display the total magnetic moment along **B** relative to Nm_B. Moreover, the estimates would be worthy of great confidence on our part, for the statistical fluctuations would be small. Problem 5 offers routes to substantiating these claims.

The positive values in equation (5.13) and in figure 5.3 show that the magnetic moment tends to be aligned along **B**. This is, indeed, the origin of the adjective "paramagnetic:" working in the laboratory with a macroscopic sample, one finds that the total magnetic moment is aligned *parallel* to the external magnetic field.

5.4   The partition function technique

We have seen one example of how to use the canonical probability distribution to estimate a physical quantity. For many quantities of interest, such as energy or pressure, the partition function provides an especially efficient way to make the estimate. The present section develops the technique.

Energy

To estimate the energy, we start by writing

$$\langle E \rangle = \sum_j E_j P(\Psi_j) = \frac{1}{Z} \sum_j E_j \exp(-E_j/kT). \tag{5.15}$$

We can express the summand in terms of the derivative of the exponential with respect to T. Using the chain rule, we find

$$\frac{\partial \exp(-E_j/kT)}{\partial T} = \frac{1}{kT^2} E_j \exp(-E_j/kT).$$

Multiply the equation by kT^2 and then sum all over all j to find the intermediate result

$$kT^2 \frac{\partial Z}{\partial T} = \sum_j E_j \exp(-E_j/kT).$$

Using this relation to simplify equation (5.15) yields

$$\langle E \rangle = \frac{1}{Z} kT^2 \frac{\partial Z}{\partial T}$$

$$= kT^2 \frac{\partial \ln Z}{\partial T}. \tag{5.16}$$

Thus, to estimate the energy, all that one needs is the temperature dependence of $\ln Z$.

Pressure

Section 4.1 showed that the energy eigenvalue for a single particle depends upon the volume V of the container. Some dependence on volume will always occur for a confined system, and that dependence enables us to compute the pressure, as follows.

Focus attention on a specific state Ψ_j and slowly expand the volume by an infinitesimal amount ΔV; do that adiabatically. The system exerts a pressure on the walls and does work during the expansion; simultaneously, the system's energy decreases. Energy conservation requires that the amount of work done and the change in the system's energy sum to zero:

$$\left(\begin{array}{c} \text{pressure} \\ \text{in state } \Psi_j \end{array} \right) \Delta V + \frac{\partial E_j}{\partial V} \Delta V = 0, \tag{5.17}$$

and so the pressure in state Ψ_j is given by $-\partial E_j/\partial V$. Then the canonical probability distribution estimates the pressure P as

$$P = \sum_j -\frac{\partial E_j}{\partial V} \frac{\exp(-E_j/kT)}{Z}. \tag{5.18}$$

[The letters p and P necessarily get heavy use in physics. In this book, the symbol P for pressure carries no argument (except in a few homework problems). In contrast, the symbol $P(\Psi_j)$ for a probability always has an argument or a subscript accompanying the letter. This convention, together with context, should keep the meanings clear.]

Again the partition function provides a succinct re-expression. Start with the relation

$$\frac{\partial \exp(-E_j/kT)}{\partial V} = -\frac{1}{kT}\frac{\partial E_j}{\partial V}\exp(-E_j/kT);$$

multiply by kT and sum over j; insert into equation (5.18), thereby deriving the result

$$P = \frac{kT}{Z}\frac{\partial Z}{\partial V}$$
$$= kT\frac{\partial \ln Z}{\partial V}. \tag{5.19}$$

Now all that one needs is the volume dependence of $\ln Z$.

A few more words in justification of this approach to calculating the pressure are in order. Early in our education, we learn to think of gas pressure as being caused by molecular impacts with the container walls. This is an excellent—and true—picture; without it, much of the real physics in the lab would remain a mystery. It was this image of "little spheres in irregular motion" that we used successfully in the kinetic theory calculation in section 1.2. The desirability of a different point of view—a work–energy analysis of pressure—arises for two reasons. (1) Quantum theory does not lend itself to visualizing trajectories of molecules. (2) One may want to include mutual interactions among the molecules, but intermolecular forces are difficult to handle in a framework based on the kinetic theory's view of a gas. The average effects of intermolecular forces, as they modify a macroscopic quantity like pressure, can be taken into account more accurately and more consistently if one adopts the present work–energy approach. Energy arguments tend to hide the details of a process, but sometimes that is a great help.

Other external parameters

If you return to section 5.3 and evaluate the expression $kT\partial \ln Z/\partial B$, you will find that it gives the system's magnetic moment (along the direction of the external magnetic field **B**):

$$\left\langle \begin{array}{c} \text{magnetic moment} \\ \text{along } \mathbf{B} \end{array} \right\rangle = kT\frac{\partial \ln Z}{\partial B}. \tag{5.20}$$

The relationship between moment and $\ln Z$ is true for any magnetic system, no matter how large the individual spins may be, how many spins there are, or what mutual interactions they may undergo. (References to a derivation are provided at the end of the chapter.) A reason for such a relationship can be found, as follows.

In deriving the expression for pressure, we studied the response of an energy eigenvalue to the volume, which is an external parameter. An energy eigenvalue may depend on other external parameters, such as magnetic or electric fields. In general, if one investigates the response of an energy eigenvalue to a infinitesimal change in an external parameter, something identifiable and useful emerges. This result is physically reasonable, for an external parameter provides the experimentalist with an opportunity to "prod" the system in a gentle and controllable fashion. The response is bound to reflect the characteristic properties of the system. For a further example, using an external electric field in place of the magnetic field will generate an estimate of the system's electric dipole moment.

5.5 The energy range δE

In section 4.2, I promised to specify more precisely the energy range δE that appears in the quantum multiplicity:

$$S = k \ln[D(E)\delta E].$$ (5.21)

Now that we have some familiarity with the partition function, a definite specification is possible (at least for a macroscopic system). When expressed in terms of the density of states, the partition function is

$$Z = \sum_j \exp(-E_j/kT)$$

$$= \int e^{-E/kT} D(E)\, dE.$$ (5.22)

For a macroscopic system, the density of states $D(E)$ is typically a rapidly rising function of the energy. The Boltzmann factor is certainly a rapidly decreasing function of E. Figure 5.4 displays these properties and show that the product—the integrand—will be sharply peaked.

Where along the energy axis does the peak lie? To answer that question, we examine the energy estimate $\langle E \rangle$ when it is expressed in terms of $D(E)$ and Z:

$$\langle E \rangle = \sum_j E_j P(\Psi_j) = \frac{\int E e^{-E/kT} D(E)\, dE}{Z}$$

$$\cong E_{\text{at peak}}.$$ (5.23)

Because the product $e^{-E/kT} D(E)$ is sharply peaked, we may evaluate the first factor in the integrand at the peak; the remaining integral cancels with the denominator. Thus the peak must occur at or near $\langle E \rangle$, and vice versa. This concludes the preliminaries.

The procedure for defining δE rests on two qualitative observations:

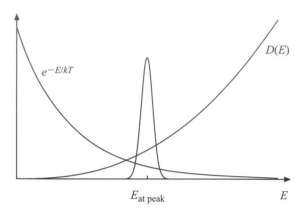

Figure 5.4 Sketches of the Boltzmann factor $\exp(-E/kT)$, the density of states $D(E)$, and their product for a typical macroscopic system. The curves are qualitatively faithful but are plotted on different vertical scales. As the total energy E grows, there are many more distinct ways to distribute energy among the particles, and so the number of distinct quantum states grows rapidly. [For this reason, the behavior of $D(E)$ differs greatly from the gentle rise of $D(\varepsilon)$, the single-particle density of states.]

1. the significant range of energies is set by the width of the sharp peak in figure 5.4;
2. the area under the peak is equal to its height times its width.

The term "width" is used qualitatively here.

Now we define the energy range δE as the width of the peak in figure 5.4 in a certain precise sense. Returning to equation (5.22), we write

$$\int e^{-E/kT} D(E)\, dE = [e^{-E/kT} D(E)]|_{E=\langle E\rangle} \delta E$$

$$= e^{-\langle E\rangle/kT} D(\langle E\rangle)\delta E. \tag{5.24}$$

The integral on the left-hand side gives the area under the curve. On the right-hand side is the product of height and width. Because $E_{\text{at peak}}$ and $\langle E\rangle$ are virtually the same numerically, it matters little whether we evaluate the integrand at its literal peak or at $E = \langle E\rangle$. The latter has some advantages. Equation (5.24) is *not* an approximation. Rather, the equation defines the value of δE and says that, in essence, δE is the width of the peak in figure 5.4.

To fix δE by the requirement (5.24) is plausible and even natural. But is it the uniquely correct choice? The answer is yes, but only in chapter 14 will we have the framework to show that.

Computing entropy efficiently

A convenient expression for the entropy follows readily from (5.24). First take the logarithm of both sides:

$$\ln Z = -\frac{\langle E \rangle}{kT} + \ln[D(\langle E \rangle)\delta E].$$

Multiply by k and rearrange to find

$$S = k\ln[D(\langle E \rangle)\delta E] = \frac{\langle E \rangle}{T} + k\ln Z. \tag{5.25}$$

In section 4.2 and in equation (5.21), the density of states was evaluated at "the system's energy E." The sharply peaked curve in figure 5.4 is a probability distribution for the system's energy E (aside from a normalizing factor). The sharpness of the peak means that, for all practical purposes, we may use the estimate $\langle E \rangle$ and "the system's energy E" interchangeably. Such an interchange was used in equation (5.25) and will be used again, from time to time.

Why is the expression for entropy in terms of $\langle E \rangle /T$ and $\ln Z$ "convenient"? Because computations often start with a calculation of the partition function. Then $\langle E \rangle$ can be gotten from $\ln Z$ by differentiation, and the entire expression for S can be evaluated readily. The next section illustrates this procedure.

5.6 The ideal gas, treated semi-classically

By now we have built a lot of theoretical machinery. This section applies it to an equally substantial problem: a monatomic ideal gas of N identical atoms, where N is realistically large, say, 10^{20}. Recall that the adjective "ideal" means "no mutual forces among the atoms." Nonetheless, coping with all N atoms at once makes the problem challenging—but also rewarding. The phrase "treated semi-classically" means that the calculation falls somewhere between a fully quantum mechanical treatment and a purely classical analysis. We start out with a rigorous quantum framework and then make approximations that are valid when the gas departs only slightly from classical behavior. (A more precise definition of "semi-classical" is provided at the end of section 9.6.)

The partition function

The partition function Z,

$$Z = \sum_j \exp(-E_j/kT), \tag{5.26}$$

is a sum of Boltzmann factors, one factor for each state of the N-particle system. Each energy E_j is a sum of single-particle energies:

$$E_j = \varepsilon_\alpha(1) + \varepsilon_\beta(2) + \varepsilon_\gamma(3) + \cdots . \tag{5.27}$$

A subscript like α specifies the single-particle state, for example, the triplet of integers $\{n_x, n_y, n_z\}$. The argument, such as (1), identifies the specific atom, here atom #1.

The factorization property of an exponential, $e^{a+b} = e^a \times e^b$, means that when we insert the structure (5.27) into equation (5.26), the result may be written as

$$Z = \sum_{\text{states } \Psi_j} e^{-\varepsilon_\alpha(1)/kT} \times e^{-\varepsilon_\beta(2)/kT} \times \cdots. \tag{5.28}$$

(By the way, if you find the steps difficult to follow, try working out an example: take $N = 2$ and suppose that there are only two or three single-particle states. Then you can compute everything explicitly.)

Now, to perform the sum, hold all Greek subscripts except α fixed and sum over all possible values of α. Then do the same with β, and so on. That process generates the intermediate result

$$Z \overset{?}{=} \left(\sum_\alpha e^{-\varepsilon_\alpha(1)/kT} \right) \times \left(\sum_\beta e^{-\varepsilon_\beta(2)/kT} \right) \times \cdots. \tag{5.29}$$

It is a product of N numerically equal factors, each of which is the partition function for a single atom. The latter will be denoted by Z_1:

$$Z_1 \equiv \sum_\alpha e^{-\varepsilon_\alpha/kT}. \tag{5.30}$$

I placed a question mark over the equal sign because the summing procedure needs to be checked. If all the atoms were distinguishable from one another and if there were no such rule as the Pauli exclusion principle, then the intermediate result would be physically correct. But now imagine multiplying out the factors in (5.29) and recombining the exponentials, so that each exponent is a sum of N energies. Consider a combination of energies such that all the associated single-particle states are different, for example,

$$\varepsilon_3(1) + \varepsilon_2(2) + \cdots,$$

in contrast to

$$\varepsilon_3(1) + \varepsilon_3(2) + \cdots,$$

which is also included in the expanded form of (5.29). The "all different" combinations of single-particle states are acceptable because the associated states Ψ_j are quantum mechanically acceptable. Nonetheless, there is a difficulty: the indistinguishability of identical particles means that, in (5.29), energies like

$$\varepsilon_3(1) + \varepsilon_2(2) + \cdots$$

and

$$\varepsilon_2(1) + \varepsilon_3(2) + \cdots$$

are associated with the same quantum state Ψ_j. (In each case, one atom is in state φ_3, and another is in state φ_2.) So we have over-counted actual quantum states by the number of permutations among N different state subscripts, which is $N!$ in value.

If only all-different combinations of single-particle states occurred in the expanded form of (5.29), then we could correct exactly by dividing by $N!$. Other combinations, however, do arise. Either they do not belong in a quantum treatment (because, for example, they violate the Pauli principle), or they are acceptable but are not over-counted by as much as $N!$. The spirit of a semi-classical treatment is to ignore such errors and distinctions. As we will see later, a semi-classical picture can be valid only if the temperature is not too low (in a sense to be specified). When the temperature is adequately high, the exponentials in equation (5.28) do not cut off the sum until enormously many single-particle states have been included. Consequently, both Z_1 and Z are extremely large numbers. Moreover, for our purposes, it suffices to know the logarithm of the partition function. A logarithm is remarkably insensitive to substantial changes in its argument. For example, $\log(10 \times 10^{100}) \cong \log(10^{100})$, to within 1 percent. Because the partition function is so large and because we need only its logarithm, we may ignore corrections and proceed as though only all-different combinations occurred. Thus we awake from this combinatorial nightmare with the approximation

$$Z_{\text{semi-classical}} = \frac{(Z_1)^N}{N!}. \tag{5.31}$$

Later, in chapter 8, the same result will emerge—but more transparently—as the semi-classical limit of another quantum calculation. A citation to a detailed yet intuitive justification of the semi-classical approximation is provided in the further reading list at the end of the chapter.

To compute Z_1, we may call upon the density of states that we derived in section 4.1. The calculation there was for a spinless particle, and that restriction holds henceforth in this section. Thus

$$Z_1 = \sum_{a} e^{-\varepsilon_a/kT} = \int e^{-\varepsilon/kT} D(\varepsilon)\, d\varepsilon$$

$$= \int_0^\infty e^{-\varepsilon/kT} \frac{2\pi(2m)^{3/2}}{h^3} V\varepsilon^{1/2}\, d\varepsilon \tag{5.32}$$

$$= \frac{(2\pi mkT)^{3/2}}{h^3} V.$$

The substitution $\varepsilon = x^2$ and then reference to an integral tabulated in appendix A takes one from the penultimate line to the final expression. Perhaps surprisingly, the factors in the last expression lend themselves to a nice physical interpretation, as follows.

In classical theory, the average kinetic energy of a free atom is $\frac{3}{2}kT$, and so the typical momentum is of order $(3mkT)^{1/2}$. Thus the factor $(2\pi mkT)^{1/2}$ is approximately the classical momentum. The de Broglie wavelength is h divided by the momentum. So it is natural to define a *thermal de Broglie wavelength* λ_{th} by

$$\lambda_{\text{th}} \equiv \frac{h}{\sqrt{2\pi mkT}}. \tag{5.33}$$

Then the single-particle partition function may be expressed as

$$Z_1 = \frac{V}{\lambda_{\text{th}}^3},$$

(5.34)

a dimensionless ratio. The semi-classical approximation for Z becomes

$$Z = \frac{(V/\lambda_{\text{th}}^3)^N}{N!}.$$

(5.35)

Energy, pressure, and entropy

The total energy, the pressure, and the entropy now follow easily. To display the dependence of $\ln Z$ on temperature and volume, write the logarithm of equation (5.35) as

$$\ln Z = N \ln V + \tfrac{3}{2}N \ln T + \text{const.}$$

(5.36)

Appeal to equation (5.16) gives

$$\langle E \rangle = kT^2 \frac{\partial \ln Z}{\partial T} = \tfrac{3}{2}NkT,$$

(5.37)

a comforting check on our calculations.

Next, for the pressure P, equation (5.19) yields the estimate

$$P = kT \frac{\partial \ln Z}{\partial V} = \frac{N}{V} kT.$$

(5.38)

Checks again.

Lastly, for the entropy, equation (5.25) implies

$$S = \frac{\langle E \rangle}{T} + k \ln Z = k \left[\tfrac{3}{2}N + \ln \frac{(V/\lambda_{\text{th}}^3)^N}{N!} \right].$$

(5.39)

The dependence on N is easier to understand if we first use Stirling's approximation for $\ln N!$ as follows:

$$\ln N! \cong N \ln N - N = N \ln(N/e).$$

(5.40)

(For a derivation of this approximation, see appendix A.) Then the entropy can be written compactly as

$$S = kN \ln \left[\frac{(V/N)}{\lambda_{\text{th}}^3} e^{5/2} \right].$$

(5.41)

The dependence on N is worth noting. In the argument of the logarithm, V/N gives the volume per particle in the system; its reciprocal is the number density, N/V. If one doubles the system by doubling V and N simultaneously, then the particle density remains constant, and so does the entire logarithm. The factor of N that precedes the logarithm will double, and so the entropy will double if one doubles the system at constant number density and temperature. This linear scaling is a general property of entropy for macroscopic systems.

Scaling

Further interpretation of the expression for entropy comes after the next subsection. Scaling, however, merits a few more sentences here. Imagine two identical macroscopic systems, each in thermal equilibrium. Two identical blocks of ice provide an example. Put the two systems together and in contact. Some quantities, such as temperature T and number density N/V, remain the same. Quantities that remain the same when the system is scaled up in this fashion are called *intensive*. Some other quantities, such as internal energy $\langle E \rangle$ and volume V, double (provided that surface effects are negligible). Quantities that double when the system is doubled are called *extensive*. In the previous paragraph, the analysis for entropy indicates that entropy S is an extensive variable. In later chapters, knowing how various quantities behave under scaling will be helpful and occasionally will be essential.

Range of validity

Under what physical circumstances are the semi-classical results valid? Let us assume, provisionally, that the results are indeed valid and see what physical conditions consistency requires. A typical atom has a momentum of magnitude $p = h/\lambda_{\mathrm{th}}$. In a semi-classical picture, the atom's momentum must be reasonably well-defined, not greatly uncertain. If we demand that the uncertainty Δp in momentum be at least an order of magnitude less than p, we get the inequality

$$\Delta p \leqslant \frac{1}{10} p = \frac{1}{10} \frac{h}{\lambda_{\mathrm{th}}}. \tag{5.42}$$

(Note that Δp is the quantum mechanical uncertainty, *not* the variation from atom to atom that arises from even a classical distribution of velocities in thermal equilibrium.) The momentum uncertainty is constrained also by the Heisenberg uncertainty principle:

$$\Delta x \Delta p \geqslant \frac{h}{4\pi}, \tag{5.43}$$

where Δx is the size of the wave packet (half a diameter, say). The Heisenberg constraint implies

$$\Delta p \geqslant \frac{1}{\Delta x} \frac{h}{4\pi}. \tag{5.44}$$

The inequalities (5.42) and (5.44) squeeze Δp between two bounds:

$$\frac{1}{10} \frac{h}{\lambda_{\mathrm{th}}} \geqslant \Delta p \geqslant \frac{1}{\Delta x} \frac{h}{4\pi}. \tag{5.45}$$

Most useful for us is that the two extremes provide a lower bound for Δx:

$$\Delta x \geqslant \frac{10}{4\pi}\lambda_{\text{th}} \cong \lambda_{\text{th}}. \tag{5.46}$$

In short, the minimum size of the wave packet is of order λ_{th}, and we may take λ_{th} as the typical size.

Semi-classical reasoning can be valid only if the wave packets for the atoms do not overlap—at least most of the time—because "overlap" corresponds to collision of two atoms and also leads to quantum correlations that are not included in classical reasoning. So, the next query is this: what is the typical separation between the atoms? The volume of space, free from neighbors, around a typical atom is one-Nth of the total volume: V/N. Thus the typical separation between atoms is $(V/N)^{1/3}$. Figure 5.5 displays the situation. The semi-classical analysis can be valid only if λ_{th} is much smaller than the typical inter-particle separation:

$$\boxed{\lambda_{\text{th}} \ll \left(\frac{V}{N}\right)^{1/3}.} \tag{5.47}$$

Unorthodox though it may be, I think of scattered, puffy white clouds drifting in an otherwise blue sky. That's fine for classical physics. When the clouds spread out and overlap much of the time, then we are in for trouble—classically—and must retreat to the shelter of quantum theory.

The condition (5.47) is not restricted to atoms. It can be applied to any gaseous system in which mutual interactions are negligible; just put in the appropriate value for the mass m that is hidden in the thermal de Broglie wavelength. Table 5.2 gives some examples.

The table indicates that

- for gases at the temperature and density encountered in a typical room, a classical treatment should be satisfactory;
- for liquids, the situation varies considerably, but is definitely against a classical treatment for liquid helium; and
- for conduction electrons in a solid, a quantum treatment is mandatory.

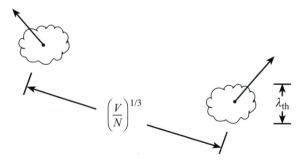

Figure 5.5 The geometry for assessing whether the semi-classical analysis is valid.

Table 5.2 *Data for determining whether a semi-classical analysis may be valid. Copper provides one conduction electron per atom. The particle number densities in liquid nitrogen and liquid helium are like those in water, in order of magnitude, and so also is the number density of conduction electrons in a metal. The mass of an "air molecule" is the average of diatomic nitrogen and oxygen, when weighted by the relative number densities.*

System	Mass (kg)	T (K)	λ_{th} (meter)	$(V/N)^{1/3}$ (meter)	$(V/N)^{1/3}/\lambda_{th}$
Air at room conditions	4.8×10^{-26}	300	1.9×10^{-11}	3.4×10^{-9}	180
Liquid nitrogen	4.7×10^{-26}	77	3.8×10^{-11}	3.9×10^{-10}	10
Liquid helium (^{4}He)	6.6×10^{-27}	4	4.4×10^{-10}	3.7×10^{-10}	0.86
Conduction electrons in copper	9.1×10^{-31}	300	4.3×10^{-9}	2.3×10^{-10}	0.053

More about entropy

For a further interpretation of the entropy expression, we return to equation (5.39) and write it as

$$S = k \ln \left[\frac{(V/e^{-3/2}\lambda_{th}^3)^N}{N!} \right]. \tag{5.48}$$

The factor λ_{th}^3, the cube of the thermal de Broglie wavelength, is the volume over which a typical wave packet extends (in order of magnitude, at least). The ratio of the container volume V to that wave-packet volume is an estimate of the (large) number of such wave-packet volumes in the entire container. (In the spirit of this interpretation, the factor $e^{-3/2}$ is so close to unity that we may ignore it.) In different words, the ratio V/λ_{th}^3 is an estimate of the number of ways in which a single atom can be put into the container when we keep in mind the "spread-out" nature of a quantum mechanical description. For distinguishable atoms, the number of distinct arrangements for the entire gas is the product of the number of arrangements for each atom. Thus we have an interpretation of why the ratio V/λ_{th}^3 appears raised to the Nth power. Moreover, the numerator of the logarithm's argument has precisely the dependence on volume V, temperature T, and number N that we reasoned out in chapter 2, when we first developed the idea of multiplicity.

What about the $N!$ in the denominator? Atoms of a single species are not distinguishable entities; rather, they are all the same and hence are indistinguishable, one from another. Placing atom #1 in the upper right corner of the container and atom #2 in the lower left corner is no different from placing atom #2 in the upper right and #1 in the lower left. And so on for other permutations of the atoms among the N positions that they occupy. The estimate in the numerator must be corrected by division with $N!$, the number of such permutations.

Partition functions

In this chapter we have met two specific instances of the partition function. In section 5.3, we noted (in passing) the partition function for a single, spatially fixed paramagnet whose spin is $\frac{1}{2}\hbar$. In the present section, we constructed an entirely different partition function—because we were dealing with N atoms of a semi-classical ideal gas. The general procedure is always the same (at least in principle): sum the Boltzmann factors for all the energy states of the entire system. How one does that and what the final form looks like vary from one physical system to another. Expect great variation in what the final partition function looks like. Table 5.3 provides a finding chart for most of the partition functions that appear in this book.

The classical analog

The quantum canonical probability distribution has an analog in classical physics. Chapter 13 develops the classical version, and you are now prepared for it. Of all the other chapters, only chapters 14 and 15 make essential use of chapter 13; so you can pick up chapter 13 anytime prior to those two chapters.

The chapters that immediately follow this one apply the quantum canonical probability distribution to a host of physical systems and also develop other valuable concepts, notably the chemical potential and the free energies.

Table 5.3 *The protean partition function. The partition function takes on different explicit forms when the general definition is evaluated in different contexts. What follows is a listing of most forms that appear in* Thermal Physics. *The first phrase is always to be understood as "The particles are . . .".*

Free to move in two or three dimensions; semi-classical approximation; no mutual interactions. Location: section 5.6 (for three dimensions) and section 7.4 (for two dimensions).

Spatially localized; one but only one particle at each site; multiple single-particle states per site; no mutual interactions. Location: chapter 5, problem 5.

Spatially localized; more sites than particles; one single-particle state per site; no mutual interactions. Location: chapter 7, problem 4.

Merely a single harmonic oscillator. Location: section 6.1.

Photons. Location: chapter 10, problem 6.

Diatomic molecules. Location: section 11.3.

In a liquid; semi-classical approximation; very simple model. Location: section 12.3.

Free to move in three dimensions; semi-classical approximation; mutual interactions included (via the van der Waals approximation). Location: section 12.9.

In a solid. Location: chapter 12, problems 2 and 3.

5.7   Theoretical threads

But before we turn to those chapters, let us pause for a glance at the big picture. This is an appropriate point for distinguishing among three theoretical threads in the fabric of thermal physics.

1. *Thermodynamics* is a strictly macroscopic and strictly deterministic theory. It uses the ideas of energy, temperature, and entropy, but it makes no assumptions about the microscopic—or atomic—structure of matter and radiation. Entropy itself is conceived macroscopically: it is the state function that is constructed by integrating equation (2.26) when the equal sign holds and by using the procedure codified in section 3.4. No microscopic interpretation of entropy need be given, and probabilities play no role.

2. *Statistical mechanics* is typified by chapter 5: probabilities are explicit and essential, and a detailed, microscopic mechanical theory (either quantum or classical) is used. The canonical probability distribution is the quintessential expression of the statistical mechanical approach.

For a macroscopic property of a macroscopic system, statistical mechanics will offer an estimate: usually the expectation value estimate or the most probable value. For the same property, thermodynamics will offer either a definite, unquestioned value or merely a relationship to other properties. Where both theories offer specific values, the best estimate from statistical mechanics is to be matched up with the single value asserted by thermodynamics. (In chapter 7, we will see in detail an example of this matching procedure.)

3. Finally, *kinetic theory* is typified by the analysis in section 1.2: a "billiard ball" view of atomic processes. Probabilities and averages are essential, and so kinetic theory may be taken as a subdivision of statistical mechanics.

In this book, the three threads are woven together, which is the way most physicists actually use them. Applications to real physical systems—in the lab, the atmosphere, or the galaxy—are difficult, and a skilled practitioner adopts whichever theoretical approach offers the easiest calculation or the most insight.

5.8   Essentials

1. For a system in thermal equilibrium, the probability that its energy eigenstate Ψ_j is the appropriate state to use in assessing properties is given by the *canonical probability distribution*:

$$P(\Psi_j) = \frac{\exp(-E_j/kT)}{Z},$$

where the numerator is called the *Boltzmann factor* and where Z is the *partition function*:

$$Z \equiv \sum_{j'} \exp(-E_{j'}/kT).$$

2. The logarithm of the partition function readily provides estimates of total energy and pressure:

$$\langle E \rangle = kT^2 \frac{\partial \ln Z}{\partial T},$$

$$P = kT \frac{\partial \ln Z}{\partial V}.$$

3. The system's entropy takes the form

$$S = \frac{\langle E \rangle}{T} + k \ln Z.$$

4. For a semi-classical ideal gas, the following relations hold:

$$Z_{\text{semi-classical}} = \frac{(Z_1)^N}{N!};$$

$$Z_1 \equiv \sum_{\alpha} e^{-\varepsilon_{\alpha}/kT} = \frac{V}{\lambda_{\text{th}}^3},$$

where λ_{th} denotes the *thermal de Broglie wavelength*:

$$\lambda_{\text{th}} \equiv \frac{h}{\sqrt{2\pi mkT}}.$$

The first equality is valid provided the thermal de Broglie wavelength is small relative to the average inter-particle separation:

$$\lambda_{\text{th}} \ll \left(\frac{V}{N}\right)^{1/3}.$$

(The second equality is valid under weaker conditions as well.)

5. Under the conditions of item 4, the entropy takes the form

$$S = kN \ln \left[\frac{(V/N)}{\lambda_{\text{th}}^3} e^{5/2}\right].$$

6. *Scaling.* Quantities that remain the same when one scales up the system are called *intensive.* (Examples are temperature T and number density N/V.) Quantities that double when one doubles the system are called *extensive.* (Examples are energy $\langle E \rangle$, volume V, and entropy S.)

7. Although the general definition of the partition function is always that given in item 1, the explicit functional form depends on the physical system and will vary widely.

Further reading

For an exposition of the degree of belief school of probability, here are some notable authors. John Maynard Keynes laid out his view of "probability" in his *Treatise on Probability* (Macmillan, London, 1921). Sir Harold Jeffreys wrote about probability in *Scientific Inference*, 2nd edition (Cambridge University Press, New York, 1957) and in *Theory of Probability*, 3rd edition (Oxford University Press, New York, 1967). Erwin Schrödinger presented his views in two papers: *Proc. R. Irish Acad.* **51**, section A, 51–66 and 141–6 (1947). Richard T. Cox's seminal paper appeared in *Am. J. Phys.* **14**, 1–13 (1946) and was expanded into *The Algebra of Probable Inference* (The Johns Hopkins Press, Baltimore, 1961).

Edwin T. Jaynes's writings on probability theory span at least four decades. A perspective is provided by his article, "A backward look to the future," in *Physics and Probability: Essays in Honor of Edwin T. Jaynes*, edited by W. T. Grandy, Jr., and P. W. Milonni (Cambridge University Press, New York, 1993). The volume contains also a bibliography of Jaynes's papers and reviews. From among them, I would recommend especially the article "Probability theory as logic" in *Maximum Entropy and Bayesian Methods*, edited by P. F. Fougère (Kluwer, Dordrecht, 1990). A baker's dozen of Jaynes's most notable papers were collected by R. D. Rosenkrantz, editor, in *E. T. Jaynes: Papers on Probability, Statistics, and Statistical Physics* (Reidel, Dordrecht, 1983).

One should remember also Pierre Simon, Marquis de Laplace, and his *Philosophical Essay on Probabilities* (Dover, New York, 1951).

The statistician I. J. Good ably discusses "Kinds of probability" in his article by that name: *Science* **129**, 443–7 (20 February 1959). He concludes that "although there are at least five kinds of probability, we can get along with just one kind."

J. Willard Gibbs introduced the "ensemble" into thermal physics in his *Elementary Principles in Statistical Mechanics* (reprinted by Dover, New York, 1960), pp. vii, 5, 16, 17, and 163. A classic of the frequency school is Richard von Mises's *Mathematical Theory of Probability and Statistics* (Academic Press, New York, 1964).

No bibliography on probability should fail to include Russell Maloney's short story, "Inflexible logic," which can be found in *A Subtreasury of American Humor*, edited by E. B. White and Katharine S. White (Random House, New York, 1941).

A detailed derivation of the general connection between magnetic moment and ln Z is provided in Ralph Baierlein, *Atoms and Information Theory* (W. H. Freeman, New York, 1971), pp. 200 and 468–72.

The accuracy of the semi-classical approximation for Z is discussed in Ralph Baierlein, "The fraction of 'all different' combinations: justifying the semi-classical partition function," *Am. J. Phys.* **65**, 314–16 (1997).

Problems

1. *Two-state system.* Section 5.3 provided an example of a *two-state system*. To simplify the analysis even further, specify that the system's two energy eigenstates have energies 0 and ε_0, where ε_0 denotes a small positive energy. Calculate and sketch (as functions of temperature) the following quantities:

(a) partition function, (b) energy estimate $\langle \varepsilon \rangle$, (c) heat capacity, and (d) entropy.

2. *Three-state system.* The nucleus of the nitrogen isotope ^{14}N acts, in some ways, like a spinning, oblate sphere of positive charge. The nucleus has a spin of $1\hbar$ and an equatorial bulge; the latter produces an electric quadrupole moment. Consider such a nucleus to be spatially fixed but free to take on various orientations relative to an external inhomogenous electric field (whose direction at the nucleus we take to be the z-axis). The nucleus has three energy eigenstates, each with a definite value for the projection s_z of the spin along the field direction. The spin orientations and the associated energies are the following: spin up ($s_z = 1\hbar$), energy $= \varepsilon_0$; spin "sideways" ($s_z = 0$), energy $= 0$; spin down ($s_z = -1\hbar$), energy $= \varepsilon_0$ (again). Here ε_0 denotes a small positive energy.

(a) In thermal equilibrium at temperature T, what is the probability of finding the nucleus with spin up? In what limit would this be $1/3$?
(b) Calculate the energy estimate $\langle \varepsilon \rangle$ in terms of ε_0, T, et cetera. Sketch $\langle \varepsilon \rangle$ as a function of T, and do the same for the associated heat capacity.
(c) What value does the estimate $\langle s_z \rangle$ have? Give a qualitative reason for your numerical result.

3. *Excited hydrogen.* The energy levels of atomic hydrogen are given by the expression $\varepsilon_n = -13.6/n^2$ electron volts, where n denotes the principal quantum number. Consider the atomic hydrogen in a stellar atmosphere, and specify that the ambient temperature is 7,000 K. For atomic hydrogen, what numerical values do the following probability ratios have?

(a) $\dfrac{\text{probability that electron has } n = 2 \text{ and no orbital angular momentum}}{\text{probability that electron has } n = 1}$.

(b) $\dfrac{\substack{\text{probability that electron has } n = 2 \\ \text{(without specification of orbital angular momentum)}}}{\text{probability that electron has } n = 1}$.

4. *Magnetic susceptibility.* The magnetic moment per unit volume is called the *magnetization* and, as a vector, is denoted by **M**. For paramagnets in an external magnetic field **B**, the vector **M** is usually parallel to **B**, and we use M to denote the component of **M** along **B**. The *magnetic susceptibility* is the rate of change of M with B: $\partial M / \partial B$.

The following questions pertain to the ideal, spin $\frac{1}{2}\hbar$ paramagnets of section 5.3.

(a) What is the magnetization as a function of B and T? Sketch M as a function of $m_B B/kT$.
(b) What is the magnetic susceptibility? (Form the partial derivative at constant temperature.)
(c) What is the limiting expression for the susceptibility when $m_B B \ll kT$? (That dependence on temperature is called *Curie's law*, after the French physicist Pierre Curie, who undertook a comprehensive experimental comparison of paramagnetism, ferromagnetism, and diamagnetism.)

5. *Spatially localized particles.* If particles are spatially localized, one to each site, and if the particles do not interact with one another, then the partition function Z for N particles is

$$Z = (Z_1)^N, \tag{1}$$

where Z_1 denotes the partition function of a single particle at a single site. To derive this result, one can reason as in section 5.6 and note that, because there is always one but only one particle per site, neither over-counting nor improper combinations arise.

Henceforth, consider paramagnetic particles with spin $\frac{1}{2}\hbar$, as in section 5.3.

(a) Confirm equation (1) when $N = 2$ by constructing the four states of the two-particle system and by forming the partition function directly as a sum over all energy eigenstates.
(b) Return to general N and estimate the total energy $\langle E \rangle$. Afterwards, you can use the relation

$$\langle E \rangle = -\langle \text{total moment along } \mathbf{B} \rangle B$$

to extract $\langle \text{total moment along } \mathbf{B} \rangle$.
(c) To assess the reliability of the estimate for $\langle E \rangle$, first confirm that the relation

$$(\Delta E)^2 \equiv \langle (E - \langle E \rangle)^2 \rangle = kT^2 \frac{\partial \langle E \rangle}{\partial T} \tag{2}$$

holds for the canonical probability distribution *in general*. Then examine the ratio $\Delta E/|\langle E \rangle|$ for the present system. Concentrate on how the ratio depends on N, the number of particles.
(d) What can you infer about the reliability of the estimate "$\langle \text{total moment along } \mathbf{B} \rangle$"?
(e) What can you infer *in general* about the sign of the heat capacity, when computed under conditions of fixed external parameters?

6. *Truncating the series.* In section 5.2, I truncated a Taylor series abruptly; the questions below help you to justify that step.

To simplify the analysis, specify that the reservoir is a monatomic classical ideal gas of N spinless atoms.

To specify an energy eigenstate, we will need to specify N triplets $\{n_x, n_y, n_z\}$ in the spirit of section 4.1 There will be one quantum state per unit volume in an enlarged mathematical space of $3N$ dimensions. (That statement would be literally correct if the N particles were distinguishable. The actual indistinguishability has no qualitative effect on the subsequent reasoning, and so we ignore it.) The energy E will continue to be proportional to n^2, where n is the "radius" vector in that space. The integral in the analog of equation (4.6), however, will contain the factor $n^{3N-1} dn$, for a "volume" in a space of $3N$ dimensions must go as n^{3N}. The density of states must have the form

$$D(E) = \text{constant} \times E^{f(N)}.$$

(a) Determine the exponent $f(N)$; do not bother to evaluate the constant.
(b) Check your value for the exponent $f(N)$ by computing $\langle E \rangle$ from the partition function.
(c) To study the reservoir's entropy, write the entropy as follows:

$$S_{\text{res}}(E_{\text{tot}} - E_j) = kf(N) \ln\left(1 - \frac{E_j}{E_{\text{tot}}}\right) + \text{part independent of } E_j. \tag{1}$$

Be sure to confirm this form, starting with the definition in equation (4.15). Because the reservoir is much larger than the sample, $E_{\text{tot}} \cong N \times (\frac{3}{2}kT)$, and this approximation gets better as N gets larger. Write out the first three nonzero terms in the Taylor series of the logarithm (about the value 1 for the argument). Then study the limit of equation (1) as N goes to infinity. Do you find that the only surviving dependence on E_j is the single term that we calculated in section 5.2?

7. *The energy range δE.* Equation (5.24) gave a precise definition of the energy range δE that appears in the expressions for multiplicity and entropy. The present exercise investigates the size of δE relative to the energy estimate $\langle E \rangle$.

Specify that the density of states for the entire system has the form

$$D(E) = g(N, V)E^{(3N/2)-1}$$

for the N atoms of a monatomic ideal gas, treated semi-classically. The prefactor $g(N, V)$ is a function of N and the volume V only.

(a) Evaluate the partition function Z exactly in terms of $g(N, V)$, N, and kT. For simplicity, take N to be an even integer.
(b) Work out the explicit expression for the energy range δE as it is defined by equation (5.24). To get a useful final result, you will need Stirling's approximation for the factorial of a large integer.
(c) Calculate the energy estimate $\langle E \rangle$. What implication can you draw from the size of δE relative to $\langle E \rangle$?

8. *More on approximating a sum by an integral.* In section 5.6, we used a density of single-particle states to replace a sum over single-particle states with an integral, namely, the sum for Z_1. That procedure is, of course, an approximation.

(a) What inequality among the quantities kT, V, and $h^2/2m$ must be satisfied if the approximation is to be a good one? (For starters, you might reason that quantization cannot affect details if the thermal wave packets are small relative to the linear size of the container. Beyond that, you can require that the Boltzmann factor change relatively little from one state to the next, at least until the factor itself becomes insignificantly small.)

(b) How does your inequality compare with the inequality that must be satisfied if a semi-classical analysis is to be adequate for the entire N-particle gas? Which inequality imposes the more stringent requirement?

Later in this text, we will replace other sums by integrals. The replacements will be good approximations under realistic physical conditions. Nonetheless, you should be aware that an approximation is being made and that, in principle, a check should be made each time.

6 Photons and Phonons

6.1 The big picture

6.2 Electromagnetic waves and photons

6.3 Radiative flux

6.4 Entropy and evolution (optional)

6.5 Sound waves and phonons

6.6 Essentials

The oscillations of electromagnetic fields and of crystal lattices have much in common. Waves form a basis for analyzing both—provided one incorporates quantum theory judiciously. This chapter addresses electromagnetic waves and sound waves—when in therrmal equilibrium—in a unified fashion. The canonical probability distribution plays a vital role, for it links the thermal and quantum aspects of the waves.

6.1 The big picture

When describing waves in thermal equilibrium, either electromagnetic waves or sound waves, there are three essential elements.

1. Normal modes. Electromagnetic waves in a metallic cavity and sound waves in a crystal can possess standing wave patterns, sometimes called *normal modes*. The standing waves have a discrete set of frequencies.

2. Quantized energy. Quantum theory restricts the energy in any given standing wave mode to be of the form

$$\begin{pmatrix} \text{energy } \varepsilon \text{ in mode} \\ \text{of frequency } \nu \end{pmatrix} = nh\nu + \text{constant}, \tag{6.1}$$

where n is zero or a positive integer: $n = 0, 1, 2, 3, \dots$. If we can estimate the value of that integer, then we can calculate the energy and also several other quantities of physical interest. We denote the expectation value estimate by $\bar{n}(\nu)$. [The estimate could be denoted by $\langle n(\nu) \rangle$, but that would be a clumsy notation.] The literal number n specifies, for electromagnetism, the number of photons in the mode and, for sound waves, the number of phonons.

In general, any wave that satisfies a linear wave equation (with time derivatives of second order) is mathematically analogous to a simple harmonic oscillator. The energy form in equation (6.1) is what quantum theory predicts for the energy of a simple harmonic oscillator of frequency ν.

3. Density of modes. A sum over the set of distinct standing wave modes may often be approximated by an integral with a *density of modes*, the number of modes per unit frequency interval.

Computing $\bar{n}(\nu)$

The easiest way to compute $\bar{n}(\nu)$ is indirect, as follows. Choose the zero of energy so that the constant in (6.1) is zero, and then form an expectation value with the canonical probability distribution:

$$\langle \varepsilon \rangle = \bar{n}(\nu) h\nu. \tag{6.2}$$

Equation (5.16) enables us to calculate $\langle \varepsilon \rangle$ as

$$\langle \varepsilon \rangle = \frac{kT^2}{Z} \frac{\partial Z}{\partial T}, \tag{6.3}$$

where Z is the partition function for the mode. The required sum is

$$Z = \sum_{n=0}^{\infty} e^{-nh\nu/kT} = \sum_{n=0}^{\infty} (e^{-h\nu/kT})^n = \frac{1}{1 - e^{-h\nu/kT}}. \tag{6.4}$$

The second equality shows that the sum is a geometric series, which can be summed in closed form.

Insert the explicit form for Z into (6.3) and find

$$\langle \varepsilon \rangle = \frac{1}{e^{h\nu/kT} - 1} h\nu. \tag{6.5}$$

Comparison with (6.2) implies

$$\left(\begin{array}{c} \text{estimated number of photons or} \\ \text{phonons in mode with frequency } \nu \end{array} \right) \equiv \bar{n}(\nu) = \frac{1}{e^{h\nu/kT} - 1}. \tag{6.6}$$

The higher the frequency at fixed temperature T, the smaller $\bar{n}(\nu)$ is.

Let's check the expression for $\bar{n}(\nu)$ against a reasonable expectation. If one increases the temperature of a solid, one expects the atoms to jiggle more vigorously (around their equilibrium sites in a lattice). Speaking classically, one would say that a sound wave mode of fixed frequency will have larger amplitude. That statement should translate into a larger number of phonons in the mode. In short, the function $\bar{n}(\nu)$ should increase when T increases. To test that proposition, note that the exponent in (6.6) will decrease when T increases; the exponential will decrease; the denominator as a whole will decrease; and so the function $\bar{n}(\nu)$ will increase. It checks.

That completes the general preparation. Now we turn to specific applications.

6.2   Electromagnetic waves and photons

The first application is to electromagnetic waves in thermal equilibrium.

Density of modes

In section 4.1 we computed a density of states for a single spinless particle, and much of that analysis can be carried over to photons. To determine the density of electromagnetic modes, consider a cubical cavity with perfectly reflecting metal walls. First we determine the discrete frequencies of the standing waves. The argument that an integral number of half wavelengths must fit along a length L remains valid. [Whatever boundary conditions need to be met at one wall will also need to be met at the opposite wall, a distance L away. Only if the wave makes an integral number of half cycles in distance L will that requirement be satisfied. Thus $L/(\lambda_x/2) = n_x$ and similarly for the other directions.] The momentum components $\{p_x, p_y, p_z\}$ of a photon must have magnitudes given by the triplet $(h/2L) \times \{n_x, n_y, n_z\}$. The momentum magnitude is

$$p = \frac{h}{2L}(n_x^2 + n_y^2 + n_z^2)^{1/2}. \tag{6.7}$$

Moreover, a photon's momentum magnitude p is connected to its energy by the relation

$$p = \frac{\text{energy}}{c} = \frac{h\nu}{c}. \tag{6.8}$$

Comparison of equations (6.7) and (6.8) implies that the frequency of a wave mode is given by

$$\nu = \frac{c}{2L}(n_x^2 + n_y^2 + n_z^2)^{1/2}. \tag{6.9}$$

Planck's constant h has canceled out, suggesting that this is a purely classical result. Indeed it is, but the excursion via quantum theory provides an easy route.

Any triplet of integers in the positive octant of figure 4.2 specifies a mode and its frequency—almost. Electromagnetic waves are transverse waves, and so for each triplet $\{n_x, n_y, n_z\}$, two orthogonal polarizations (specified by the direction of the electric field) are possible. Thus, if $A(\nu)$ is any smooth function of the frequency, the analog of equation (4.6) becomes

$$\sum_{\text{EM modes}} A(\nu) = 2 \times \sum_{\text{triplets } \{n_x, n_y, n_z\}} A(\nu) = \frac{2}{8}\int A(\nu) 4\pi n^2 \, dn$$

$$= \frac{2}{8}\int A(\nu) 4\pi \left(\frac{2L}{c}\right)^3 \nu^2 \, d\nu. \tag{6.10}$$

The factor of 2 in the first line incorporates the two polarizations, and the sum goes over the positive triplets only. In the step to an integral, the variable n is the radius in the mathematical space of triplets $\{n_x, n_y, n_z\}$. (Note. The n here is entirely distinct from the n of section 6.1. Unfortunately, the letter n gets heavy use in physics.) Equation (6.9) indicates that $v = (c/2L)n$; that relation was used to arrive at the last line. From equation (6.10) we extract the density of electromagnetic modes $D_{EM}(v)$ as

$$D_{EM}(v) = \frac{8\pi}{c^3} V v^2. \tag{6.11}$$

Lest there be ambiguity, let me express things this way:

$$\left(\begin{array}{c} \text{number of EM modes in the} \\ \text{frequency range } v \text{ to } v + dv \end{array} \right) \equiv D_{EM}(v)\, dv$$

$$= \frac{8\pi}{c^3} V v^2\, dv. \tag{6.12}$$

A density of modes, such as $D_{EM}(v)$, is quite analogous to a density of single-particle energy eigenstates $D(\varepsilon)$. Each counts the number of discrete entities (modes or energy eigenstates) per unit range of some continuous argument (frequency v or single-particle energy ε). Using a capital letter D provides a good mnemonic for "density." Then the argument and, at times, a subscript distinguish one kind of density from the others.

Note, however, that a wave mode of frequency v may have energy equal to $0hv$, $1hv$, $2hv$, etc. To know how much energy the mode has, one needs to know how many photons are associated with the mode (either precisely or in an average sense). We turn to this issue in the next subsection.

Energy

An estimate of the total electromagnetic energy follows readily. Equation (6.6) gives us an expression for $\bar{n}(v)$, the estimated number of photons in a mode of frequency v. Since each photon has energy hv, the estimated energy in a mode is $hv\bar{n}(v)$. Replacing a sum over all modes with an integral, we find that the total energy is

$$\langle E \rangle = \int hv\bar{n}(v) D_{EM}(v)\, dv$$

$$= \int_0^\infty \left(\frac{8\pi h}{c^3} \frac{v^3}{e^{hv/kT} - 1} \right) V\, dv. \tag{6.13}$$

In the integrand, the factor in parentheses has a useful interpretation. Note that it is multiplied by the volume V and by a frequency interval dv. Thus the function in parentheses is the estimated energy per unit volume and per unit frequency range:

$$\left(\begin{array}{c} \text{spatial energy density} \\ \text{per unit frequency interval} \end{array} \right) = \frac{8\pi h}{c^3} \frac{v^3}{e^{hv/kT} - 1}. \tag{6.14}$$

This is Max Planck's great result: *the Planck distribution*. Figure 6.1 displays the distribution as a function of frequency for two values of the temperature.

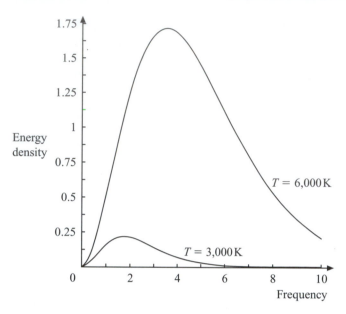

Figure 6.1 The Planck distribution. The estimated energy per unit volume and per unit frequency interval, in units of 10^{-15} J/(m^3 · Hz), is plotted against frequency, in units of 10^{14} Hz. Changing the temperature by a mere factor of 2 makes a tremendous difference.

When $h\nu$ is small relative to kT, the denominator is approximately $h\nu/kT$ and cancels one factor of ν in the numerator and the factor of h; so the Planck distribution rises as $kT\nu^2$ for small ν and does not depend on h. That portion of the distribution had been derived by classical methods before Planck's discovery in 1900. At high frequency, such that $h\nu \gg kT$, the distribution falls as $h\nu^3 \exp(-h\nu/kT)$, an effect that is thoroughly quantum mechanical.

Now we return to equation (6.13). The temperature dependence of $\langle E \rangle$ can be made more evident by a change of integration variable to the dimensionless variable $x = h\nu/kT$. The change produces the form

$$\langle E \rangle = \left(\frac{8\pi V k^4}{c^3 h^3}\right) T^4 \times \int_0^\infty \frac{x^3}{e^x - 1}\, dx.$$

The remaining integral is tabulated in appendix A; its value is $\pi^4/15$. Thus

$$\langle E \rangle = \left(\frac{8\pi^5 k^4}{15 c^3 h^3}\right) V T^4. \tag{6.15}$$

The total electromagnetic energy is proportional to the fourth power of the temperature, a strong dependence.

A simple argument provides insight into the T^4 dependence. As displayed in equation (6.6), the estimated number of photons per mode, $\bar{n}(\nu)$, drops exponentially to insignificance as soon as $h\nu$ exceeds $3kT$ or so. Thus the numerically significant sum (or integral) over modes extends only to photon momenta of order $3kT/c$. In the mathematical space of the triplets $\{n_x, n_y, n_z\}$, that implies an octant "volume" proportional to $(kT/c)^3$. In that octant, the photons that contribute significantly to the energy have individual energies of order kT, and $\bar{n}(\nu)$ is of order unity for those modes. So we can estimate the total energy by the product

$$h\nu \times \bar{n}(\nu) \times \left(\begin{array}{c}\text{number of}\\\text{significant modes}\end{array}\right) \propto kT \times 1 \times \left(\frac{kT}{c}\right)^3.$$

The right-hand side shows that the product is proportional to $(kT)^4/c^3$, and that proportionality reproduces the dependence of $\langle E \rangle$ on k, T, and c. The dependence on temperature—namely, T^4—is the crucial property to understand and to remember.

Radiation pressure

Electromagnetic radiation can exert a pressure, an effect calculated already by James Clerk Maxwell. In section 5.4, we found that the pressure exerted by a system in quantum state Ψ_j could be computed as $-\partial E_j/\partial V$. According to equation (6.9), the frequency of an electromagnetic mode scales with volume as $V^{-1/3}$, and so the energy $h\nu$ scales that way, too. The energy E_j of any state Ψ_j of the entire radiation system must have the form

$$E_j = \left(\begin{array}{c}\text{constant dependent}\\\text{on index } j\end{array}\right) \times V^{-1/3}, \tag{6.16}$$

whence

$$-\frac{\partial E_j}{\partial V} = \tfrac{1}{3}\frac{E_j}{V}. \tag{6.17}$$

Then, for the pressure P, equation (5.18) implies

$$P = \sum_j \tfrac{1}{3}\frac{E_j}{V} P(\Psi_j)$$

$$= \tfrac{1}{3}\frac{\langle E \rangle}{V}. \tag{6.18}$$

Because $\langle E \rangle$ itself is proportional to V, the radiation pressure is independent of the volume, being a function of temperature only: $P \propto T^4$.

Total number of photons

An estimate of the total number of photons $\langle N \rangle$ will prove useful and is readily set up:

$$\langle N \rangle = \int \bar{n}(\nu) D_{\text{EM}}(\nu)\, d\nu = \int_0^\infty \frac{8\pi}{c^3} V \frac{\nu^2}{e^{h\nu/kT} - 1}\, d\nu$$

$$= 8\pi V \left(\frac{kT}{ch}\right)^3 \times 2.404. \tag{6.19}$$

Again, the definite integral can be cast into dimensionless form, and the dimensionless version is tabulated in appendix A. The final expression for $\langle N \rangle$ has at least two uses, as follows.

1. Energy of typical photon. The ratio of total energy $\langle E \rangle$ to total number of photons provides an estimate of the typical photon energy. From equations (6.15) and (6.19), one finds

$$\frac{\langle E \rangle}{\langle N \rangle} = 2.7kT, \tag{6.20}$$

and so the typical photon has an energy of approximately $3kT$.

The peak in the Planck distribution produces a similar estimate. Differentiate the distribution with respect to ν; then set the derivative to zero. The process generates a transcendental equation in $h\nu/kT$:

$$\frac{h\nu}{kT} = 3(1 - e^{-h\nu/kT}). \tag{6.21}$$

Guessing that $h\nu/kT$ is close to 3 and inserting that number into the right-hand side yields 2.85; inserting the latter number on the right yields 2.83; and further iteration leaves the right-hand side at 2.82, when rounded to two decimals. Thus the Planck distribution yields

$$(h\nu)_{\text{at peak}} = 2.82kT, \tag{6.22}$$

in satisfying agreement with our other estimate for the energy of a typical photon.

The relationship,

$$\nu_{\text{at peak}} = 2.82 \frac{k}{h} T, \tag{6.23}$$

is one version of *Wien's displacement law*: the peak frequency shifts linearly with changes in temperature T. Problem 3 develops another version of Wien's law.

2. Constancy during slow adiabatic change. If one expands the cavity slowly and adiabatically, it is plausible that the total number of photons remains constant. If that is so, then the product $T^3 V$ will remain constant:

$$T^3 V = \text{constant} \tag{6.24}$$

during a slow adiabatic expansion. This relationship is in fact true; we can prove it quickly by an entropy argument, as follows.

For the photon gas, the heat capacity at constant volume is

$$C_V = \left(\frac{\partial \langle E \rangle}{\partial T} \right)_V = \left(\frac{32\pi^5 k^4}{15 c^3 h^3} \right) T^3 V. \tag{6.25}$$

Imagine slowly heating the photon gas to temperature T by heating the metal walls of the cavity; start at absolute zero, where there are no photons, and hence there is no entropy. Then

$$S = \int_0^T \frac{C_V(T')}{T'} \, dT' = \tfrac{1}{3}\left(\frac{32\pi^5 k^4}{15c^3 h^3}\right) T^3 V. \tag{6.26}$$

The entropy is proportional to $T^3 V$. In a slow adiabatic process, the entropy remains constant, and therefore $T^3 V$ does.

6.3 Radiative flux

Hot objects radiate electromagnetic waves. After all, that is how the heating coils in a kitchen oven fill the oven with infrared radiation. In this section, we explore the topic of *radiative flux*, meaning the flow of electromagnetic energy.

First, a definition is in order. A surface is called perfectly *black* if it absorbs all the radiation incident upon it.

Consider now an entirely closed cavity whose interior walls are black and are held at temperature T. The cavity becomes filled with radiation and has an energy density given by $\langle E \rangle / V$, where $\langle E \rangle$ is given by equation (6.15). Next, make a small hole in one wall, as indicated in figure 6.2, but pretend that the tube and its termination are absent. The cavity radiates through the hole. If all electromagnetic waves moved strictly leftward and rightward, then—in a short time interval Δt—the energy flux through the hole (of area A) would be half the energy in a cylinder extending rightward from the hole a distance $c\Delta t$ and having cross-sectional area A. Thus a rough estimate of the energy flux (in joules per second and per square meter) is

$$\text{(radiative flux)} \cong \tfrac{1}{2} \frac{\langle E \rangle}{V} \frac{c\Delta t A}{\Delta t A} = \tfrac{1}{2} \frac{\langle E \rangle}{V} c, \tag{6.27}$$

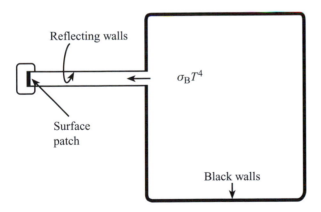

Figure 6.2 Blackbody radiation. The interior walls are perfectly black and are held at a temperature T. Thus the interior constitutes a "blackbody." A narrow tube, whose interior walls are perfectly reflecting, connects a hole in the cavity wall with a patch of surface, initially taken to be perfectly black. The tube itself is made of insulating material; hence the surface patch is thermally isolated except for the radiative flux through the tube (both leftward and rightward). Further exterior insulation isolates the surface patch from the environment.

where $\langle E \rangle / V$ is the spatial energy density in the cavity. A more accurate assessment notes that the distribution of photon velocities is isotropic and that radiation can pass through the hole obliquely. Such a calculation replaces the $1/2$ by $1/4$ and yields

$$\left(\begin{array}{c} \text{energy flux from} \\ \text{hole in cavity} \end{array} \right) = \tfrac{1}{4} \frac{\langle E \rangle}{V} c$$

$$= \tfrac{1}{4} \left(\frac{8 \pi^5 k^4}{15 c^3 h^3} \right) T^4 c = \sigma_B T^4. \tag{6.28}$$

The symbol σ_B denotes the *Stefan–Boltzmann constant*, defined by

$$\sigma_B \equiv \frac{2 \pi^5 k^4}{15 c^2 h^3} = 5.67 \times 10^{-8} \ \mathrm{W/(m^2 \cdot K^4)}. \tag{6.29}$$

Next we pay attention to the tube with perfectly reflecting walls and the thermally isolated patch of black surface at its left-hand end. The black surface will come to thermal equilibrium with the cavity walls and will acquire the same temperature that the walls have. In thermal equilibrium, the surface must radiate as much energy as it absorbs. After glancing back at equation (6.28), we conclude that

$$\left(\begin{array}{c} \text{energy flux from} \\ \text{perfectly } black \text{ surface} \end{array} \right) = \sigma_B T^4. \tag{6.30}$$

In short, a *black surface* radiates precisely as does a *hole* in the cavity. The relationship (6.30) is called the *Stefan–Boltzmann law*.

Perfect absorbers are difficult to find. Even lampblack (soot) absorbs only 95 percent in the visible range of electromagnetic waves. So we consider a situation where the surface absorbs the *fraction a* and reflects the fraction $1 - a$ of the incident radiation. The coefficient a is called the *absorptivity*. In thermal equilibrium at temperature T, energy balance holds in the sense that

$$\left(\begin{array}{c} \text{flux emitted} \\ \text{by surface} \end{array} \right) + \left(\begin{array}{c} \text{flux reflected} \\ \text{by surface} \end{array} \right) = \left(\begin{array}{c} \text{flux incident} \\ \text{from hole} \end{array} \right), \tag{6.31}$$

that is, the sum of the rightward fluxes through the tube equals the leftward flux. Solving for the emitted flux, we find

$$\left(\begin{array}{c} \text{flux emitted by surface} \\ \text{with absorptivity } a \end{array} \right) = \left(\begin{array}{c} \text{flux incident} \\ \text{from hole} \end{array} \right) - \left(\begin{array}{c} \text{flux reflected} \\ \text{by surface} \end{array} \right)$$

$$= \sigma_B T^4 - (1 - a) \sigma_B T^4$$

$$= a \sigma_B T^4. \tag{6.32}$$

At first sight, this equation just says that the emitted flux equals the absorbed flux. But the non-black surface will emit at the rate given in (6.32) even in the absence of an incident flux; such emission is a consequence of the surface's nonzero temperature T and the atomic characteristics of its constituents. In short, equation (6.32) generalizes equation (6.30), and it tells us that a good absorber (with large coefficient a) is also a good emitter.

For purposes of tabulation, the *emissivity* of a surface is defined by the relation

$$\left(\begin{array}{c} \text{actual} \\ \text{emitted flux} \end{array} \right) = (\text{emissivity}) \times \left(\begin{array}{c} \text{flux that a black} \\ \text{surface would emit} \end{array} \right). \tag{6.33}$$

Comparing equations (6.32) and (6.33) tells us that the emissivity equals the absorptivity. The latter is often the tabulated quantity, but we may use one for the other numerically whenever convenience recommends it. Table 6.1 provides a selection of emissivities.

Absorption and emission as a function of frequency

Note that we could carry through a similar analysis with just a portion of the Planck spectrum. Imagine placing in the tube a "reflection filter" that passes a narrow spectral band, v to $v + dv$, and reflects all else on both of its sides. Insert the filter after thermal equilibrium has been reached. Then all frequencies outside the pass band remain in equilibrium; consequently, the energy fluxes in the pass band must also remain in equilibrium, and to them we apply our preceding analysis.

Thus, if the absorptivity is a function of frequency, $a = a(v)$, then

Table 6.1 *Some emissivities (for total radiation) and the temperature range to which they apply. For the temperatures listed here, the peak in the Planck spectrum lies in the infrared. (The peak will lie in the frequency range where we see—the visible range— only if the radiation source has a temperature comparable to that of the solar surface, which is 5,800 K.)*

Material	Temperature (°C)	Emissivity
Gold, polished	200–600	0.02–0.03
Silver, clean and polished	200–600	0.02–0.03
Aluminum, polished	50–500	0.04–0.06
Lead, oxidized	200	0.05
Chromium, polished	50	0.1
Graphite	0–3,600	0.7–0.8
Varnish, glossy black sprayed on iron	20	0.87
Asbestos board	20	0.96
Lampblack (carbon)	20–400	0.96
Varnish, dull black	40–100	0.96–0.98

Source: CRC Handbook of Chemistry and Physics, 71st edn, edited by David R. Lide (Chemical Rubber Publishing Company, Boston, 1992).

$$\left(\begin{array}{c}\text{energy flux from surface}\\\text{in range } v \text{ to } v + dv\end{array}\right) = a(v) \times \left(\begin{array}{c}\text{flux from hole in cavity}\\\text{in same spectral range}\end{array}\right)$$

$$= a(v)\sigma_B \frac{15}{\pi^4}\left(\frac{h}{k}\right)^4 \frac{v^3}{e^{hv/kT} - 1}\, dv. \qquad (6.34)$$

Reasoning by analogy with equation (6.28), one sets "flux from hole in cavity in spectral range v to $v + dv$" equal to $c/4$ times the corresponding spatial energy density, which is the Planck distribution, given in equation (6.14). If the coefficient $a(v)$ is a constant, then integration of (6.34) over the frequency range from zero to infinity reproduces equation (6.32).

But, in fact, the absorptivity sometimes varies strongly with frequency. Figure 6.3 shows three metal cans, each filled initially with half a liter of almost boiling water (at temperature 370 K). One can has the original shiny metallic surface; another was painted flat white; and the last was painted flat black. To the eye—and hence in the visible portion of the electromagnetic spectrum—the shiny and the white cans are good reflectors and therefore poor emitters. The black can is a good absorber and hence a good emitter. One might expect the black can to cool relatively rapidly and the two other cans to lag behind. When the experiment is done, however, both black and white cans cool rapidly; only the shiny can lags.

Why? We see the cans in the visible and judge the absorptivity of each in that frequency domain: $a(v_{\text{visible}})$. For the cooling, however, what matters is $a(v_{\text{peak when } T=370\text{ K}})$, the absorptivity at the peak in the Planck spectrum for an emitter at the cans' actual temperature. How different are the frequencies? And the associated absorptivities? The proportionality between the frequency at the peak and the temperature of the emitter enables us to make the first comparison readily.

The wavelength of orange light is $\lambda_{\text{orange}} = 6 \times 10^{-7}$ meter, and so visible light has a frequency of order $v_{\text{orange}} = c/\lambda_{\text{orange}} = 5 \times 10^{14}$ Hz, a convenient round number. Most of our visible light arises from the solar surface, whose temperature is approximately 6,000 K. For our purposes, the peak in the solar spectrum (at the sun and as a function of frequency) is close enough to the visible that we can associate v_{orange} with

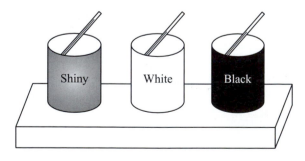

Figure 6.3 Is a good reflector necessarily a poor emitter? Comparing the emission rates of three hot surfaces: shiny metal, flat white, and flat black. Richard A. Bartels reported the experiment in *Am. J. Phys.* **58**, 244–8 (1990).

the solar peak. The cans have a temperature of only 370 K to start with and cool to 350 K or so in the experiment. According to equation (6.22), the peak in the Planck spectrum is shifted downward by the factor $370/6{,}000 = 0.06$, and so the peak frequency for the cans is approximately $0.06 \times (5 \times 10^{14}) = 3 \times 10^{13}$ Hz. That frequency lies well into the infrared region of the spectrum. In the infrared, the absorptivity of flat white and black paint are virtually the same (roughly 95 percent), and so the white can should cool just as rapidly as the black can, which it does. The shiny metallic surface remains a good reflector and poor emitter at infrared frequencies; hence that can lags.

Cosmic background radiation

According to standard cosmological theory, the early universe contained electromagnetic radiation in thermal equilibrium with a hot plasma of primarily protons and electrons. As the universe expanded, it cooled. When the temperature dropped to approximately 3,000 K, the electrons and protons combined to form neutral hydrogen. (There had been a dynamic equilibrium: hydrogen being formed and then dissociating. Such a process continued at still lower temperatures. Nonetheless, as chapter 11 will demonstrate, the balance shifts from mostly dissociated to mostly combined over a surprisingly narrow interval of temperature.) The radiation had a Planck distribution, as exhibited in (6.14), with the temperature of the combination era. As the universe continued to expand, the radiation underwent adiabatic expansion, dropping further in temperature as the volume increased; equation (6.26) provides the quantitative relationship. By today, the temperature has descended to approximately 3 K, but the radiation—in its distribution of energy as a function of frequency—has preserved the shape of the Planck distribution. (You can think of all wavelengths as having been expanded by the same factor; such a scaling leaves invariant the shape of the energy distribution.)

Figure 6.4 shows the cosmic spectrum as it was measured in 1990.

Planck's determination of h and k

In section 1.2, I remarked that Max Planck determined the numerical values of the "Naturconstanten" h and k by comparing his theory with existing data on radiation. Now we are in a position to understand his route. The energy flux from a blackbody was known as a function of temperature; so equations (6.28) and (6.29) imply that the Stefan–Boltzmann constant was known experimentally:

$$\sigma_{\mathrm{B}} \equiv \frac{2\pi^5 k^4}{15 c^2 h^3} = 5.67 \times 10^{-8} \text{ watts}/(\text{m}^2 \cdot \text{K}^4), \qquad (6.35)$$

where the modern value is quoted here. Because the speed of light c was well known, equation (6.35) provided a value for the combination k^4/h^3.

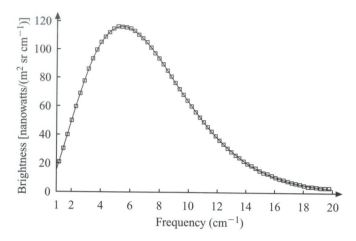

Figure 6.4 Cosmic background radiation. Measurements by a spectrometer aboard a satellite, the Cosmic Background Explorer, produced the data "squares" that display energy flux versus frequency. The continuous curve is a fit to the data that uses the Planck distribution and only one free parameter: the temperature T. Optimizing the fit yields a temperature of $T = 2.735 \pm 0.06$ K for what is clearly a blackbody spectrum. The authors cited the frequency as ν/c, where $c = 3 \times 10^{10}$ cm/s, and so the apparent units of "frequency" are cm^{-1}. Plotted vertically is the energy flux (in nanowatts per square meter) per unit solid angle (in steradians: sr) and per unit frequency interval (in cm^{-1}). [*Source*: J. C. Mather *et al.*, *Astrophys. J.* **354**, L37–L40 (1990).]

For a second experimental relation connecting h and k, we could take equation (6.22), which has the content

$$\frac{k}{h} = \frac{\nu_{\text{at peak}}}{2.82\,T}, \tag{6.36}$$

where $\nu_{\text{at peak}}$ denotes the frequency at which the spatial energy density per unit frequency interval has its maximum. An experimental determination of $\nu_{\text{at peak}}/T$ would fix the ratio k/h, and then both h and k could be extracted from equations (6.35) and (6.36).

Equivalent information about a peak value was available to Planck. Otto Lummer and Ernst Pringsheim had determined an experimental maximum in the quantity "spatial energy density per unit wavelength interval." (Problem 3 suggests a route to the detailed theoretical expression.) Their measurement gave Planck a value for the ratio k/h, equivalent to that provided by equation (6.36), and so Planck could extract both h and k. His values were within 3 percent of today's values.

6.4 Entropy and evolution (optional)

This section poses a paradox and then resolves it. We begin with two eras in the Earth's history. Bacteria arose in warm ocean waters approximately 3.5 billion years ago. Dinosaurs roamed the lush, tree-studded valley of the Connecticut River 200 million years ago. A macrostate with complex animals like dinosaurs surely has smaller multiplicity than a macrostate with just bacteria. It looks as though the Earth's entropy *decreased* in the time interval between the formation of bacteria and the age of dinosaurs. In short, history suggests

$$\Delta S_{\text{Earth}} < 0 \tag{6.37}$$

for the time interval under consideration. But how can that be?

The Earth absorbs sunlight on its illuminated side and thus absorbs energy. The energy arrives as visible light (for the most part). Over periods of geologic time, the Earth has not gotten dramatically hotter, and so it must be radiating away energy at about the same rate that it receives energy. The Earth does that by emitting infrared radiation. Indeed, between a scorching summer afternoon and dawn the next day, the air temperature can drop by 15 degrees Celsius as infrared radiation streams out through a clear night sky. (Note. The greenhouse effect and global warming are matters of extremely serious concern. They are, however, very recent on a geologic time scale: a matter of 100 years or so relative to billions of years. According to the geologic record, the Earth's temperature fluctuates, even when averaged over thousands or millions of years. Almost surely, however, the average temperature has varied by less than several tens of degrees Celsius since life formed.) Thus, to first approximation, the energy flow satisfies the equation

$$\begin{pmatrix} \text{energy in from sun} \\ \text{with the visible} \\ \text{radiation} \end{pmatrix} = \begin{pmatrix} \text{energy out from} \\ \text{Earth with the} \\ \text{infrared radiation} \end{pmatrix}. \tag{6.38}$$

In section 6.2, we calculated the entropy per unit volume, S/V, associated with blackbody radiation. If we approximate both the radiation from the sun and the radiation leaving the Earth as blackbody radiation, then we may say that the radiation has not only an energy density $\langle E \rangle / V$ but also an entropy density S/V. The heating and cooling in our general relationship

$$\Delta S \geq \frac{q}{T}, \tag{6.39}$$

can be expressed here in terms of the entropy of incoming and outgoing radiation:

$$\Delta S_{\text{Earth}} \geq \left[\begin{pmatrix} \text{entropy in from} \\ \text{sun with the} \\ \text{visible radiation} \end{pmatrix} - \begin{pmatrix} \text{entropy out from} \\ \text{Earth with the} \\ \text{infrared radiation} \end{pmatrix} \right]. \tag{6.40}$$

On the right-hand side, which term dominates? The visible radiation is characteristic of the sun's surface, whose temperature is roughly 6,000 K. The infrared radiation is characteristic of the Earth's surface temperature, roughly 300 K. The temperatures differ by a factor of 20. Recall that the energy of a typical blackbody photon is $2.7kT$, and so the energy of a typical infrared photon is only $1/20$ that of a photon of the visible light. Thus, for every photon arriving from the hot solar surface, the Earth emits approximately 20 infrared photons. That increase in number suggests higher multiplicity and higher entropy.

Indeed, equations (6.19) and (6.26) imply the relationship

$$\frac{S}{\langle N \rangle} = 3.60k, \tag{6.41}$$

that is, the entropy per photon in blackbody radiation has the constant value $3.60k$, independent of temperature. Equation (6.38) told us that the rates of energy flow are equal. Because 20 photons leave the Earth for every photon that arrives, the constant value of the entropy per photon in blackbody radiation suggests that 20 times as much entropy leaves as arrives. [In fact, the entropy ratio exceeds 20. A patch of Earth radiates into a hemisphere of sky, but the light that reaches us from the sun forms nearly parallel rays. Thus the outgoing radiation has a greater directional multiplicity than the incoming radiation. The quotient in equation (6.41) refers to pure blackbody radiation, in which radiation travels equally in all directions.] The difference in square brackets in equation (6.40) is dominated by the infrared term (the right-hand term), and so the difference is negative. In short, thermodynamics implies

$$\Delta S_{\text{Earth}} \geqslant (\text{some negative value}). \tag{6.42}$$

In a manner of speaking, the Earth exports more entropy than it imports, and so—despite rapid processes on Earth, such as volcanic eruptions—the net value of ΔS_{Earth} is *not* required to be greater than zero.

Figure 6.5 displays the implications for ΔS_{Earth} made by history and by thermodynamics. Because the right-hand side of equation (6.42) is negative, it is possible for ΔS_{Earth} to be both greater than the right-hand side and yet negative. Without the sun and without the consequent influx and outflux of radiation, thermodynamics would have implied $\Delta S_{\text{Earth}} \geqslant 0$, and hence dinosaurs could not have developed from bacteria.

Conclusions

1. History and thermodynamics are consistent.
2. The sun was essential to the development of our life—alone from a thermodynamic point of view.

To be sure, the thermodynamic analysis does not imply that life must arise, nor does it replace the insights of Charles Darwin. Rather, the thermodynamic calculation shows that life is permitted to arise and to evolve.

6.5 Sound waves and phonons

Recall the general structure of this chapter, thus far: an introductory section on the thermal physics of waves and then several sections devoted to electromagnetic waves. Now the chapter turns to sound waves in a solid. We continue to rely on the unified

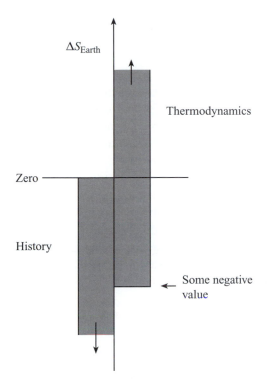

Figure 6.5 What history and thermodynamics individually imply for the entropy change of the Earth. Each shaded bar with an arrow (for continuation) represents the *range of possible values* for ΔS_{Earth} according to history or thermodynamics. The time interval over which the change in entropy occurs stretches from the emergence of bacteria to the age of dinosaurs; what is plotted is ΔS_{Earth} for the interval from 3.5×10^9 years ago to 200×10^6 years ago. For the implied ranges of ΔS_{Earth}, there is a region of overlap.

description in section 6.1, but we apply that analysis to the coherent vibration of atoms in a crystal.

In a solid, the interaction of a given atom with its neighbors confines the atom to the immediate vicinity of some site in the lattice structure. (Actually, some wandering occurs occasionally, but we suppress that possibility.) The atoms can, however, vibrate about their equilibrium positions in the lattice, and there is energy, both kinetic and potential, associated with such motion and mutual interaction. Our objective is to estimate the heat capacity of a solid, especially as a function of temperature, on the basis of this picture.

A hierarchy of methods, successively increasing in complexity, exists for coping with the mutual interactions of the atoms. The essential ingredient of the simplest model is this: the position-dependent potential energy for any given atom is calculated as though all the other atoms were at precisely their equilibrium positions. This approximation decouples the vibrational motions of the various atoms from one another. Albert Einstein worked out the consequences of that picture, called the

Einstein model, in 1906; it was the first application of quantum theory to solid state physics. Problem 16 explores the Einstein model, but here we go on to a more realistic picture.

For the moment, we can think of the atoms as coupled together by Hooke's law springs between neighboring pairs of atoms. (There is an amusing consistency here, for Hooke's law comes precisely from the actual behavior of interatomic forces in a solid.) In one dimension, an instantaneous picture might look like part (a) of figure 6.6. The sinusoidal curve indicates the amount of displacement from the equilibrium sites, each of which is indicated by a short vertical line.

Because the force exerted on a given atom by its neighbors now depends on *their* displacement as well as its own, the theory admits the possibility of *coherent* vibrations, neighboring atoms vibrating with definite amplitude and phase relations between them. We have opened the door to sound waves in the solid. The sketch displays a portion of such a wave, one wavelength's worth, to be precise. The frequency of vibration will be related to the wavelength by the speed of sound c_s in the solid. The heart of the *Debye model* consists of an astute association of two notions: coherently vibrating atoms and sound waves.

The longest wavelength will be of the order of the crystal's diameter. The associated frequency is relatively low, and we take it to be zero without loss of accuracy.

The shortest wavelength arises when adjacent atoms vibrate in opposite directions: out of phase by 180 degrees. Part (b) of figure 6.6 illustrates this situation. Thus

$$\lambda_{min} = 2 \times (\text{interatomic spacing}), \tag{6.43}$$

and so the maximum frequency is given by

$$\nu_{max} = \frac{c_s}{\lambda_{min}}$$

$$= \frac{c_s}{2 \times (\text{interatomic spacing})} = \frac{c_s}{2a}. \tag{6.44}$$

(a)

(b)

Figure 6.6 Coherent vibrations as sound waves. (a) A typical longitudinal wave. (b) The wave with smallest wavelength and highest frequency.

The symbol a denotes the interatomic spacing. This result for the maximum frequency, derived in one dimension, continues to hold in three dimensions, at least in order of magnitude, as we shall see.

Density of modes

The two parts of figure 6.6 show atomic displacements parallel to the direction in which a traveling wave would propagate. Such displacements constitute a *longitudinal* wave. The atoms may also vibrate perpendicular to the propagation direction. Such displacements produce a *transverse* wave in the solid, just as the electric and magnetic fields make an electromagnetic wave a transverse wave. Two distinct polarizations exhaust the transverse possibilities.

In an actual crystal and for a given wavelength, the frequencies of longitudinal and transverse waves usually differ, and hence so do the associated speeds of sound. The simplest version of the Debye model ignores this difference. It takes the speed of sound to be a constant, independent of wavelength, the longitudinal-versus-transverse distinction, and the direction of propagation. The presence of longitudinal modes, in addition to the two transverse modes, increases the density of modes by the ratio $3/2$ (relative to the electromagnetic density of modes). Thus the density of modes $D_{\text{Debye}}(\nu)$ for the simplest Debye model is

$$D_{\text{Debye}}(\nu) = \frac{3}{2} \times \frac{8\pi}{c_s^3} V\nu^2. \tag{6.45}$$

If the N atoms that form the crystal were entirely uncoupled from one another, each could vibrate independently in three orthogonal directions, and so there would be $3N$ distinct modes. Coupling, which leads to coherent vibration, cannot change the total number of modes. Thus, to determine the maximum frequency ν_{max}, stipulate that the integral of the density of modes from zero to ν_{max} yields $3N$ modes:

$$\int_0^{\nu_{\text{max}}} D_{\text{Debye}}(\nu)\, d\nu = 3N. \tag{6.46}$$

Thus

$$\int_0^{\nu_{\text{max}}} \frac{3}{2} \times \frac{8\pi}{c_s^3} V\nu^2\, d\nu = \frac{4\pi V}{c_s^3} \nu_{\text{max}}^3 = 3N, \tag{6.47}$$

whence

$$\nu_{\text{max}} = \left(\frac{3}{4\pi}\right)^{1/3} \frac{c_s}{(V/N)^{1/3}} = \left(\frac{3}{4\pi}\right)^{1/3} \frac{c_s}{a}. \tag{6.48}$$

The root $(V/N)^{1/3}$ gives the average interatomic spacing a. The present result agrees with equation (6.44) both in its dependence on crystal parameters and also in order of magnitude.

The maximum frequency defines a characteristic temperature, the *Debye temperature* θ_D, by the relation

$$k\theta_D \equiv h\nu_{max}. \tag{6.49}$$

Equation (6.48) enables one to express the Debye temperature in terms of two crystal parameters, the sound speed and the average interatomic spacing:

$$\theta_D = \left(\frac{3}{4\pi}\right)^{1/3} \frac{hc_s}{ka}. \tag{6.50}$$

The usefulness of the Debye temperature will become apparent soon.

Energy and heat capacity

As noted in section 6.1, quantum theory restricts the energy of a sound wave mode to increments of magnitude $h\nu$. With each such increment, one says that an additional *phonon* is present. The estimated number of phonons $\bar{n}(\nu)$ is given by (6.6).

If the coherent vibrations are responsible for the crystal's total energy, then the expression for $\langle E \rangle$ takes the form

$$\langle E \rangle = \int_0^{\nu_{max}} h\nu\bar{n}(\nu)D_{Debye}(\nu)\, d\nu$$

$$= \int_0^{\nu_{max}} \frac{12\pi h}{c_s^3} V \frac{\nu^3}{e^{h\nu/kT} - 1}\, d\nu. \tag{6.51}$$

For explicit evaluation, we consider two regimes of temperature: high and low.

High temperature: $T \gg \theta_D$

When the inequality $T \gg \theta_D$ holds, the exponent $h\nu/kT$ in (6.51) is always much less than 1, for its maximum value is

$$\frac{h\nu_{max}}{kT} = \frac{\theta_D}{T} \ll 1. \tag{6.52}$$

To simplify the integral, we can expand the exponential and retain just the first two terms:

$$\langle E \rangle = \frac{12\pi h}{c_s^3} V \int_0^{\nu_{max}} \frac{\nu^3}{1 + (h\nu/kT) + \cdots - 1}\, d\nu$$

$$= kT \times \frac{12\pi V}{c_s^3} \int_0^{\nu_{max}} \nu^2\, d\nu$$

$$= kT \times 3N, \tag{6.53}$$

where the last step follows from (6.47).

The heat capacity C_V is simply

$$C_V = \left(\frac{\partial \langle E \rangle}{\partial T}\right)_V = 3Nk. \tag{6.54}$$

Note that, in this high temperature regime, C_V is independent of the crystal parameters c_s and a.

Low temperature: $T \ll \theta_D$

When the inequality $T \ll \theta_D$ holds, expansion of the exponential in (6.51) for merely a few terms is not a valid route. Rather, the integral is best expressed first in terms of the dimensionless variable $x = h\nu/kT$:

$$\langle E \rangle = \frac{12\pi hV}{c_s^3}\left(\frac{kT}{h}\right)^4 \times \int_0^{\theta_D/T} \frac{x^3}{e^x - 1}\, dx. \tag{6.55}$$

The upper limit of integration is $x_{max} = h\nu_{max}/kT = \theta_D/T$. Because the physical temperature T is much less than the Debye temperature, the upper limit of integration is large relative to 1. Consequently, the integrand is exponentially small at the upper limit, and we may—for convenience and without loss of accuracy—extend the upper limit to infinity; then the integral takes the value $\pi^4/15$, as tabulated in appendix A. The energy is proportional to T^4, and the heat capacity C_V becomes

$$C_V = \left(\frac{\partial \langle E \rangle}{\partial T}\right)_V = \left(\frac{48\pi^5 k^4}{15 c_s^3 h^3}\right) T^3 V. \tag{6.56}$$

This low temperature expression for C_V, the *Debye T^3 law*, is just $3/2$ times the electromagnetic result, once the speed of sound has replaced the speed of light. Basically, when the inequality $T \ll \theta_D$ holds, the estimated phonon number $\bar{n}(\nu)$ becomes much smaller than 1 well before the frequency approaches ν_{max}, and so the existence of a cut-off to the integration at ν_{max} is irrelevant. The outcome must be like the electromagnetic result, except for the $3/2$ to account for the longitudinal waves.

Figure 6.7 shows the excellent agreement between the Debye model and experiment for solid argon at low temperature. The general run of the heat capacity is illustrated in figure 6.8; that general run must be calculated from the derivative of $\langle E \rangle$ as given in equation (6.51) or (6.55) and then computed numerically. At temperatures greater than the Debye temperature, C_V approaches the value $3Nk$, as we calculated earlier. Later, in chapter 13, we will see that $3Nk$ is the value that classical physics would predict for C_V (at every temperature, alas). For the low temperature end of the graph, note that equation (6.50) enables one to express the factor $Vk^3/c_s^3 h^3$ in (6.56) in terms of N and θ_D:

$$C_V = \frac{12\pi^4}{5} Nk \left(\frac{T}{\theta_D}\right)^3. \tag{6.57}$$

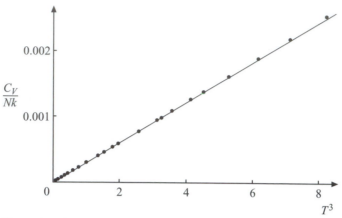

Figure 6.7 The heat capacity of solid argon at low temperatures: 0.44 K to 2.02 K. The slope of the C_V versus T^3 line is well fit with $\theta_D = 92$ K. Then the upper end of the temperature range corresponds to $T/\theta_D = 2/92 = 0.022$, and so the Debye approximation should be excellent, as indeed it is. [*Source*: Leonard Finegold and Norman E. Phillips, *Phys. Rev.* **177**, 1383–91 (1969).]

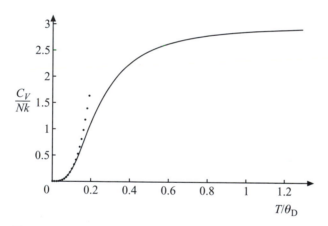

Figure 6.8 The full run of the Debye theory's heat capacity as a function of temperature. The dotted curve is a strictly T^3 curve whose third derivative at the origin agrees with the Debye theory's third derivative. Evidently the T^3 behavior is restricted to temperatures below $T \cong 0.1\theta_D$. Indeed, when $T \leqslant 0.1\theta_D$, the T^3 form differs from the exact value by 1 percent or less. At the other limit, the heat capacity has reached 95 percent of its asymptotic value when $T/\theta_D = 1$, specifically, $C_V/Nk = (0.952) \times 3$ at $T = \theta_D$.

When the temperature is low, specifically, when $T \leqslant 0.1\theta_D$, so that the T^3 behavior holds, then the heat capacity is substantially smaller than its classical limit. That conclusion is evident in figure 6.8, also.

Table 6.2 gives a selection of Debye temperatures. Crystals formed from the noble gases are loosely bound; so the maximum frequency ν_{max} should be small and,

Table 6.2 *Some elements and their experimental Debye temperatures, taken from experiments in the range where the T^3 law holds. Most of the experiments were done at liquid helium temperatures and below: 4.2 K and lower.*

	θ_D (K)		θ_D (K)
Neon	75	Carbon (diamond)	2,230
Argon	93	Iron	467
Lithium	344	Cobalt	445
Sodium	158	Copper	343
Potassium	91	Zinc	327
Rubidium	56	Silver	225
Cesium	38	Gold	165
Silicon	640	Boron	1,250

Source: *AIP Handbook*, 3rd edn, edited by D. E. Gray (McGraw-Hill, New York, 1972).

correspondingly, θ_D should be small. The sequence of alkali metals, lithium to cesium, goes toward less reactive elements and hence toward weaker binding, lower maximum frequency, and lower θ_D. Diamond is extremely hard; hence high frequencies should be expected, and a high Debye temperature should follow.

Some history

In 1819, the French physicists Pierre-Louis Dulong and Alexis-Thérèse Petit announced a remarkable finding: the heat capacity per atom was very nearly the same for the dozen metals (plus sulfur) that they had studied. Their units were different from ours, but their result was equivalent to the statement that $C_V/N \cong 3k$ at the temperatures they used. Table 6.3 shows some modern values for C_P/Nk at 298 K, equivalent to 25 °C. (For a solid, which expands little, $C_P \cong C_V$, and C_P is what Dulong and Petit actually measured.) The elements iron through gold in the second column are ones that Dulong and Petit investigated. They determined heat capacities by surrounding the bulb of a mercury thermometer with a hot powdered sample and subsequently monitoring the cooling rate. The initial temperatures lay in the range 300 to 500 K. Their experimental values for C_P/N had comparably small variations from an average, and so their conclusion was amply warranted. In their honor, the high temperature limit for electrically insulating solids, $C_V/N \to 3k$, is called the *Dulong and Petit value*. (Later, in chapter 9, we will see that conduction electrons in an electrically conducting solid contribute separately and additionally to the heat capacity; so one needs to distinguish between electrically insulating and electrically conducting solids.)

Einstein had been impressed by Planck's theory of blackbody radiation. In 1905, he suggested a particulate nature for light, what came to be called the photon aspect of

Table 6.3 *Heat capacity per atom at 298 K for selected solids. The tabulated numbers are C_P/Nk, a dimensionless quantity.*

	C_P/Nk		C_P/Nk
Lithium	2.93	Silicon	2.37
Sodium	3.39	Iron	3.01
Potassium	3.54	Cobalt	3.20
Rubidium	3.70	Copper	2.94
Cesium	3.81	Zinc	3.04
Carbon (diamond)	0.749	Silver	3.07
Boron	1.33	Gold	3.05

Source: *CRC Handbook of Chemistry and Physics*, 71st edn, edited by David R. Lide (Chemical Rubber Publishing Company, Boston, 1992).

the wave–particle duality. A year later, Einstein reasoned that the discreteness in energy should carry over to atomic vibrations in a crystal. That extension, he said, could resolve some difficulties in understanding the thermal and optical properties of certain solids. Among other items, Einstein had in mind the heat capacity (per atom) of carbon, boron, and silicon. As table 6.3 shows, those heat capacities fall well below the Dulong and Petit value. For diamond, data in the range $222 \leqslant T \leqslant 1258$ K were available to Einstein. At the lowest temperature, $C_P/Nk = 0.385$, a mere 13 percent of the Dulong and Petit value. Einstein's model, with only a single vibrational frequency (rather than the spectrum of the Debye theory), did well over that range of temperatures (although a systematic deviation began to appear in the lowest 15 percent of the range). The Debye T^3 behavior is clearly evident only when $T \leqslant 0.1\theta_D$, and table 6.2 indicates that the threshold for diamond is $0.1\theta_D \cong 220$ K, right where the data available to Einstein ended.

By 1912, experiments in Walther Nernst's lab had pushed heat capacity measurements down to temperatures as low as 23 K. For Einstein's theory, the evidence was unambiguous: the theory failed at such low temperatures. Unerringly, Einstein had put his finger on the key element—discrete energies—but a single vibrational frequency was too crude an approximation to represent a solid at very low temperature. In a comprehensive paper published in 1912, Peter Debye proposed the theory that this section developed: a spectrum of frequencies, derived from the picture of sound waves. Debye even calculated the speed of longitudinal and transverse waves in terms of basic crystal parameters such as the compressibility and the shear coefficient. The T^3 behavior when $T \leqslant 0.1\theta_D$ was the theory's novel prediction. Of the data that Debye could marshal in support of his new theory, perhaps the most convincing was the data for copper. Figure 6.9 reconstructs a figure from Debye's paper. The lowest temperature, 23.5 K, corresponds to $T/\theta_D = 23.5/309 = 0.076$. It lies within the region where the T^3 behavior is evident (theoretically), and Debye was pleased with the fit to the data, as well he might be.

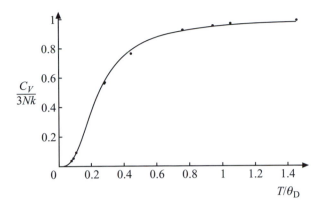

Figure 6.9 The ratio $C_V/3Nk$ versus T/θ_D for copper, as presented by Debye, who chose $\theta_D = 309$ K. The experiments provided C_P; Debye corrected to C_V before plotting. (The maximum correction was 3.6 percent.) [*Source*: Peter Debye, *Ann. Phys. (Leipzig)* **39**, 789–839 (1912).]

6.6   Essentials

1. The unified description of electromagnetic waves and sound waves (in a crystal) centers on normal modes and the relationship

$$\left(\begin{array}{l} \text{estimated number of photons or} \\ \text{phonons in mode with frequency } \nu \end{array} \right) \equiv \bar{n}(\nu) = \frac{1}{e^{\,h\nu/kT} - 1}.$$

Thermal equilibrium is presupposed, and the canonical probability distribution provides the foundation for the derivation.

2. The estimated energy of a mode is given by

$$\langle \varepsilon \rangle = \frac{1}{e^{\,h\nu/kT} - 1}\, h\nu.$$

3. A sum over all electromagnetic modes may be replaced with an integral:

$$\sum_{\text{EM modes}} A(\nu) = \int A(\nu) D_{\text{EM}}(\nu)\, d\nu,$$

where

$$\left(\begin{array}{l} \text{number of EM modes in the} \\ \text{frequency range } \nu \text{ to } \nu + d\nu \end{array} \right) \equiv D_{\text{EM}}(\nu)\, d\nu = \frac{8\pi}{c^3}\, V\nu^2\, d\nu.$$

4. The *Planck distribution* is the relationship

$$\left(\begin{array}{l} \text{spatial energy density} \\ \text{per unit frequency interval} \end{array} \right) = \frac{8\pi h}{c^3}\, \frac{\nu^3}{e^{\,h\nu/kT} - 1}.$$

5. In thermal equilibrium, a gas of photons has energy and pressure as follows:

$$\langle E \rangle = \left(\frac{8\pi^5 k^4}{15 c^3 h^3} \right) V T^4,$$

$$P = \frac{1}{3} \frac{\langle E \rangle}{V}.$$

6. The peak in the Planck distribution occurs at

$$(h\nu)_{\text{at peak}} = 2.82 kT.$$

Thus the energy of a typical photon is approximately $3kT$. The frequency at the peak shifts linearly with changes in temperature; this is one version of *Wien's displacement law*.

7. A body at temperature T emits radiation according to the *Stefan–Boltzmann law*:

$$\left(\begin{array}{c} \text{flux emitted by surface} \\ \text{with absorptivity } a \end{array} \right) = a\sigma_{\text{B}} T^4.$$

The absorptivity a lies in the range $0 \leqslant a \leqslant 1$, being 1 for a perfect absorber, that is, for a black body.

8. The *Debye theory* of crystalline lattice vibrations is based on sound waves and normal modes. In the simplest version, the number of modes per unit frequency interval, $D_{\text{Debye}}(\nu)$, differs from the electromagnetic analog in three ways: (1) multiplication by $3/2$ because a crystal has longitudinal waves in addition to transverse waves; (2) replacement of the speed of light by the (suitably averaged) speed of sound, and (3) a cut-off at a maximum frequency, ν_{max}, because a crystal with N atoms has only $3N$ independent modes.

9. The maximum frequency leads to a characteristic temperature, the *Debye temperature* θ_{D}:

$$k\theta_{\text{D}} \equiv h\nu_{\text{max}}.$$

10. When the temperature is low, specifically, when $T \leqslant 0.1\theta_{\text{D}}$, the heat capacity of the lattice is

$$C_V = \frac{12\pi^4}{5} Nk \left(\frac{T}{\theta_{\text{D}}} \right)^3.$$

This heat capacity varies as T^3 and is much smaller than the classical value of Dulong and Petit: $3Nk$.

Further reading

The cosmic blackbody radiation is ably surveyed by R. B. Partridge in his book, *3K: The Cosmic Microwave Background Radiation* (Cambridge University Press, New York, 1995).

Problems

1. *Contributions to the pressure within the sun.* The major contributors to the pressure inside the sun are (1) atoms, ions, and electrons (collectively called "particles" here) and (2) photons. Calculate the contribution from each under the conditions specified below; also, form the ratio of radiation pressure to particle pressure. Take the particles to act like a classical ideal gas.

(a) Photosphere (the solar surface): $T = 5,800$ K. Particle number density: $N/V = 10^{23}$ particles per cubic meter.

(b) Center: $T = 1.5 \times 10^7$ K. Particle number density: $N/V = 5 \times 10^{31}$ m^{-3}.

2. *Expanding radiation.*

(a) Consider radiation in thermal equilibrium at an initial temperature T_i. The radiation is allowed to expand slowly and adiabatically so that it occupies a volume which is larger by the factor $x_{v.e.f.}$, where "v.e.f." denotes "volume expansion factor." The peak in the energy distribution (as a function of frequency) shifts. What is the ratio

$$(\nu_{\text{at peak, final}})/(\nu_{\text{at peak, initial}})?$$

(b) In the early universe, free electrons and nuclei combined to form neutral hydrogen and helium when the temperature dropped to approximately 3,000 K. Neutral matter scatters radiation much less than charged particles do; therefore, the then-existing blackbody radiation could subsequently expand freely, uncoupled to matter (to good approximation). By what factor has the "volume of the universe" increased since the ionization in space dropped to essentially zero?

3. *Spatial energy density per unit wavelength interval.* Wavelength λ and frequency ν are related by the speed of light c. Therefore, to a narrow frequency range ν to $\nu + \Delta\nu$, there corresponds a unique wavelength range λ to $\lambda + \Delta\lambda$.

(a) Determine the ratio $\Delta\lambda/\Delta\nu$.

Then reason that the energy in corresponding intervals must be equal:

$$\left(\begin{array}{c}\text{spatial energy density} \\ \text{per unit wavelength interval}\end{array}\right) \times |\Delta\lambda| = \left(\begin{array}{c}\text{spatial energy density} \\ \text{per unit frequency interval}\end{array}\right) \times |\Delta\nu|.$$

(b) Determine the "spatial energy density per unit wavelength interval" as a function of λ, c, h, k, and T.

(c) For fixed temperature T, what value of λ yields the maximum in the density? Express $\lambda_{at\ max}$ in terms of c, h, k, T, and a dimensionless number. Then compare $\lambda_{at\ max}$ with the quantity $\nu_{at\ peak}$ that is provided by equation (6.22).

(d) Based upon your analysis of spatial energy density per unit wavelength interval, provide a "wavelength version" of Wien's displacement law, complete with its own numerical coefficient.

(e) Take the sun to have a surface temperature of 5,800 K. What is the associated value of $\lambda_{at\ max}$? To what color of light does it correspond? (For your mental data bank, note the "Convenient typical values" in appendix A.)

4. *Radiation results by dimensional analysis.*

(a) For radiation in thermal equilibrium, the spatial energy density $\langle E \rangle / V$ must depend on merely kT, h, c, and a dimensionless constant of proportionality. By considering the dimensions (or the units) of kT, h, and c, determine how $\langle E \rangle / V$ depends on those three quantities.

(b) Extend your reasoning to the energy flux from a small hole in a cavity that contains blackbody radiation.

5. *Carnot cycle with photons.* Figure 3.2 displayed the classic Carnot cycle; take the "gas" to be a gas of photons in thermal equilibrium.

(a) For the isothermal expansion from V_1 to V_2, compare the magnitudes of the work done by the gas and the energy absorbed from the hot reservoir. How does the relationship here compare with that for a monatomic classical ideal gas?

(b) Calculate the total work done by the photons in one cycle (in terms of V_1, V_2, and the two temperatures T_{hot} and T_{cold}). Do this by evaluating separately the work done in each of the four segments of the cycle and then combining the contributions (with due regard for sign).

(c) Calculate the efficiency and compare it with the Carnot efficiency that we derived as a general result.

6. The entropy of a certain (realistic) physical system is $S = AE^{3/4}$ as a function of the energy E at fixed volume. The constant A has the value $A = 1.26 \times 10^{-4}$ J$^{1/4}$/K. When the system's energy is $E = 3.3 \times 10^{-6}$ J, what is the system's temperature?

7. *An older route to the bulk properties of blackbody radiation.*

(a) Derive the general relationship

$$P = -\left(\frac{\partial \langle E \rangle}{\partial V}\right)_T + T\left(\frac{\partial P}{\partial T}\right)_V. \tag{1}$$

A good route can be constructed from equations (5.16) and (5.19).

(b) Check equation (1) against the ideal gas law.

(c) Specify that

$$\langle E \rangle = f(T)V \text{ and } P = \text{const} \times \langle E \rangle/V,$$

that is, the total energy scales linearly with the volume (at fixed temperature), and the pressure is proportional to the energy density. Use equation (1) to determine the function $f(T)$ as far as you can.

(d) When the radiation is isotropic, already classical electromagnetism implies that $P = \frac{1}{3}\langle E \rangle/V$. For thermal radiation, what conclusion can you draw about $\langle E \rangle/V$ as a function of temperature?

8. *Kitchen oven.* The goal is to raise the temperature of 1 cm^3 of water by transferring energy from a cavity filled with blackbody radiation. The water is to change from 299 K to 300 K. (To adequate accuracy, the heat capacity of water is 4.2 J/gram · K, and 1 cm^3 of water has a mass of 1 gram.) The radiation is initially at "cookie baking temperature," 350 °F, equivalent to 450 K (and the heating coils have been turned off).

(a) If the radiation is to drop in temperature by no more than 10 K, how large must the cavity be? (Provisionally, take the energy of the radiation to be the sole energy source.)

(b) If the cavity has a volume of 0.1 cubic meter (which is typical of a kitchen oven), how long would you need to have the heating coils turned on to heat the water? (The voltage is typically 240 volts, and the current is limited to 20 amperes, say.)

9. *The factor of 1/4 in the radiative flux.* Adapt the methods of section 1.2 and problem 1.2 to compute the energy per second and per unit area delivered to a cavity wall (or to a hole). Aim to emerge with the form in the first line of equation (6.28). If you need the average of $\cos\theta$ over a hemisphere, try integration with spherical polar coordinates.

10. *Light bulb.* A light bulb filament is made of 12 cm of tungsten wire; the wire's radius is 1.0×10^{-4} meter. Take the emissivity of tungsten to be 0.4 at the bulb's operating temperature.

(a) What is that temperature when the bulb consumes a power of 67 watts?

(b) Suppose you want to shift the peak in the emitted spectrum to a frequency that is higher by 10 percent. By what percentage must you increase the power supplied to the bulb?

11. *Energy balance for the Earth.* Take the sun to radiate like a blackbody with a surface temperature of 5,800 K. Relevant distances and sizes are provided in appendix A.

(a) Calculate the flux of solar radiant energy at the Earth's orbital distance (in watts per square meter).

(b) Take the Earth to be in a thermal steady state, radiating as much energy (averaged over the day) as it receives from the sun. Estimate the average surface temperature of the Earth. Spell out the approximations that you make.

12. *Radiation shield: black surfaces.* Two large plane surfaces, at fixed temperatures T_{hot} and T_{cold}, face each other; a narrow evacuated gap separates the two black surfaces.

(a) Determine the net radiant energy flux from the hotter to the colder surface (in watts per square meter).
(b) Specify now that a thin metallic sheet, black on both sides, is inserted between the original two surfaces. When the sheet has reached thermal equilibrium, (1) what is the temperature of the metallic sheet and (2) what is the new net radiant energy flux (in terms of T_{hot} and T_{cold})?
(c) If n such black metallic sheets are inserted, what is the net radiant energy flux?

13. *Radiation shield: reflective surfaces.* The context is the same as in problem 12 except that all surfaces are partially reflective and have the same absorptivity a, where $a < 1$.

(a), (b), (c) Repeat parts (a), (b), and (c) of problem 12.
(d) Specify that liquid helium at 4.2 K fills a spherical tank of radius 30 cm and is insulated by a thin evacuated shell that contains 60 layers of aluminized plastic, all sheets slightly separated from one another. Take the absorptivity of the aluminum coating to be $a = 0.02$. The outer surface is at room temperature. If the only significant energy influx is through the radiation shield, how long will it take before all the helium has evaporated? [A value for the energy required to evaporate one atom of ^{4}He (the latent heat of vaporization, L_{vap}) is provided by table 12.2, and the number density N/V can be extracted from table 5.2.]

14. *Carnot cycle on a spaceship.* A Carnot engine is to supply the electrical power (denoted *Pwr* and measured in units of joules per second) needed on a small spaceship (that is far from the sun and hence far from solar energy). The temperature T_{hot} of the hot reservoir is fixed (by some reaction, say). The energy that enters the cold reservoir is radiated into empty space at the rate $\sigma_B T_{cold}^4 \times A$, where A denotes the area of a black surface. A steady state is to hold for that reservoir, that is, the surface radiates as much energy per second as the engine dumps into the cold reservoir.

(a) Determine the minimum area A_{min} as a function of the fixed parameters *Pwr* and T_{hot} (together with the Stefan–Boltzmann constant σ_B).
(b) When the engine operates at the minimum area, what is its efficiency?
(c) Could you improve the efficiency by changing the size of the area A? Explain your response.

15. *Simulating a laser.* The typical helium–neon laser in an instructional lab produces a nearly monochromatic red beam of wavelength 632.8 nanometers and a total power of 0.5 milliwatt. The wavelength spread of the beam is 10^{-3} nanometers, and the beam emerges as a (narrow) cone of light with half-angle (angle from axis of cone to cone itself) of 5×10^{-4} radian. Imagine approximating the laser beam by filtering appropriately the radiation from 1 mm^2 of a blackbody.

(a) What temperature would be required for the body? Explain your reasoning and any approximations.
(b) At what frequency would the radiation spectrum—before filtering—have its maximum? What would be the energy (in eV) of a photon associated with the peak frequency?
(c) In what essential characteristic would the blackbody simulation fail to match the laser?

16. *Einstein model.* Specify that all N atoms in a crystal vibrate independently of one another and with the same frequency ν, a fixed value. Define the *Einstein temperature* θ_E, a characteristic temperature, by the equation $k\theta_E \equiv h\nu$.

(a) Calculate $C_V/3Nk$ as a function of temperature. Graph the behavior as a function of the dimensionless variable T/θ_E.
(b) In 1906, Einstein had available the data for diamond displayed in table 6.4. Graph both data and theoretical curve as a function of T/θ_E. (There is no need to distinguish between $C_V/3Nk$ and $C_P/3Nk$.) You will have to adopt a good value for the Einstein temperature θ_E of diamond. You can do that by forcing agreement at one temperature or by some more sophisticated method. Cite your value for θ_E. How well did Einstein's theory explain the data? [Einstein himself noted that refinements were in order, such as a slow change in the frequency ν as the material contracted with decreasing temperature. For details, you can consult A. Einstein, *Ann. Phys. (Leipzig)* **22**, 180–90 (1907).]

Table 6.4 *Data on $C_P/3Nk$ for diamond as a function of temperature (as of 1906).*

T (K)	$C_P/3Nk$	T (K)	$C_P/3Nk$
222.4	0.1278	413.0	0.4463
262.4	0.1922	479.2	0.5501
283.7	0.2271	520.0	0.6089
306.4	0.2653	879.7	0.8871
331.3	0.3082	1,079.7	0.9034
358.5	0.3552	1,258.0	0.9235

Source: A. Einstein, *Ann. Phys. (Leipzig)* **22**, 180–90 (1907).

17. *Debye model with two distinct speeds.* Suppose one preserved the distinction between longitudinal and transverse waves in a crystal, assigning them speeds c_l and c_t. How would the results of the simplest Debye theory be modified?

For copper, the measured speeds are $c_l = 4,760$ m/s and $c_t = 2,325$ m/s. The average interatomic spacing is $a = 2.28 \times 10^{-10}$ meter. How does your predicted value of θ_D compare with the measured value?

18. *A high temperature expansion.*

(a) For the Debye theory, section 6.5 provided the first term in both the energy $\langle E \rangle$ and the heat capacity C_V when the temperature is large relative to the Debye temperature. Extend the calculation by computing, for both $\langle E \rangle$ and C_V, the *second non-vanishing* term in an expansion in powers of θ_D/T. You can expand the integrand in a power series, which will also simplify the integration. Take a sufficient number of terms at each stage so that your final results will be complete, that is, they will contain all contributions at the desired power of θ_D/T.

(b) When $\theta_D/T = 0.5$, by what percentage do $\langle E \rangle$ and C_V differ from the values given by the first term only?

(c) Graph $C_V/3Nk$ versus T/θ_D for the range $0.6 \leqslant T/\theta_D \leqslant 3$. Are your results consistent with the graphs in section 6.5?

19. Calculate the entropy change of the system in the following circumstances. Initial and final states are states of thermal equilibrium. Provide numerical answers.

(a) A monatomic classical ideal gas is initially at $T_i = 300$ K and volume V_i. The gas is allowed to expand into a vacuum so that $V_f = 3V_i$. Then the gas is heated to a final temperature $T_f = 400$ K. Take the number of atoms to be $N = 10^{20}$.

(b) A diamond has been in liquid helium (at $T_i = 4.2$ K) and is transferred to liquid nitrogen (at $T_f = 77$ K). The diamond's heat capacity while in the helium is $C = 10^{-6}$ J/K.

20. *A different density of modes.* Suppose a solid consisting of N atoms has a density of wave modes given by

$$\left(\begin{array}{c} \text{number of wave modes in the} \\ \text{frequency range } \nu \text{ to } \nu + d\nu \end{array} \right) = b\nu^4 \, d\nu,$$

where b denotes a constant (with appropriate dimensions) and where the exponent is indeed 4. The frequencies range from zero to some cut-off.

Calculate the total energy $\langle E \rangle$ and the heat capacity in the limits of both low and high temperature. Computing the leading term in each will suffice. Explain what you mean (and what you think I mean) by "low temperature" and "high temperature." Express your results exclusively in terms of N, h, k, T, and the constant b.

21. A small sphere of silver, painted black, is set adrift in outer space. It has a radius $r_0 = 1$ cm, contains $N = 2.45 \times 10^{23}$ silver atoms, and has an initial temperature $T_i = 300$ K.

(a) How long does it take the sphere to cool to 250 K?
(b) Later, how long does it take to cool from 20 K to 10 K?
(c) What do you expect the sphere's final temperature will be (that is, the asymptotic value)? Why?

22. A block of boron is placed in contact with a block of solid neon. The blocks are otherwise isolated, and no chemical reaction occurs. Additional data are given below. When thermal equilibrium has been reached, what is the final temperature?

Block	Number of atoms	Initial temperature
Boron	$N_B = 1.21 \times 10^{24}$	$T_B = 120$ K
Neon	$N_{Ne} = 1.0 \times 10^{23}$	$T_{Ne} = 90$ K

If you need to approximate, explain and justify your approximations.

23. Consider a thin film of solid material—only one atom thick—deposited on an inert substrate. The atoms may vibrate parallel to the surface but not perpendicular to it. Take the material to be a two-dimensional solid (so that sound waves travel only parallel to the surface). Adapt the Debye theory to this context, and use it for *both* part (a) *and* part (b) below.

A total of N atoms form the solid, which is spread over an area A. The speed of sound c_s is a constant.

(a) Calculate the heat capacity C_A (at constant area) at high temperature.
(b) Calculate the heat capacity C_A at low temperature.
(c) State *explicit* criteria for "high" and "low" temperature.

Express all your final answers in terms of N, c_s, A, h, k, T, and pure numbers.

7 The Chemical Potential

7.1 Discovering the chemical potential
7.2 Minimum free energy
7.3 A lemma for computing μ
7.4 Adsorption
7.5 Essentials

The preface noted that "the book's conceptual core consists of four linked elements: entropy and the Second Law of Thermodynamics, the canonical probability distribution, the partition function, and the chemical potential." By now, three of those items are familiar. The present chapter introduces the last item and, for illustration, works out a typical application. The chemical potential plays a significant role in most of the succeeding chapters.

7.1 Discovering the chemical potential

The density of the Earth's atmosphere decreases with height. The concentration gradient—a greater concentration lower down—tends to make molecules diffuse upward. Gravity, however, pulls on the molecules, tending to make them diffuse downward. The two effects are in balance, canceling each other, at least on an average over short times or small volumes. Succinctly stated, the atmosphere is in equilibrium with respect to diffusion.

In general, how does thermal physics describe such a diffusive equilibrium? In this section, we calculate how gas in thermal equilibrium is distributed in height. Certain derivatives emerge and play a decisive role. The underlying purpose of the section is to discover those derivatives and the method that employs them. We will find a quantity that measures the tendency of particles to diffuse.

Figure 7.1 sets the scene. Two volumes, vertically thin in comparison with their horizontal extent, are separated in height by a distance H. A narrow tube connects the upper volume V_u to the lower volume V_l. A total number N_{total} of helium atoms are in thermal equilibrium at temperature T; we treat them as a semi-classical ideal gas What value should we anticipate for the number N_u of atoms in the upper volume, especially in comparison with the number N_l in the lower volume?

We need the probability $P(N_l, N_u)$ that there are N_l atoms in the lower volume and N_u in the upper. The canonical probability distribution gives us that probability as a sum over the corresponding states Ψ_j:

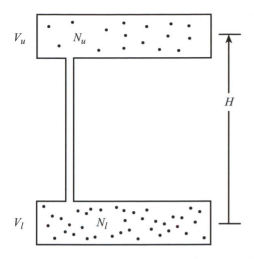

Figure 7.1 The context. The narrow tube allows atoms to diffuse from one region to the other, but otherwise we may ignore it.

$$P(N_l, N_u) = \sum_{\substack{\text{states } \Psi_j \text{ with } N_l \text{ in } V_l \\ \text{and } N_u \text{ in } V_u}} \frac{\exp(-E_j/kT)}{Z} \equiv \frac{Z(N_l, N_u)}{Z}. \tag{7.1}$$

The second equality merely defines the symbol $Z(N_l, N_u)$ as the sum of the appropriate Boltzmann factors. In analogy with our analysis in section 5.6, an energy E_j will have the form

$$E_j = \varepsilon_\alpha(1) + \varepsilon_\beta(2) + \cdots + \varepsilon_\gamma(N_l) \quad + \quad \varepsilon_a'(1') + \varepsilon_b'(2') + \cdots + \varepsilon_c'(N_u).$$

The single-particle states with energy ε_α describe an atom in the lower volume V_l; the states with energy ε_a' describe an atom in the upper volume V_u and are shifted in energy by the gravitational potential energy mgH, where g is the force per unit mass exerted by the Earth's gravitational field. Further reference to section 5.6 tells us that the sum in (7.1) must yield the expression

$$Z(N_l, N_u) = \frac{(V_l/\lambda_{\text{th}}^3)^{N_l}}{N_l!} \times \frac{[(V_u/\lambda_{\text{th}}^3)e^{-mgH/kT}]^{N_u}}{N_u!}. \tag{7.2}$$

The numerators alone correspond to relations (5.29) and (5.34). But we must correct for over-counting. Every set of N_l different single-particle states of the lower volume has been included $N_l!$ times in the numerator; to correct, we divide by $N_l!$. Similar reasoning applies to the upper volume.

Because $Z(N_l, N_u)$ factors so nicely into something dependent on N_l and something similar dependent on N_u, we write it as

$$Z(N_l, N_u) = Z_l(N_l) \times Z_u(N_u), \tag{7.3}$$

where the factors themselves are defined by (7.2). Each factor is equal to the partition function that the corresponding volume has when it contains the indicated number of atoms.

Common experience suggests that, given our specifically macroscopic system, the probability distribution $P(N_l, N_u)$ will have a single peak and a sharp one at that, as illustrated in figure 7.2. An efficient way to find the maximum in $P(N_l, N_u)$ is to look for the maximum in the logarithm of the numerator in equation (7.1). That logarithm is

$$\ln Z(N_l, N_u) = \ln Z_l(N_l) + \ln Z_u(N_{total} - N_l).$$

The number N_u in the upper volume must be the total number N_{total} minus the number of atoms in the lower volume. The derivative with respect to N_l is

$$\frac{d\ln Z(N_l, N_u)}{dN_l} = \frac{\partial \ln Z_l}{\partial N_l} + \frac{\partial \ln Z_u}{\partial N_u}\frac{\partial(N_{total} - N_l)}{\partial N_l}.$$

Partial derivatives appear on the right-hand side because variables such as temperature and volume are to be held constant while one differentiates with respect to the number of atoms. The last derivative on the right equals -1. To locate the maximum, set the full rate of change of $\ln Z(N_l, N_u)$ equal to zero. Thus the maximum in the probability distribution arises when N_l and N_u have values such that

$$\frac{\partial \ln Z_l}{\partial N_l} = \frac{\partial \ln Z_u}{\partial N_u}. \tag{7.4}$$

The equality of two derivatives provides the criterion for the most probable situation. This is the key result. The remainder of the section fills in details, reformulates the criterion, and generalizes it.

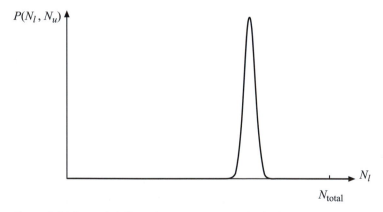

Figure 7.2 The probability of finding N_l atoms in the lower volume and $N_u = (N_{total} - N_l)$ atoms in the upper volume. No drawing can do justice to the sharpness of the peak. Suppose each volume is one liter and that the total number of atoms is $N_{total} = 5 \times 10^{22}$ (as would be true for air under room conditions). If mgH/kT were sufficiently large to place the peak as far off-center as it is shown, then the peak's full width at half maximum would be 5×10^{-12} of the full abscissa (from zero to N_{total}). If the full printed range is 7 centimeters, then the peak's full width at half maximum would be approximately one-hundredth of the diameter of a hydrogen atom.

To compute the required partial derivatives, we regard them as shorthand for a finite difference associated with adding one atom. Thus, for the lower volume, we have

$$\frac{\partial \ln Z_l}{\partial N_l} = \frac{\Delta \ln Z_l}{\Delta N_l} = \frac{\ln Z_l(N_l) - \ln Z_l(N_l - 1)}{1}. \tag{7.5}$$

Then the explicit form,

$$\ln Z_l(N_l) = N_l \ln(V_l/\lambda_{\mathrm{th}}^3) - \ln N_l!,$$

and the decomposition,

$$\ln N_l! = \ln[N_l \times (N_l - 1)!] = \ln N_l + \ln (N_l - 1)!,$$

imply

$$\frac{\partial \ln Z_l}{\partial N_l} = \ln(V_l/\lambda_{\mathrm{th}}^3) - \ln N_l$$

$$= \ln \left(\frac{V_l}{N_l} \frac{1}{\lambda_{\mathrm{th}}^3} \right). \tag{7.6}$$

A similar expression holds for the upper volume. Now equation (7.4) becomes

$$\ln \left(\frac{V_l}{N_l} \frac{1}{\lambda_{\mathrm{th}}^3} \right) = \ln \left(\frac{V_u}{N_u} \frac{e^{-mgH/kT}}{\lambda_{\mathrm{th}}^3} \right).$$

The arguments of the logarithms must be equal, and so we can solve for the number density N_u/V_u in the upper volume:

$$\frac{N_u}{V_u} = \frac{N_l}{V_l} e^{-mgH/kT}. \tag{7.7}$$

The number density drops exponentially with height, a result with which you may be familiar already from a purely macroscopic treatment of the "isothermal atmosphere." (Problem 1 outlines such a route.)

Reformulation and generalization

Now we start to reformulate our major result, equation (7.4). (Why reformulate? To connect with functions that are defined in thermodynamics as well as in statistical mechanics.) Section 5.5 established a connection among $\langle E \rangle$, T, S, and $\ln Z$, namely,

$$S = \frac{\langle E \rangle}{T} + k \ln Z.$$

Let us rearrange this equation as

$$\langle E \rangle - TS = -kT \ln Z. \tag{7.8}$$

The combination on the left-hand side is called the *Helmholtz free energy* and is denoted by F:

$$F \equiv \langle E \rangle - TS. \tag{7.9}$$

Thus equation (7.8) tells us that the Helmholtz free energy equals $-kT$ times the logarithm of the partition function:

$$F = -kT \ln Z. \tag{7.10}$$

In chapter 5, we found that we could compute the energy $\langle E \rangle$ and the pressure P by differentiating $\ln Z$ appropriately. The same must be true of the Helmholtz free energy, and therein lies some of its extensive usefulness. Most of the development is deferred to chapter 10.

Right now, the merit of equation (7.10) lies in this: we can express $\ln Z$ (which is unique to statistical mechanics) in terms of the Helmholtz free energy, which is defined in thermodynamics as well as in statistical mechanics. Thus we write

$$F_l(N_l) = -kT \ln Z_l(N_l),$$

$$F_u(N_u) = -kT \ln Z_u(N_u).$$

Then (7.4) takes the form

$$\frac{\partial F_l}{\partial N_l} = \frac{\partial F_u}{\partial N_u} \tag{7.11}$$

and provides the criterion for the most probable distribution of atoms.

Equations (7.4) and (7.11) locate the peak in $P(N_l, N_u)$. But how wide is that peak? A calculation, outlined in problem 3, shows that

$$\frac{\left(\begin{array}{c} \text{full width of peak} \\ \text{at half maximum} \end{array} \right)}{N_{\text{total}}} = \text{Order of } \frac{1}{\sqrt{N_{\text{total}}}}. \tag{7.12}$$

When $N_{\text{total}} = 10^{22}$ or so, the peak is extremely sharp. Thermodynamics ignores the peak's width; it uses, as the sole and actual numbers N_l and N_u, the values that the canonical probability distribution says are merely the most probable. That is eminently practical, though one should file the distinction away in the back of one's mind. At the moment, the following conclusion suffices: in thermal equilibrium, the physical system is pretty sure to show values for N_l and N_u extremely near the most probable values.

Chemical potential

Equation (7.11) provides a criterion for diffusive equilibrium: certain derivatives of F must be equal. That makes the derivatives so useful that they deserve a separate name and symbol. The relation

$$\mu(T,\,V,\,N) \equiv \left(\frac{\partial F}{\partial N}\right)_{T,V} = F(T,\,V,\,N) - F(T,\,V,\,N-1) \tag{7.13}$$

defines the *chemical potential* μ for the species (be it electrons, atoms, or molecules) whose number is denoted by N. If more than one species is present, then the numbers of all the other species are to be kept fixed while the derivative is computed. Similarly, if there are other external parameters, such as a magnetic field, they are to be kept fixed also.

Equation (7.13) indicates that the chemical potential may be computed by applying calculus to a function of N or by forming the finite difference associated with adding one particle. When N is large, these two methods yield results that are the same for all practical purposes. Convenience alone determines the choice of method.

J. Willard Gibbs introduced the chemical potential μ in the 1870s. He called it simply the "potential," but so many "potentials" appear in physics that the adjective "chemical" was later added. In chapter 11, we will see that the chemical potential plays a major role in the quantitative description of chemical equilibrium. That role alone justifies the name.

For some examples of the chemical potential, we return to the context of figure 7.1. From the definition of μ and from equations (7.6) and (7.10), the chemical potential for the helium atoms in the lower volume follows as

$$\mu_l = \frac{\partial F_l}{\partial N_l} = kT \ln\left(\frac{N_l}{V_l}\lambda_{\text{th}}^3\right). \tag{7.14}$$

The minus sign that appears in equation (7.10) was used to invert the logarithm's argument.

Equation (7.2) tells us that Z_u differs from Z_l principally by an exponential in the gravitational potential energy. Thus the chemical potential for the atoms in the upper volume is

$$\mu_u = \frac{\partial F_u}{\partial N_u} = -kT \ln\left(\frac{V_u}{N_u}\frac{e^{-mgH/kT}}{\lambda_{\text{th}}^3}\right)$$

$$= mgH + kT \ln\left(\frac{N_u}{V_u}\lambda_{\text{th}}^3\right). \tag{7.15}$$

Figure 7.3 displays the run of the two chemical potentials. Suppose we found the gaseous system with the number N_l significantly below its "equilibrium" or most probable value. Almost surely atoms would diffuse through the connecting tube from V_u to V_l and would increase N_l toward $(N_l)_{\text{most probable}}$. Atoms would diffuse from a region where the chemical potential is μ_u to a place where it is μ_l, that is, they would diffuse to smaller chemical potential. In Newtonian mechanics, a literal force pushes a particle in the direction of lower potential energy. In thermal physics, diffusion "pushes" particles toward lower chemical potential.

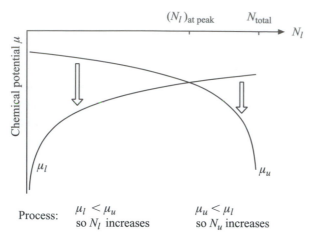

Figure 7.3 Graphs of the two chemical potentials as functions of N_l and $N_u = (N_{total} - N_l)$. The arrows symbolize the direction of particle diffusion relative to the graphed values of the chemical potentials (and *not* relative to the local vertical). When N_l is less than its most probable value, particles diffuse toward the lower volume and its smaller chemical potential; when N_l is greater than its most probable value, diffusion is toward the upper volume and its smaller chemical potential.

As a system heads toward its most probable configuration, particles diffuse from high chemical potential to low. At equilibrium, as it is understood in thermodynamics, the chemical potential is uniform in space: μ has everywhere the same numerical value. In the context of figure 7.1 and its two volumes, the statement means

$$\mu_l = \mu_u, \tag{7.16}$$

which is precisely the content of equation (7.11) but expressed in the language of the chemical potential.

Next comes a profound parallel. Take a tablespoon from the kitchen drawer and place one end in a pot of hot soup that is bubbling on the stove. The spoon's other end soon warms up, indeed even becomes too hot to hold. Energy diffuses (by conduction) from the boiling hot end to the room temperature end. Just as temperature tells us how energy diffuses, so the chemical potential tells us how particles diffuse: from high chemical potential toward low. The meaning of temperature, as stated earlier, is "hotness measured on a definite scale." Analogously, the meaning of the chemical potential is "the tendency of particles to diffuse, as measured on a definite scale." (The word "tendency" is rarely used in technical writing, but it seems to be the best word to describe the physics here.) When the chemical potential has the same numerical value everywhere, the tendency of particles to diffuse is the same every-where, and so no net diffusion occurs. Before such equality, particles migrate away from regions where they have the greatest tendency to diffuse.

With this understanding of the chemical potential in mind, let us return to the examples of this section. Equation (7.14) shows that the chemical potential grows when the concentration of particles, N/V, grows. As one might expect, a high concentration increases the tendency to diffuse.

The chemical potential for the upper volume, μ_u, has the term mgH. That expression represents the additional gravitational potential energy of an atom (relative to the energy of an atom in the lower volume). The increase in energy arises from the downward gravitational force. In turn, that force tends to make atoms diffuse downward, away from the upper volume. Accordingly, the chemical potential μ_u is larger by mgH to reflect that diffusive tendency.

Understanding the chemical potential usually takes a long time to achieve. In chapter 10 we will return to that project and gain some insight into the numerical values that the chemical potential takes.

The present section developed many new concepts and properties. Table 7.1 provides a synopsis.

We will understand the chemical potential best by seeing how it is used. Thus another, and different, example is in order. Before we turn to that, however, we note a corollary and establish a useful lemma.

7.2 Minimum free energy

For the corollary, we return to the canonical probability distribution and its full spectrum of values for N_l and N_u . With the aid of equation (7.10), we can form a composite Helmholtz free energy by writing

Table 7.1 *A listing of the major steps and conclusions as we developed the chemical potential. Statements are given in abbreviated form, and some conditions have been omitted.*

Construct $P(N_l, N_u)$ from the canonical probability distribution.

Locate the maximum in its logarithm.

Define the Helmholtz free energy: $F \equiv \langle E \rangle - TS$.

Relate the Helmholtz free energy to $\ln Z$: $F = -kT \ln Z$.

Define the chemical potential: $\mu(T, V, N) \equiv \left(\dfrac{\partial F}{\partial N} \right)_{T,V}$.

Recast the criterion for locating the maximum in $P(N_l, N_u)$ in terms of the chemical potential: $\mu_l = \mu_u$.

Note generalization: in the most probable situation, which is the only situation that thermodynamics considers, the chemical potential μ is uniform in space.

Before the most probable situation becomes established, particles diffuse toward lower chemical potential.

Succinctly stated, the chemical potential measures the tendency of particles to diffuse.

$$F(N_l, N_u) \equiv -kT \ln Z(N_l, N_u) = F_l(N_l) + F_u(N_u). \tag{7.17}$$

Note especially the minus sign. Where $P(N_l, N_u)$ and hence $Z(N_l, N_u)$ are relatively large and positive, $F(N_l, N_u)$ will be large and negative. Figure 7.4 illustrates this. At the maximum for $P(N_l, N_u)$ and hence for $Z(N_l, N_u)$, the function $F(N_l, N_u)$ will have its minimum. Thus we find that the composite Helmholtz free energy has a *minimum* at what thermodynamics calls the equilibrium values of N_l and N_u. This is a general property of the Helmholtz free energy (at fixed positive temperature T and fixed external parameters):

> In the most probable situation, which is the only situation that thermodynamics considers, the Helmholtz free energy achieves a minimum.

Chapter 10 will derive this conclusion directly from the Second Law of Thermodynamics and thus will confirm its generality. Right now, the minimum property is a bonus from our route to discovering the chemical potential. The method used in chapter 10 is powerful but abstract, and so it is good to see the minimum property emerge here in transparent detail.

7.3   A lemma for computing μ

Whenever the partition function has the structure

$$Z = \frac{(Z_1)^N}{N!} \tag{7.18}$$

and if Z_1 is *independent* of N, then the definition of the chemical potential in (7.13) implies that μ will have the form $\mu = -kT \ln(Z_1/N)$. The result is derived as follows.

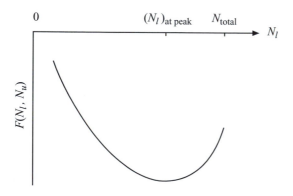

Figure 7.4 The run of $F(N_l, N_u)$. Because of the constraint $N_u = (N_{\text{total}} - N_l)$, the composite Helmholtz free energy is effectively a function of only one particle number, which we may take as N_l. The composite Helmholtz free energy has a minimum at what thermodynamics calls the equilibrium values of N_l and N_u.

The given structure implies

$$F(N) = -kT \ln Z = -kT(N \ln Z_1 - \ln N!).$$

Recall that

$$\ln N! = \ln[N \times (N-1)!] = \ln N + \ln(N-1)!.$$

Then

$$\mu = \left(\frac{\partial F}{\partial N}\right)_{T,V} = F(N) - F(N-1) = -kT(\ln Z_1 - \ln N)$$

$$= -kT \ln(Z_1/N). \tag{7.19}$$

The condition that Z_1 be independent of N is usually met. Nonetheless, one should always check. If the condition is not met, then one can go back to the very definition of the chemical potential and compute μ for the specific case at hand.

Now we turn to another physical situation where the chemical potential comes into play.

7.4 Adsorption

Suppose we put N_{total} atoms of argon into an initially evacuated glass container of volume V and surface area A. The walls are held at the low temperature T by contact (on the outside) with liquid nitrogen. How many argon atoms become adsorbed onto the walls, that is, become attached to the walls? And how many atoms remain in the gas phase? To answer these questions, we compute the chemical potential for each phase and then equate them.

For the N_{gas} atoms that remain in the gas phase, we adapt equation (7.14) and write the chemical potential of that phase as

$$\mu_{\text{gas}} = -kT \ln\left(\frac{V}{N_{\text{gas}}} \frac{1}{\lambda_{\text{th}}^3}\right). \tag{7.20}$$

This form is valid for any spinless, monatomic, semi-classical ideal gas (that is free to move in three dimensions).

Two-dimensional theory: the featureless plane

The adsorbed atoms are bound to the surface through attraction by the atoms that constitute the walls. For an electrical insulator like glass, the attraction arises from an electric dipole moment that is induced in an argon atom by fluctuating electric dipole moments in the silicon dioxide of the walls. The induction process is actually mutual, but the details are not essential. Figure 7.5 sketches the potential energy of an argon

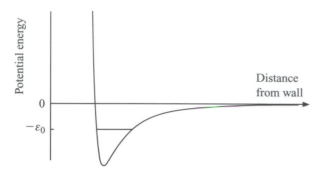

Figure 7.5 The potential energy of an argon atom near a wall. The attractive part (where the slope is positive) arises from mutual electric dipole interactions. The repulsive part (where the slope is negative) arises from electrostatic repulsion and the Pauli exclusion principle. The horizontal line represents the energy $-\varepsilon_0$ of an atom trapped in the potential well (that is, trapped so far as motion perpendicular to the surface is concerned). The distance from the wall to the potential minimum is approximately equal to the radius of an argon atom.

atom as a function of distance from the glass. We suppose that the potential well has only one bound state. The state's energy is $-\varepsilon_0$, where ε_0 is positive and of order 0.05 electron volt. (The zero of energy corresponds to an atom that is at rest far from the surface and hence is part of the gas phase.) The energy $-\varepsilon_0$ incorporates not only the potential energy but also the kinetic energy of the tiny vibrational motion perpendicular to the surface that arises even while the atom is trapped in the well. In the simplest theory, the atom remains free to move around *on* the surface, which is taken to be a perfectly flat and uniform plane. We treat the total area A as though it were an $x-y$ plane of square shape, $L_A \times L_A$, where L_A denotes an edge length.

Single-particle partition function Z_1 for adsorbed atom

Now we can turn to calculating the single-particle partition function Z_1. The analysis of section 4.1 carries over, and so we can write

$$\begin{pmatrix} \text{energy of single-particle} \\ \text{surface state} \end{pmatrix} = \begin{pmatrix} \text{kinetic energy of} \\ \text{motion on surface} \end{pmatrix}$$

$$+ \begin{pmatrix} \text{kinetic and potential energy} \\ \text{of motion perpendicular to surface} \end{pmatrix}$$

$$= \frac{h^2}{8m}\frac{1}{L_A^2}(n_x^2 + n_y^2) - \varepsilon_0. \tag{7.21}$$

The pair of integers $\{n_x, n_y\}$ specifies a point in the $x-y$ plane of figure 4.2. The single-particle partition function Z_1 is

$$Z_1 = \sum_{n_x=1}^{\infty} \sum_{n_y=1}^{\infty} \exp\left[-\frac{h^2}{8m}\frac{1}{L_A^2 kT}(n_x^2 + n_y^2) - \frac{(-\varepsilon_0)}{kT}\right]. \tag{7.22}$$

The double sum goes over all points in the positive quadrant of figure 4.2. Because there is one state per unit area in that mathematical plane, we convert the sum to an integral with differential area $\frac{1}{4} \times 2\pi n \, dn$, that is, $1/4$ of the area in an annulus of circumference $2\pi n$ and radial width dn:

$$Z_1 = e^{+\varepsilon_0/kT} \int_{n=0}^{\infty} \exp\left[-\frac{h^2}{8m} \frac{1}{L_A^2 kT} n^2 \right] \frac{1}{4} \times 2\pi n \, dn$$

$$= \frac{A}{\lambda_{\text{th}}^2} e^{\varepsilon_0/kT}. \tag{7.23}$$

The differential $n \, dn$ is the differential of the exponent (except for a multiplicative constant), and so the integral can be done as the integral of an exponential. The thermal de Broglie wavelength λ_{th} collects the constants into a succinct expression, one that is entirely analogous to the quotient V/λ_{th}^3 that we found for the gas phase in section 5.6.

Chemical potential for adsorbed atom

Section 5.6 and, in particular, equation (5.31) imply that the partition function for the adsorbed phase is

$$Z_{\text{adsorbed}}(N_{\text{adsorbed}}) = \frac{(Z_1)^{N_{\text{adsorbed}}}}{N_{\text{adsorbed}}!}. \tag{7.24}$$

The expression has precisely the structure specified in the lemma of section 7.3, and Z_1 is independent of N. Consequently, the lemma implies that the chemical potential for the adsorbed phase is

$$\mu_{\text{adsorbed}} = -kT \ln(Z_1/N_{\text{adsorbed}})$$

$$= -kT \ln\left(\frac{A}{N_{\text{adsorbed}}} \frac{e^{\varepsilon_0/kT}}{\lambda_{\text{th}}^2} \right). \tag{7.25}$$

Thermodynamic equilibrium

To describe equilibrium, we equate the chemical potentials of the adsorbed and gaseous phases:

$$\mu_{\text{adsorbed}} = \mu_{\text{gas}}. \tag{7.26}$$

That equation takes the explicit form

$$-kT \ln\left(\frac{A}{N_{\text{adsorbed}}} \frac{e^{\varepsilon_0/kT}}{\lambda_{\text{th}}^2} \right) = -kT \ln\left(\frac{V}{N_{\text{gas}}} \frac{1}{\lambda_{\text{th}}^3} \right).$$

The arguments of the logarithms must be equal, and so we can solve for the number density in the gas phase:

$$\frac{N_{\text{gas}}}{V} = \frac{N_{\text{adsorbed}}}{A\lambda_{\text{th}}} e^{-\varepsilon_0/kT}. \tag{7.27}$$

This equation and the constraint,

$$N_{\text{gas}} + N_{\text{adsorbed}} = N_{\text{total}}, \tag{7.28}$$

provide two equations in the two unknowns. Thus, implicitly, the equations answer the questions about how many atoms are adsorbed and how many remain in the gas phase.

People who study adsorption often assess the adsorptive coverage N_{adsorbed}/A by measuring the vapor pressure P_{gas}. If we multiply equation (7.27) by kT and presume that the ideal gas law holds, then we find

$$P_{\text{gas}} = \frac{N_{\text{adsorbed}}}{A} \frac{kT}{\lambda_{\text{th}}} e^{-\varepsilon_0/kT}. \tag{7.29}$$

The implication is that, at fixed temperature, the coverage N_{adsorbed}/A is directly proportional to the vapor pressure.

The proportionality in (7.29) arises when we take the wall surface to be a featureless plane and treat the adsorbed atoms as a semi-classical ideal gas restricted to two-dimensional motion on that plane. This picture forms one end of a spectrum of ways to treat the surface's effect on the argon atoms. If the surface has a pronounced periodic structure, as would graphite or mica, an atom may be restricted to discrete sites on the surface, being both unable to move and unwilling to share the site with another atom. That picture, developed by the American chemist Irving Langmuir in 1918, leads to a different relationship between vapor pressure and coverage; problem 4 shows a route to Langmuir's relationship. In the summary of his paper [*J. Am. Chem. Soc.* **40**, 1361– 403 (1918)], Langmuir commented, "No single equation other than purely thermo-dynamic ones [such as equality of chemical potentials for the phases] should be expected to cover all cases of adsorption any more than a single equation should represent equilibrium pressures for all chemical reactions." Rather, said Langmuir, one should examine various limiting cases, as we did in adopting the featureless plane.

In the next two chapters, the chemical potential plays a supporting role. It is certainly *not* the dominant concept, but it is ever-present. For those reasons, this chapter introduced the chemical potential. A lot more remains to be done—both in developing formal relationships and also in achieving that elusive goal, an intuitive understanding—but those projects are deferred to chapter 10.

7.5   Essentials

1. The *Helmholtz free energy*, denoted by F, is defined by

$$F \equiv \langle E \rangle - TS$$

and may be computed from the logarithm of the partition function as

$$F = -kT \ln Z.$$

2. The *chemical potential* μ is defined by

$$\mu(T, V, N) \equiv \left(\frac{\partial F}{\partial N}\right)_{T,V} = F(T, V, N) - F(T, V, N-1).$$

The chemical potential measures the tendency of particles to diffuse.

3. In the most probable situation, which is the only situation that thermodynamics considers,

(a) the chemical potential μ is uniform in space, and
(b) the Helmholtz free energy achieves a minimum.

Before the most probable situation becomes established, particles diffuse toward lower chemical potential.

4. If a system can be divided meaningfully into two subsystems, such as adsorbed atoms and atoms in a vapor, then the chemical potentials of the two subsystems are equal at thermal equilibrium. [This statement is just a corollary of point (a) in item 3.]

5. Whenever the partition function has the structure

$$Z = \frac{(Z_1)^N}{N!}$$

and if Z_1 is *independent* of N, then the chemical potential has the form

$$\mu = -kT \ln(Z_1/N).$$

6. Item 5 implies

$$\mu_{gas} = -kT \ln\left(\frac{V}{N_{gas}} \frac{1}{\lambda_{th}^3}\right)$$

for any spinless, monatomic, semi-classical ideal gas (that is free to move in three dimensions).

Further reading

For adsorption, J. G. Dash provides a thorough, yet readable, discourse in his *Films on Solid Surfaces: The Physics and Chemistry of Physical Adsorption* (Academic Press, New York, 1975).

G. Cook and R. H. Dickerson offer help in "understanding the chemical potential" in their article by that name: *Am. J. Phys.* **63**, 737–42 (1995). Charles Kittel argues for using the chemical potential as a major analytical tool in his paper, "The way of the chemical potential," *Am. J. Phys.* **35**, 483–7 (1967).

Problems

1. *The isothermal atmosphere.* Consider a tall column of gaseous nitrogen standing in the Earth's gravitational field; its height is several kilometers, say. The gas is in thermal equilibrium at a uniform temperature T, but the local number density $n(z)$, the number of molecules per cubic meter at height z, will be greater near the bottom.

(a) Note figure 7.6 and then explain why the equation

$$P(z) = P(z + \Delta z) + mgn(z)\Delta z$$

must hold. Start with Newton's second law applied to the gas in the height interval z to $z + \Delta z$.

(b) Convert the equation to a differential equation and solve it for $P(z)$, given that the pressure is $P(0)$ at the base. Recall that the ideal gas law may be used to re-express $n(z)$ in terms of more convenient quantities. Take g to be a constant.

(c) Describe the behavior with height of the local number density. At what height has the number density dropped to one-half of its value at the surface? Provide both algebraic and numerical answers.

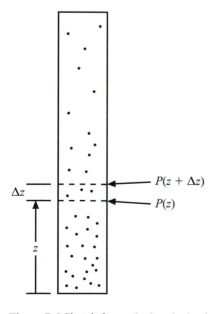

Figure 7.6 Sketch for analyzing the isothermal atmosphere.

2. *Isothermal atmosphere via the chemical potential.* Let $n(z)$ denote the local number density, that is, the number of atoms per cubic meter at height z. Adapt the relationship in (7.15) to fill in the two blanks in the expression below for the chemical potential:

$$\mu(z) = mg\_ + kT \ln[\_].$$

(a) How does the chemical potential $\mu(z)$ vary with concentration $n(z)$?
(b) At equilibrium, how must $n(z)$ vary with z? Explain your reasoning.

3. *Relative width.* Continue the heuristic calculation of section 7.1 and determine the relative width of the peak in the probability distribution $P(N_l, N_u)$. In short, work out the details for equation (7.12). For a good route, expand the logarithm of $P(N_l, N_u)$ to second order around the peak. For simplicity, take the two volumes to be equal.

4. *The Langmuir model.* As atoms from the vapor become bound to the surface, they are restricted to discrete sites and are *not* able to move about on the surface. Specify that sites of individual area A_1 cover, like tiles, the entire surface area A, so that there are a total of $N_s = A/A_1$ sites. Only one atom may occupy any given site. Suppose further that an atom on a site has energy $-\varepsilon_0$ relative to an atom at rest in the gas phase. (The energy $-\varepsilon_0$ incorporates both the binding to the surface and any motion confined to the immediate vicinity of the site. In short, each site has only one quantum state.) All other conditions are as specified in section 7.4.

One further preliminary may be useful. Suppose you have a checkerboard with N_s squares and a handful of pebbles from the beach, N of them, each pebble slightly different in shape or color. In how many distinct ways can you place the pebbles on the checkerboard, one to a square? You may place the first pebble in any one of N_s locations. The second pebble may go into any one of the remaining $N_s - 1$ free locations, giving you $N_s \times (N_s - 1)$ possibilities so far. And so it goes until you place the Nth pebble in one of the remaining $N_s - (N - 1)$ open locations. Thus we find that, given N_s sites,

$$\begin{pmatrix} \text{number of distinct arrangements} \\ \text{of } N \text{ distinguishable pebbles} \end{pmatrix} = N_s \times (N_s - 1) \times \cdots \times [N_s - (N - 1)]$$

$$= \frac{N_s!}{(N_s - N)!}.$$

(a) Explain why the partition function for the adsorbed atoms is

$$Z_{\text{adsorbed}}(N_{\text{ad}}) = \frac{N_s!}{N_{\text{ad}}!(N_s - N_{\text{ad}})!} e^{N_{\text{ad}}\varepsilon_0/kT},$$

where N_{ad} denotes the number of adsorbed atoms.
(b) Calculate the chemical potential of the adsorbed atoms. Then express the vapor pressure in terms of the ratio N_{ad}/N_s, kT, ε_0, and λ_{th}.

Langmuir found the relationship to hold quite well for hydrogen, oxygen, argon, nitrogen, carbon monoxide, methane, and carbon dioxide when adsorbed on both mica and glass (so long as the coverage remained below a complete monomolecular layer). The run of coverage versus pressure at fixed temperature is called the *Langmuir isotherm*.

5. *Liquid–vapor equilibrium*. A liquid is substantially incompressible, and so we may approximate its volume by the product Nv_0, where N is the number of molecules and v_0 is a fixed volume of approximately the size of a single molecule. The short-range attractive forces between molecules (which give cohesion to the liquid) "bind" a molecule into the liquid by an energy $-\varepsilon_0$ relative to the gas phase, where ε_0 is a positive constant. Otherwise, treat the molecules as forming two semi-classical ideal gases, one the vapor, the other the "liquid gas." Use this information to estimate the vapor pressure over a liquid at temperature T. (Imagine squirting liquid into a totally evacuated volume. Some liquid will evaporate, and an equilibrium between vapor and remaining liquid will be established.)

6. *Semi-classical ideal gas in two dimensions*. Extend the analysis in section 7.4 to calculate (a) the energy $\langle E \rangle$, (b) the heat capacity at constant area C_A, and (c) the entropy S of N atoms confined to two-dimensional motion on a featureless plane.

7. *Centrifuge*. Specify that a dilute gas fills a hollow cylinder whose outer curved wall rotates with angular velocity ω about the cylinder's axis. The gas—when the molecular motion is averaged over a small volume (1 cm^3, say)—rotates with the same angular velocity. The goal is to calculate the radial dependence of the number density $n(r)$ when the gas is in thermal equilibrium at temperature T. Two routes (at least) are open to you, as follows.

Route 1. In order that the gas move in circular motion, a net force, directed radially inward, must act on each little volume. Only a pressure gradient can supply that force. A pressure gradient implies (here) a gradient in the number density. Newton's second law will provide a differential equation for the number density.

Route 2. Transform to a reference frame rotating with angular velocity ω. In this non-inertial frame, the gas is at rest (macroscopically), but each atom is subject to an apparent force of magnitude $mr\omega^2$, where m is the atomic mass; that fictitious force is directed radially outward. There is an associated effective potential energy, and so one can reason with the chemical potential.

(a) Determine the radial dependence of the number density $n(r)$ as a function of r, m, ω, and T.
(b) If the gas were a mixture of two uranium isotopes, ^{235}U and ^{238}U, both as the gas uranium hexafluoride, would you expect the ratio of the number densities for the two isotopes to be the same everywhere? If no, what implications does this have?
(c) At which radius, if any, does the number density have the same value that it had when $\omega = 0$? (Only one isotope is present. Denote the maximum radius by r_{max}. You may invoke "slow rotation," but—if you do—then specify that condition in dimensionless fashion.)

8. Return to the context of problem 1. Specify now that the gravitational force per unit mass falls off inversely as the square of the distance from the Earth's center. (a) If the isothermal column extended to a height of only three Earth radii, how would the number density vary?

(b) If the column stretched indefinitely far from the Earth's surface, would thermal equilibrium be possible? What consequences for the Earth's atmosphere do you infer?

9. *Adiabatic relation in two dimensions.* An ideal gas of N spinless atoms is confined to two-dimensional motion on a featureless plane of area A. (Desorption from the plane is no longer an option.) What combination of area A and temperature T remains constant during a slow adiabatic compression of the area? Justify your assertion with a derivation.

8 The Quantum Ideal Gas

8.1 Coping with many particles all at once

8.2 Occupation numbers

8.3 Estimating the occupation numbers

8.4 Limits: classical and semi-classical

8.5 The nearly classical ideal gas (optional)

8.6 Essentials

By now, the classical ideal gas should be a familiar system. We can turn to the full quantum theory of an ideal gas. The applications are numerous, and some will be examined in chapter 9. The present chapter develops an effective technique for working with a quantum ideal gas.

8.1 Coping with many particles all at once

Every macroscopic system consists of a great many individual particles. Coping with all of them simultaneously, even if only statistically, can be a daunting task. A flow chart will help us to see some of our accomplishments and to see what steps lie ahead. Figure 8.1 displays such a chart on the supposition that the system consists of one species of particle only, for example, N electrons or N helium atoms (of a specific isotope).

If the forces between the particles must be included, then the analysis is difficult. Moreover, there is no general algorithm for proceeding, not even by computer. To be sure, the canonical probability distribution usually is applicable and provides estimates in principle, but the actual evaluation of the partition function, say, defies a direct approach. We have, however, successfully dealt with one such difficult situation. The Debye model for a solid incorporates the inter-particle forces, for example, those between adjacent copper atoms in a crystal. The forces ensure that the motions of adjacent atoms are correlated, and the correlation leads to sound waves, the theoretical basis of the Debye model.

(Note. The Einstein model of a solid provides a rarity: a calculation that incorporates inter-particle forces and yet is easily soluble. The simplicity arises because the force on any one atom is calculated as though all the other atoms were at their equilibrium locations in a periodic lattice. Thus each atom experiences a force that arises from a known, time-independent potential energy. That splits the difficult N-particle problem into N separate one-particle problems, which are easy.)

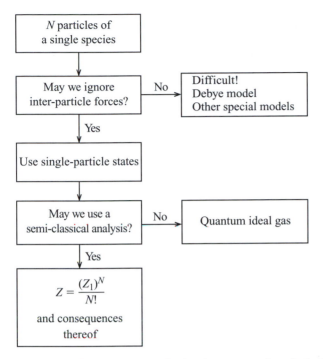

Figure 8.1 A flow chart for methods when one needs to deal with many particles simultaneously. The phrase "inter-particle forces" means the forces between whatever particles one takes as the basic entities. If atoms are the basic entities, as they would be in a solid, then "inter-particle forces" means the forces between the atoms. If the basic entities are the molecules in a gas, then "inter-particle forces" means the forces between the molecules, not the forces that bind the atoms together to make the molecules. The explicit partition function in the lowest box presumes gas-like behavior. For spatially localized particles, a different expression applies.

If we may ignore the inter-particle forces, then we may base our analysis on the single-particle quantum states (without further significant approximation). Specify a gaseous context (in contrast to spatially localized particles). The indistinguishability of identical particles presents us with the next question. When we ask "may we use a semi-classical analysis?", we are asking whether the indistinguishability can be taken into account adequately by a single simple approximation. In section 5.6, we reasoned that, when the thermal de Broglie wavelength is much smaller than the typical separation between particles, then we could adequately approximate the partition function for N indistinguishable particles by two steps.

1. Calculate the partition function as though indistinguishability imposed no restrictions, thus finding the provisional result $(Z_1)^N$, where Z_1 is the partition function for a single particle.
2. Divide by $N!$, the number of permutations among N particles (that had been regarded as distinguishable).

Often the $N!$ will not affect the estimate of a physical quantity. For example, energy, pressure, and magnetic moment are not affected. The technical reason is that such quantities can be computed by differentiating the *logarithm* of Z with respect to temperature or an external parameter; the $N!$ appears as a separate term in the logarithm and does not survive such differentiation. If a derivative is taken with respect to N or if $\ln Z$ enters without differentiation, then the $N!$ does matter. Such is the case with the chemical potential and the entropy.

If a semi-classical analysis is not adequate, then we move to the quantum ideal gas, which is the subject of this chapter.

8.2   Occupation numbers

Following Wolfgang Pauli, we divide the particles of physics into two classes.

1. Fermions. The first class consists of particles whose intrinsic angular momentum equals half an odd positive integer times $\hbar$, where $\hbar \equiv h/2\pi$; that is, the spin is $\frac{1}{2}\hbar, \frac{3}{2}\hbar, \frac{5}{2}\hbar, \ldots$. Some examples are the following: spin $\frac{1}{2}$—an electron, a proton, or a neutron; spin $\frac{3}{2}$—a nucleus of the most common lithium isotope, ^{7}Li; and spin $\frac{5}{2}$—a nucleus of the rare oxygen isotope, ^{17}O. These particles obey the Pauli exclusion principle. Enrico Fermi (in 1926) and Paul A. M. Dirac (also in 1926, but independently) worked out the statistical consequences of the exclusion principle, and so their names are intimately associated with these particles.

2. Bosons. The second class consists of particles whose intrinsic angular momentum equals zero or a positive integer times $\hbar$; that is, the spin is $0, 1\hbar, 2\hbar, \ldots$. Again, some examples are the following: spin 0—the common isotope of helium, ^{4}He, both the nucleus alone and also the entire atom; spin 1—a nucleus of the uncommon lithium isotope, ^{6}Li; and spin 2—a nucleus of a radioactive nitrogen isotope, ^{16}N. The Indian physicist Satyendra Nath Bose (in 1924) and Albert Einstein (also in 1924 and building on Bose's paper) worked out the statistical consequences of indistinguishability for such particles, and so their names are intimately linked to them. (Writing from India, Bose submitted his paper to the *Philosophical Magazine* in England. The referee's report was negative, and so the editor rejected the paper. Then Bose sent the manuscript to Einstein in Berlin, asking him—if he thought it had merit—to arrange for publication in the *Zeitschrift für Physik*. Thus it came about that Bose's manuscript was translated for publication in German by none other than Einstein himself. His "Note from the translator" praised the paper as providing "an important advance.")

Most of the examples are composite particles, for instance, a bound collection of protons and neutrons that form an atomic nucleus. A collection of fermions, regarded as a single composite particle, behaves like a fermion if the collection is composed of an odd number of fermions and acts like a boson if composed of an even number of fermions. Thus, while ^{4}He is a boson, the isotope ^{3}He has an odd number of fermions and is itself a fermion. The striking difference in behavior between ^{3}He and ^{4}He at low temperature arises from this distinction.

In the following, we consider a system consisting of N identical particles, for example, N electrons (fermions) or N atoms of ^{4}He (bosons).

Occupation number

An energy eigenstate Ψ_j of the entire N-particle system is determined as soon as we say

(a) which single-particle states are occupied and
(b) how often each of those states is occupied.

For fermions, of course, the exclusion principle prohibits multiple occupancy. In contrast, a system of bosons may have, for example, five bosons in single-particle state φ_7.

The indistinguishability of identical particles implies that we should not even try to specify which particles are in which single-particle states. It suffices to know the following set of numbers:

$$
n_\alpha(\Psi_j) \equiv \left(\begin{array}{c} \text{number of times that} \\ \text{single-particle state } \varphi_\alpha \text{ is} \\ \text{occupied in the full state } \Psi_j \end{array} \right). \tag{8.1}
$$

The number $n_\alpha(\Psi_j)$ is called the *occupation number* for the single-particle state φ_α in Ψ_j. For fermions, the occupation number is always restricted to zero or 1. For bosons, the range is $0, 1, 2, \ldots, N$. To be sure, the occupation numbers must satisfy the equation

$$
\sum_\alpha n_\alpha(\Psi_j) = N. \tag{8.2}
$$

Altogether, N single-particle states are occupied (when multiple occupancy is counted), even though we cannot say which of the N identical and indistinguishable particles occupies which single-particle state.

Estimating the total energy

To determine the energy E_j of a state Ψ_j, we add up the energies of the occupied single-particle states. Thus

$$
E_j = \sum_\alpha \varepsilon_\alpha\, n_\alpha(\Psi_j). \tag{8.3}
$$

When the physical system is in thermal equilibrium at some temperature T, we do not know which quantum state Ψ_j to use to describe the system. Nonetheless, we can estimate the total energy by applying the canonical probability distribution to the entire system of N particles. We start as in section 5.4 and then use (8.3):

$$\langle E \rangle = \sum_j E_j P(\Psi_j) = \sum_j \left[\sum_\alpha \varepsilon_\alpha \, n_\alpha(\Psi_j) \right] P(\Psi_j)$$

$$= \sum_\alpha \varepsilon_\alpha \left[\sum_j n_\alpha(\Psi_j) P(\Psi_j) \right]$$

$$= \sum_\alpha \varepsilon_\alpha \langle n_\alpha \rangle. \tag{8.4}$$

The double summation in the first line is a sum over all terms in a rectangular array; the elements are labeled by a row index α and a column index j. As specified in that line, one is to sum first over all elements in a given column and then over all columns. But one may equally well sum first over all elements in a given row and then over all rows. That order of summing is shown in the second line, and it has some advantages. The step to the last line follows from the definition

$$\langle n_\alpha \rangle \equiv \sum_j n_\alpha(\Psi_j) P(\Psi_j). \tag{8.5}$$

The function $\langle n_\alpha \rangle$ is the estimated number of particles in single-particle state φ_α. Often we will call it the *estimated occupation number*. A verbal description of equation (8.4) runs as follows: to estimate the total energy, multiply the energy of each single-particle state by the estimated number of particles that occupy the state and then add up all the contributions.

Whenever a physical quantity is well-defined already in a single-particle state, we can reduce a calculation for the entire system of N particles to a form like that for energy in (8.4). Therein lies the great usefulness of the estimated occupation numbers $\langle n_\alpha \rangle$. In the next section, we calculate them in general.

8.3   Estimating the occupation numbers

A few words about tactics are in order. First, recall that the occupation numbers for fermions are restricted to zero or 1, whereas for bosons there is no such restriction. Consequently, we handle fermions and bosons separately.

Second, equation (8.2) asserts that, for each full state Ψ_j, the sum of the occupation numbers is N. The same must be true for the sum of the estimated values:

$$\sum_\alpha \langle n_\alpha \rangle = N. \tag{8.6}$$

[To prove the claim, multiply both sides of (8.2) by $P(\Psi_j)$ and sum over j.]

Third, to keep the notation from getting out of hand, we start with $\alpha = 1$ and then generalize to other values of the index α.

Fermions

We begin with the definition of $\langle n_1 \rangle$ and then write out the probability $P(\Psi_j)$ in some detail:

$$\langle n_1 \rangle = \sum_j n_1(\Psi_j) \, P(\Psi_j)$$

$$= \frac{1}{Z(N)} \sum_{n_1} \sum_{n_2} \cdots n_1 \, e^{-(n_1\varepsilon_1 + n_2\varepsilon_2 + \cdots)/kT}$$

$$\text{subject to } n_1 + n_2 + \cdots = N. \tag{8.7}$$

I replaced a sum over j by a sum over *all sets* of occupation numbers, consistent with there being N particles in the system. The unique correspondence between a state Ψ_j and the set of occupation numbers permits this replacement. (Appendix B provides an example worked out in detail, and you may find it helpful to look at that now.) The ellipsis dots following the two summation signs indicate more summations, associated with n_3, n_4, etc.

Next we focus our attention on the first summation, that over n_1, the occupation number for state φ_1. When n_1 is taken to be zero, the contribution vanishes. When n_1 is taken to be 1, which is its only other value, we get

$$\langle n_1 \rangle = \frac{1}{Z(N)} e^{-\varepsilon_1/kT} \sum_{n_2} \cdots 1 e^{-(n_2\varepsilon_2 + \cdots)/kT}$$

$$\text{subject to } n_2 + n_3 + \cdots = N - 1. \tag{8.8}$$

When $n_1 = 1$, the sum of the other occupation numbers must be $N - 1$. For convenience later, I factored out the exponential that contains ε_1.

Now comes the unexpected move: arrange matters so that the summation once again goes over *all sets* of occupation numbers, n_1 included, but subject to the restriction that the occupation numbers add up to $N - 1$. To achieve this arrangement, insert a factor of $(1 - n_1)$:

$$\langle n_1 \rangle = \frac{1}{Z(N)} e^{-\varepsilon_1/kT} \sum_{n_1} \sum_{n_2} \cdots (1 - n_1) e^{-(n_1\varepsilon_1 + n_2\varepsilon_2 + \cdots)/kT}$$

$$\text{subject to } n_1 + n_2 + n_3 + \cdots = N - 1.$$

Let's check the legitimacy of this unanticipated step. When $n_1 = 0$, the contribution reproduces (8.8). When $n_1 = 1$, the contribution vanishes. It checks.

Moreover, we can interpret the two terms produced by the "1" and the "n_1" in the full summation, as follows:

$$\langle n_1 \rangle = \frac{e^{-\varepsilon_1/kT}}{Z(N)} [Z(N-1) - \langle n_1 \rangle_{\text{for } N-1} Z(N-1)]. \tag{8.9}$$

The "1" gives a sum of Boltzmann factors but for a system with $N-1$ particles; hence the sum produces the partition function for such a system. The sum with "n_1" is proportional to an estimated occupation number, but for a system with $N-1$ particles.

Nothing depended on the index α having the value 1; so we may generalize to

$$\langle n_\alpha \rangle = \frac{Z(N-1)}{Z(N)} e^{-\varepsilon_\alpha/kT}[1 - \langle n_\alpha \rangle_{\text{for } N-1}]. \tag{8.10}$$

No one would claim that the steps or this result are intuitively obvious, but the steps are few, the result is exact, and the canonical probability distribution has sufficed.

When the system is macroscopic and when the temperature has a realistic value, tiny fractional changes in individual $\langle n_\alpha \rangle$s will shift their sum from N to $N-1$ or vice versa. So we set

$$\langle n_\alpha \rangle_{\text{for } N-1} = \langle n_\alpha \rangle, \tag{8.11}$$

our sole approximation, and make the corresponding replacement in (8.10):

$$\langle n_\alpha \rangle = \frac{Z(N-1)}{Z(N)} e^{-\varepsilon_\alpha/kT}[1 - \langle n_\alpha \rangle]. \tag{8.12}$$

The ratio of partition functions introduces the chemical potential, as follows. From section 7.1,

$$\mu = \left(\frac{\partial F}{\partial N}\right)_{T,V} = F(N) - F(N-1)$$

$$= -kT[\ln Z(N) - \ln Z(N-1)], \tag{8.13}$$

whence

$$\frac{Z(N-1)}{Z(N)} = e^{\mu/kT}. \tag{8.14}$$

Inserting the last expression into (8.12) and solving for $\langle n_\alpha \rangle$, we emerge with the result

$$\langle n_\alpha \rangle_{\text{F}} = \frac{1}{e^{(\varepsilon_\alpha - \mu)/kT} + 1}, \tag{8.15}$$

where the subscript F emphasizes that the expression applies to fermions. The "+1" in the denominator ensures that the relation $\langle n_\alpha \rangle_{\text{F}} \leq 1$ holds, a consequence of the Pauli principle. The chemical potential itself is to be chosen (numerically) so that equation (8.6) is satisfied. [In principle, the chemical potential is already determined by the relations in equation (8.13), but we do not know either $F(N)$ or $Z(N)$ in detail, and so an indirect route is the best option.]

Bosons

The derivation for bosons follows a similar route, and we can dispatch it quickly. The definition of $\langle n_1 \rangle$ leads to an equation like (8.7), much as before. The sole difference is that each sum, such as the sum over n_1, runs over the values 0, 1, 2, ..., N. (Again, appendix B provides an explicit example.) As before, when n_1 is taken to be zero, the contribution vanishes. The significant sum over n_1 starts with $n_1 = 1$. To benefit from this circumstance, make the substitution

$$n_1 = 1 + n_1',$$

where the new summation variable n_1' ranges from 0 to $N - 1$. Thus

$$\langle n_1 \rangle = \frac{1}{Z(N)} \sum_{n_1'} \sum_{n_2} \cdots (1 + n_1') e^{-[(1+n_1')\varepsilon_1 + n_2\varepsilon_2 + \cdots]/kT}$$

subject to $n_1' + n_2 + \cdots = N - 1$.

Factoring and identifying as before, we get

$$\langle n_1 \rangle = \frac{e^{-\varepsilon_1/kT}}{Z(N)} [Z(N - 1) + \langle n_1 \rangle_{\text{for } N-1} Z(N - 1)].$$

Comparing this expression with the analogous fermion expression, equation (8.9), we find that the minus sign in the latter has become a plus sign for the bosons.

The remaining steps are just as before. The result is

$$\langle n_\alpha \rangle_{\text{B}} = \frac{1}{e^{(\varepsilon_\alpha - \mu)/kT} - 1}, \tag{8.16}$$

where the subscript B emphasizes that the expression applies to bosons. The minus sign permits the estimated occupation number to be larger than 1, indeed, much larger. The structure in (8.16) is fundamentally the same as what we found for $\bar{n}(\nu)$, the estimated number of photons in a mode of frequency ν. Photons are bosons, and so the similarity is no coincidence. Section 10.3 will elaborate on the comparison.

8.4 Limits: classical and semi-classical

Classical physics does not distinguish between fermions and bosons. When the conditions of temperature and number density are such that classical reasoning provides a good approximation, then fermions and bosons should act the same way. The ± 1 distinction in $\langle n_\alpha \rangle$ must be insignificant relative to the exponential. Thus the strong inequality

$$e^{(\varepsilon_\alpha - \mu)/kT} \gg 1 \tag{8.17}$$

must hold for all states φ_α and energies ε_α. In short, in the classical limit,

$$\langle n_\alpha \rangle = \frac{1}{e^{(\varepsilon_\alpha - \mu)/kT}} \ll 1, \tag{8.18}$$

where the strong inequality follows because the exponential must be much larger than $|\pm 1|$.

To evaluate the chemical potential that appears in (8.18), turn to the summation condition (8.6) and use the classical limit of $\langle n_\alpha \rangle$:

$$\sum_\alpha \langle n_\alpha \rangle = e^{\mu/kT} \sum_\alpha e^{-\varepsilon_\alpha/kT} = N. \tag{8.19}$$

The sum over α is the single-particle partition function, which we previously denoted by Z_1. Thus

$$e^{\mu/kT} = \frac{N}{Z_1}, \tag{8.20}$$

whence

$$\mu = -kT \ln(Z_1/N), \tag{8.21}$$

a result that reproduces what we found in section 7.3.

Moreover, equation (8.18) now takes on an appealing form:

$$\langle n_\alpha \rangle = N \frac{e^{-\varepsilon_\alpha/kT}}{Z_1}. \tag{8.22}$$

Estimates of energy or pressure made with this $\langle n_\alpha \rangle$ will have a form that looks as though there were N totally independent particles, each of them described by the canonical probability distribution for a single particle.

The semi-classical limit of the partition function

The elements are at hand to work out the limiting expression for $Z(N)$, the full partition function. Equation (8.14) relates $Z(N)$ to $Z(N-1)$:

$$Z(N) = e^{-\mu/kT} Z(N-1). \tag{8.23}$$

Substitution from (8.20) for the exponential gives

$$Z(N) = \frac{Z_1}{N} Z(N-1). \tag{8.24}$$

Here is a tidy relationship between partition functions that differ by one particle. We can use the connection repeatedly. Thus we can relate $Z(N-1)$ to $Z(N-2)$:

$$Z(N) = \frac{Z_1}{N} \frac{Z_1}{N-1} Z(N-2).$$

Proceeding inductively, we find

$$Z(N) = \frac{Z_1^N}{N!}.$$ (8.25)

Note the division by $N!$. The limiting value of the full partition function retains a trace of the indistinguishability of identical particles. Moreover, in (8.25) we have confirmation that our approximate summation in section 5.6 was done correctly. The $N!$ is preserved in the expression for the entropy, equation (5.39), and the factorial is responsible for the N in the chemical potential, as seen in equation (7.19), for example. In a vital sense, these three quantities—full partition function, entropy, and chemical potential—do *not* have a true classical limit. What we have found for them is better called a semi-classical limit.

Moreover, a little digging would show that all three quantities—full partition function, entropy, and chemical potential—retain Planck's constant h when they are expressed in their most explicit forms. The adjective "semi-classical" is used (at least in this book) to denote an expression or situation in which vestiges of indistinguishability or of Planck's constant remain.

8.5 The nearly classical ideal gas (optional)

In the next chapter, we apply our results for the quantum ideal gas to systems at low temperature and high number density, a context where the behaviors of fermions and bosons differ radically. Before doing that, however, it would be good to become more familiar with the notion of estimated occupation numbers. So this section calculates the pressure and energy of a quantum ideal gas when its behavior is close to that of a classical ideal gas. We know what to expect in lowest order, and the first-order corrections to the classical results nicely foreshadow a characteristic distinction between fermions and bosons.

First we develop a general connection between pressure and energy, valid for any non-relativistic ideal gas. Section 5.4 gave us the entirely general relationship

$$P = \sum_j -\frac{\partial E_j}{\partial V} \frac{\exp(-E_j/kT)}{Z}.$$ (8.26)

Section 4.1 showed that the single-particle energy ε_α depends on volume as $V^{-2/3}$. Implicit were the assumptions of non-relativistic motion and zero potential energy. The derivative of ε_α with respect to volume will be

$$\frac{\partial \varepsilon_\alpha}{\partial V} = -\frac{2}{3}\frac{\varepsilon_\alpha}{V}.$$ (8.27)

The energy eigenvalue E_j of the entire ideal gas will be a sum of single-particle energies, and so the derivative with respect to volume will have the same structure as in (8.27):

$$\frac{\partial E_j}{\partial V} = -\frac{2}{3}\frac{E_j}{V}.$$ (8.28)

Upon inserting this form into (8.26), we find

$$P = \frac{2}{3}\frac{\langle E \rangle}{V}$$ (8.29)

as the connection between pressure and kinetic energy for any non-relativistic ideal gas.

Now we turn to calculating $\langle E \rangle$, which equation (8.4) gave as

$$\langle E \rangle = \sum_\alpha \varepsilon_\alpha \langle n_\alpha \rangle$$ (8.30)

in terms of the estimated occupation numbers. We can compute for fermions and bosons simultaneously by adopting a $\pm$ notation and the convention that the *upper* sign always refers to fermions. Thus

$$\langle n_\alpha \rangle = \frac{1}{e^{(\varepsilon_\alpha - \mu)/kT} \pm 1}$$ (8.31)

handles both cases.

When the gas is nearly classical, the exponential is much greater than $|\pm 1|$, and we can profitably expand with the binomial theorem. Let A denote the exponential. Then the structure of the expansion is

$$\frac{1}{A \pm 1} = \frac{1}{A(1 \pm A^{-1})} = A^{-1}(1 \pm A^{-1})^{-1}$$

$$= A^{-1}(1 \mp A^{-1} + \cdots) = A^{-1} \mp A^{-2},$$

provided A is much larger than 1 and provided we truncate after the first correction term. Thus

$$\langle n_\alpha \rangle = e^{(\mu - \varepsilon_\alpha)/kT} \mp e^{2(\mu - \varepsilon_\alpha)/kT}.$$ (8.32)

The chemical potential is determined by the requirement that the sum of $\langle n_\alpha \rangle$ over all α yields N:

$$e^{\mu/kT} \sum_\alpha e^{-\varepsilon_\alpha/kT} \mp e^{2\mu/kT} \sum_\alpha e^{-2\varepsilon_\alpha/kT} = N.$$ (8.33)

To evaluate the sums, one converts them to integrals with a density of single-particle states. Section 4.1 gave us the density $D(\varepsilon)$ for a spinless particle. When a particle has a spin $s\hbar$ (and when the energy does not depend on the spin orientation), the density is augmented by the factor $(2s + 1)$:

$$D(\varepsilon) = (2s + 1)\frac{2\pi(2m)^{3/2}}{h^3} V\varepsilon^{1/2}. \tag{8.34}$$

Here is the reason: a spin of $s\hbar$ can have $2s + 1$ distinctly different orientations, and so each state of a spinless particle splits into $2s + 1$ new states but without any change in energy.

The first sum in (8.33) was evaluated in section 5.6; the second sum is similar. The outcome can be arranged as

$$e^{\mu/kT} \mp \frac{1}{2^{3/2}} e^{2\mu/kT} = \frac{1}{(2s + 1)} \frac{N}{V} \lambda_{\text{th}}^3. \tag{8.35}$$

In section 8.4, we reasoned that $e^{(\varepsilon_a - \mu)/kT}$ must be large for every value of α. This requires that $e^{-\mu/kT}$ be large and hence that $e^{\mu/kT}$ be small. Thus (8.35) implies that λ_{th} must be small relative to $(V/N)^{1/3}$, the typical inter-particle separation. In short, we recover here the criterion for classical behavior that qualitative reasoning gave us in section 5.6.

Knowing that $e^{\mu/kT}$ is small enables us to solve equation (8.35) readily—at least to the required accuracy. We merely replace $e^{2\mu/kT}$ by the square of the lowest order expression for $e^{\mu/kT}$, as follows:

$$e^{\mu/kT} = \frac{1}{(2s + 1)} \frac{N}{V} \lambda_{\text{th}}^3 \pm \frac{1}{2^{3/2}} e^{2\mu/kT}$$

$$= \frac{1}{(2s + 1)} \frac{N}{V} \lambda_{\text{th}}^3 \pm \frac{1}{2^{3/2}} \left[\frac{1}{(2s + 1)} \frac{N}{V} \lambda_{\text{th}}^3 \right]^2$$

$$= \frac{1}{(2s + 1)} \frac{N}{V} \lambda_{\text{th}}^3 \times \left[1 \pm \frac{1}{2^{3/2}} \frac{1}{(2s + 1)} \frac{N}{V} \lambda_{\text{th}}^3 \right]. \tag{8.36}$$

Return now to $\langle E \rangle$ as presented in (8.30) and use (8.32):

$$\langle E \rangle = e^{\mu/kT} \sum_{\alpha} \varepsilon_a e^{-\varepsilon_a/kT} \mp e^{2\mu/kT} \sum_{\alpha} \varepsilon_a e^{-2\varepsilon_a/kT}. \tag{8.37}$$

Again the sums can be converted to integrals with a density of states. Moreover, the exponentials that contain the chemical potential are known. To first order in $N\lambda_{\text{th}}^3/V$, the outcome is

$$\langle E \rangle = \tfrac{3}{2}NkT \left[1 \pm \frac{1}{2^{5/2}} \frac{N\lambda_{\text{th}}^3}{(2s + 1)V} \right]. \tag{8.38}$$

Several conclusions follow from this result.

First, recall that $P = \tfrac{2}{3}\langle E \rangle/V$ for any non-relativistic ideal gas, as derived earlier in this section. Thus, when one uses equation (8.38) for $\langle E \rangle$, the leading term in P is the familiar NkT/V. The correction term is positive for fermions, negative for bosons. The Pauli exclusion principle compels fermions to populate single-particle states of high energy more heavily than if no such exclusion principle held. At low temperature, the effect becomes extreme, as the next chapter shows.

Second, the average kinetic energy per particle, $\langle E \rangle / N$, differs between fermions and bosons even when the particles have the same temperature. For example, one could have a dilute gaseous mixture of ^{3}He atoms, which are fermions, and ^{4}He atoms, which are bosons. As gases coexisting in the same volume at thermal equilibrium, the gases are unquestionably at the same temperature. By using a dilute mixture of these inert gases, one can approximate nearly-classical ideal gases. Some textbooks assert that temperature is a measure of average translational kinetic energy per particle, as though such a statement provided an adequate definition of temperature. Not only is the statement inadequate; it is also incorrect. Gases that are manifestly at the same temperature can have different average translational kinetic energy per particle. The goal of the temperature notion is *not* to tell us about energy per particle. Rather, the goal is to quantify "hotness."

In turn, the attribute "hotness" arises because, when macroscopic objects from two different environments are placed in thermal contact, one object will gain energy by being heated and the other will lose energy (by being cooled). For a homely example, imagine putting a package of frozen strawberries into a pot of hot water for a quick thaw in time for dessert. Energy flows by conduction from the water to the strawberries. The direction of flow implies, by definition, that the water is "hotter" than the strawberries. Assigning temperatures of $(80 + 273)$ K to the water and $(-25 + 273)$ K to the strawberries is a way to quantify the "hotness" of each. Then one knows, for example, that shortcake at $T = (20 + 273)$ K would be hotter than the strawberries and would transfer energy to them if the berries were immediately placed on the cake.

As a third conclusion, equation (8.38) displays clearly the quantity that must be small if the classical limit is to be adequate. That is, the strong inequality

$$\lambda_{th}^3 \ll \frac{V}{N} \times (2s + 1)2^{5/2}$$

must hold. The analysis in section 5.6, which culminated in equation (5.47), gave essentially the cube root of the strong inequality displayed here.

8.6 Essentials

1. The particles of Nature divide into two classes: (1) *fermions*, whose spin is one-half of an odd positive integer times $\hbar$ and which are subject to the Pauli exclusion principle; (2) *bosons*, whose spin equals zero or a positive integer times $\hbar$ and which are exempt from the Pauli exclusion principle.

2. *Estimated occupation numbers* are defined by

$$\langle n_\alpha \rangle \equiv \sum_j n_\alpha(\Psi_j)P(\Psi_j),$$

where the *occupation numbers* themselves are defined by

$$n_\alpha(\Psi_j) \equiv \begin{pmatrix} \text{number of times that} \\ \text{single-particle state } \varphi_\alpha \text{ is} \\ \text{occupied in the full state } \Psi_j \end{pmatrix}.$$

The context is an ideal gas: no literal forces between the particles.

3. For a first glimpse of the utility of estimated occupation numbers, note that

$$\langle E \rangle = \sum_\alpha \varepsilon_\alpha \langle n_\alpha \rangle.$$

The complicated N-body problem is reduced to a sum over single-particle states.

4. The canonical probability distribution provides the following expressions for the estimated occupation numbers:

$$\langle n_\alpha \rangle_{\text{F}} = \frac{1}{e^{(\varepsilon_\alpha - \mu)/kT} + 1};$$

$$\langle n_\alpha \rangle_{\text{B}} = \frac{1}{e^{(\varepsilon_\alpha - \mu)/kT} - 1}.$$

The expressions differ by a crucial ± 1 in the denominator.

5. In both cases, the estimated occupation numbers must sum to N, the fixed total number of particles:

$$\sum_\alpha \langle n_\alpha \rangle = N.$$

If the chemical potential is not otherwise known, this constraint serves to determine it.

6. When a particle has a spin $s\hbar$ (and when the energy does not depend on the spin orientation), the density of single-particle states is augmented by the factor $(2s + 1)$:

$$D(\varepsilon) = (2s + 1)\frac{2\pi(2m)^{3/2}}{h^3} V\varepsilon^{1/2}.$$

Here is the reason: a spin of $s\hbar$ can have $2s + 1$ distinctly different orientations, and so each state of a spinless particle splits into $2s + 1$ new states but without any change in energy.

Further reading

The derivation of $\langle n_\alpha \rangle$ for fermions and bosons is based on a paper by Helmut Schmidt, *Z. Phys.* **134**, 430–1 (1953). An English translation was given by R. E. Robson, *Am. J. Phys.* **57**, 1150–1 (1989). Some papers that focus on rigor in the derivation are provided by A. R. Fraser, *Phil. Mag.* **42**, 156–64 and 165–75 (1951), and by F. Ansbacher and P. T. Landsberg, *Phys. Rev.* **96**, 1707–8 (1954).

The seminal papers on bosons and fermions were the following:

S. N. Bose, *Z. Phys.* **26**, 178–81 (1924).

Albert Einstein, *Berlin Ber.* 261–7 (1924) and 3–14 (1925).

Wolfgang Pauli, *Z. Phys.* **31**, 765–83 (1925), proposing the exclusion principle.

Enrico Fermi, *Z. Phys.* **36**, 902–12 (1926).

Paul A. M. Dirac, *Proc. R. Soc. Lond.* **A 112**, 661–77 (1926).

The papers by Bose and Einstein took account of the true indistinguishability of identical particles but placed no restrictions on multiple occupancy. (In the preceding classical physics, it had been considered meaningful to label and thereby to distinguish among identical particles, much as names distinguish between human twins, say.) After Pauli introduced the exclusion principle for electrons in atoms, Fermi and Dirac extended the principle to gas molecules. A sentence from Dirac's paper gives us some sense for the era: "The solution with antisymmetrical eigenfunctions, though, is probably the correct one for gas molecules, since it is known to be the correct one for electrons in an atom, and one would expect molecules to resemble electrons more closely than light-quanta." The recognition that even "gas molecules" come in two classes, fermions and bosons, came only later.

More accessible for readers of English is the fine chapter, "Exclusion principle and spin," by B. L. van der Waerden in *Theoretical Physics in the Twentieth Century: A Memorial Volume to Wolfgang Pauli*, edited by M. Fierz and V. F. Weisskopf (Interscience, New York, 1960).

An endearing vignette of Bose—based in part on a long interview—is provided by William A. Blanpied, "Satyendranath Bose: Co-founder of quantum statistics," *Am. J. Phys.* **40**, 1212–20 (1972).

Problems

1. *Another meaning for* $\langle n_\alpha \rangle_F$.

(a) For fermions, construct a proof that

$$\begin{pmatrix} \text{probability that single-particle} \\ \text{state } \varphi_\alpha \text{ is occupied} \end{pmatrix} = \langle n_\alpha \rangle_F.$$

Start with the probabilities $P(\Psi_j)$ or at least use them. This result extends the significance of $\langle n_\alpha \rangle_F$ from merely "the estimated number of particles in φ_α."

(b) Why does such a result *not* hold for bosons?

2. *Some effects of spin.*

(a) Use the limiting expressions in section 8.4, specifically relations (8.18) to (8.22), to compute the energy, pressure, and chemical potential of an N-particle gas in terms of T, V, and N. Specify that each particle has a spin $s\hbar$, where s may differ from 0 and $\frac{1}{2}$.

(b) In which calculated quantities does the value of s make a numerical difference (in this limiting regime)?

3. *Criterion for the classical limit.*

(a) If equations (8.18) to (8.22) are to be self-consistent, what strong inequality must hold among the quantities N, V, m, and T?

(b) What can you say about the sign and size of the chemical potential?

4. Section 8.5 outlined the calculation of pressure and energy for a nearly classical ideal gas. Compute the four integrals and otherwise fill in the omitted steps, so that you present a complete derivation of the expression for $\langle E \rangle$. Do you confirm the factor of $\pm 1/2^{5/2}$?

5. Continuing with the approximations in section 8.5, calculate $\langle n_a \rangle$ inclusive of the first terms that distinguish between fermions and bosons. That is, eliminate the chemical potential from the expression in equation (8.32). Check that you have preserved the correct value for the integral over all single-particle states. Offer a verbal interpretation of the different behaviors of $\langle n_a \rangle$ for fermions and bosons (based, in part, on the exclusion principle).

6. *Entropy.* The entropy S of a quantum ideal gas can be expressed in terms of the estimated occupation numbers:

$$S/k = \sum_{a} [\mp(1 \mp \langle n_a \rangle) \ln(1 \mp \langle n_a \rangle) - \langle n_a \rangle \ln\langle n_a \rangle],$$

where the upper sign applies for fermions and the lower for bosons.

(a) In general, how should the derivative $(\partial S/\partial T)_V$ be related to the derivative $(\partial \langle E \rangle/\partial T)_V$?

(b) Check whether the expression for S displayed above satisfies the relationship in (a). (Your steps reverse the process by which S was computed by integration—and are much easier steps.)

9 Fermions and Bosons at Low Temperature

9.1 Fermions at low temperature
9.2 Pauli paramagnetism (optional)
9.3 White dwarf stars (optional)
9.4 Bose–Einstein condensation: theory
9.5 Bose–Einstein condensation: experiments
9.6 A graphical comparison
9.7 Essentials

This chapter applies the occupation number analysis to ideal gases at low temperature. While "low" temperature will mean 3 K in one instance, in another situation already room temperature is "low" temperature. Each gaseous system has a characteristic temperature, distinct from its physical temperature T and defined by parameters such as particle mass and typical inter-particle separation. The physical temperature T is "low" whenever it is substantially smaller than the characteristic temperature. This suffices as an introduction. The subject is understood best by seeing the applications themselves, and so we proceed to them.

9.1 Fermions at low temperature

Chapter 8 provided several general results for an ideal fermion gas. The estimated occupation number $\langle n_\alpha \rangle$ has the structure

$$\langle n_\alpha \rangle = \frac{1}{e^{(\varepsilon_\alpha - \mu)/kT} + 1}. \tag{9.1}$$

The chemical potential μ is determined by the equation

$$\sum_\alpha \langle n_\alpha \rangle = N. \tag{9.2}$$

This relation says that summing the estimated occupation numbers over all single-particle states yields the fixed total number of particles in the system. The total energy $\langle E \rangle$ may be computed from the relation

$$\langle E \rangle = \sum_\alpha \varepsilon_\alpha \langle n_\alpha \rangle. \tag{9.3}$$

The sums are best evaluated by converting them to integrals with a density of single-particle states. Section 4.1 gave us the density $D(\varepsilon)$ for a spinless particle. When a particle has a spin $s\hbar$ (and when the energy does not depend on the spin orientation), the density is augmented by the factor $(2s + 1)$:

$$D(\varepsilon) = (2s + 1)\frac{2\pi(2m)^{3/2}}{h^3}V\varepsilon^{1/2}$$

$$= C\varepsilon^{1/2}, \tag{9.4}$$

where C is a constant defined by the preceding line. The most common application is to electrons, whose spin is $\frac{1}{2}\hbar$, and for which the factor $(2s + 1)$ equals 2. That value will be used henceforth. The behavior of $D(\varepsilon)$ as a function of its argument was displayed in figure 4.3, and it would be a good idea to look at that graph again now.

Conversion to integrals implies that the discrete single-particle energy spectrum is being approximated by a continuous energy spectrum. The common notation for $\langle n_\alpha \rangle$ in that context is

$$f(\varepsilon) \equiv \frac{1}{e^{(\varepsilon-\mu)/kT} + 1}. \tag{9.5}$$

The function $f(\varepsilon)$ is called the *Fermi distribution function*. The name is apt, for $f(\varepsilon)$ describes the way that fermions are "distributed" over the various single-particle states (in the sense of an estimated distribution). A shorter name for $f(\varepsilon)$ is simply the *Fermi function*. Although the notation $f(\varepsilon)$ displays only the dependence on energy ε, the Fermi function depends also on the temperature T and the chemical potential μ, which itself is a function of temperature. Remember, too, that we have introduced new notation only; the Fermi function is just the function $\langle n_\alpha \rangle$ but with different clothes on.

In section 5.6, we reasoned that a classical (or semi-classical) analysis is valid provided the thermal de Broglie wavelength is small relative to the typical inter-particle separation:

$$\frac{h}{\sqrt{2\pi mkT}} = \lambda_{\text{th}} \ll \left(\frac{V}{N}\right)^{1/3}. \tag{9.6}$$

At fixed number density, the classical limit corresponds to a temperature T high enough so that the strong inequality holds. Quantum effects dominate at the other limit, when the temperature is so low that the inequality is reversed.

Limit $T \to 0$

Let us take first the extreme situation: the limit as the temperature descends toward absolute zero. The canonical probability distribution,

$$P(\Psi_j) = \frac{\exp(-E_j/kT)}{Z},$$

implies that the system settles into its ground state. [For all states of higher energy, their probability (relative to that of the ground state, denoted by g.s.) is $\exp[-(E_j - E_{\text{g.s.}})/kT]$ and hence vanishes exponentially as $T \to 0$.] We can mentally construct the system's ground state by filling single-particle states, starting with the lowest and continuing until we have placed N fermions into single-particle states, one particle per single-particle state. The single-particle energy that marks the end of this process is called the *Fermi energy* and is denoted by ε_F. Thus the Fermi function for the system's ground state (and hence for the limit $T \to 0$) has the rectangular form shown in figure 9.1.

Return now to the structure of $f(\varepsilon)$ as displayed in equation (9.5). When the Fermi function is evaluated at $\varepsilon = \mu$, its value is always $\frac{1}{2}$. One may turn the sentence around: whenever the equation $f = \frac{1}{2}$ holds, the relation $\varepsilon = \mu$ holds. The value $\frac{1}{2}$ appears in figure 9.1 only on the (limiting) vertical line at the Fermi energy ε_F. (At temperatures above absolute zero, the Fermi function descends smoothly from 1 to zero, as we will find later. The vertical line in figure 9.1 is the low temperature limit of a curve that steepens as the temperature drops.) Because the vertical line occurs at the Fermi energy and contains the value $f = \frac{1}{2}$, the chemical potential has the Fermi energy ε_F as its limiting value:

$$\lim_{T \to 0} \mu(T) = \varepsilon_F. \tag{9.7}$$

Here is a check. If $\varepsilon > \varepsilon_F$, the limit of the exponential in $f(\varepsilon)$ is infinite, and the Fermi function goes to zero, as required by figure 9.1. If $\varepsilon < \varepsilon_F$, the exponential vanishes as $T \to 0$, and so $f(\varepsilon)$ goes to 1, again as required.

Equation (9.2) and the density of states $D(\varepsilon)$ enable us to determine ε_F as a function of the system's fundamental parameters. The general form for the continuous version of equation (9.2) is

$$\int_0^\infty f(\varepsilon)D(\varepsilon)\,d\varepsilon = N. \tag{9.8}$$

Given the rectangular shape for the Fermi function, equation (9.8) reduces to

$$\int_0^{\varepsilon_F} 1 D(\varepsilon)\,d\varepsilon = \int_0^{\varepsilon_F} C\varepsilon^{1/2}\,d\varepsilon = \tfrac{2}{3}C\varepsilon_F^{3/2} = N. \tag{9.9}$$

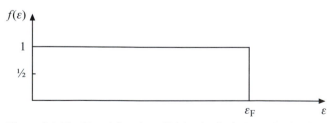

Figure 9.1 The Fermi function $f(\varepsilon)$ in the limit $T \to 0$. The distribution represents the estimated occupation numbers for the system's ground state.

Taking the constant C from (9.4) and solving for ε_F, we find

$$\varepsilon_F = \left(\frac{3}{8\pi}\right)^{2/3} \frac{h^2}{2m} \left(\frac{N}{V}\right)^{2/3}. \tag{9.10}$$

The Fermi energy is determined by the particle mass m and the number density N/V. Low mass and high number density imply large ε_F. Table 9.1 gives values for the conduction electrons in some metals. The quantity $(V/N)^{1/3}$ is the average inter-particle separation. If a particle has a de Broglie wavelength equal to that separation, then $h/(V/N)^{1/3}$ is its momentum. Moreover, the square of its momentum, divided by $2m$, is the particle's kinetic energy. Thus, in order of magnitude, the Fermi energy ε_F is the energy of a particle whose de Broglie wavelength is equal to the inter-particle separation.

As the characteristic energy for the fermion system, the Fermi energy defines a characteristic temperature, denoted T_F and called the *Fermi temperature*, by the relation

$$kT_F \equiv \varepsilon_F. \tag{9.11}$$

Table 9.1 indicates that the Fermi temperature for conduction electrons in metals is of order 5×10^4 K. Already room temperature ($\cong 300$ K) is low by comparison. Shortly we will put this observation to good use.

Table 9.1 *Fermi energies and Fermi temperatures for some metals. The integer that follows the element symbol is the number of conduction electrons contributed by each atom. The typical separation $(V/N)^{1/3}$ is given in units of 10^{-10} meter (1 angstrom). The separations are those at 20 °C [except for Na (25 °C) and K (18 °C)]. In almost all cases, the experimental Fermi temperature comes from measurements at 4 K or less.*

Element		$(V/N)^{1/3}$ $(10^{-10}$ m)	ε_F (theory) (eV)	T_F (theory) $(10^4$ K)	T_F (experimental) $(10^4$ K)
Li	1	2.78	4.70	5.45	2.5
Na	1	3.41	3.14	3.65	2.9
K	1	4.23	2.03	2.36	2.0
Rb	1	4.53	1.78	2.07	1.7
Cs	1	4.88	1.53	1.78	1.3
Cu	1	2.28	7.03	8.16	6.0
Ag	1	2.58	5.50	6.38	6.3
Au	1	2.57	5.53	6.41	5.9
Mg	2	2.26	7.11	8.25	6.3
Ca	2	2.79	4.70	5.45	2.8
Al	3	1.77	11.65	13.51	9.1

Sources: Encyclopedia of the Chemical Elements, edited by Clifford A. Hampel (Reinhold, New York, 1968) and *AIP Handbook*, 3rd edn, edited by D. E. Gray (McGraw-Hill, New York, 1972).

But first we compute the system's total energy at $T = 0$. Denoting the ground state energy by $E_{\text{g.s.}}$, we find from (9.3)

$$E_{\text{g.s.}} = \int_0^{\varepsilon_F} 1 \varepsilon D(\varepsilon)\, d\varepsilon = \tfrac{2}{5} C \varepsilon_F^{5/2}$$

$$= \tfrac{3}{5} N \varepsilon_F. \qquad (9.12)$$

The step to the second line uses equation (9.9). Single-particle states from energy zero (approximately) to energy ε_F are filled by a total of N fermions. Thus the total energy ought to be roughly N times $\tfrac{1}{2}\varepsilon_F$. The growth of the density of states with ε, produced by the factor $\varepsilon^{1/2}$ in $D(\varepsilon)$, pushes the numerical factor up to $\tfrac{3}{5}$.

Before we leave the domain of absolute zero, we should note one more relationship. The density of states, evaluated at the Fermi energy, may be written as

$$D(\varepsilon_F) = C\varepsilon_F^{1/2} = \frac{3}{2}\frac{N}{\varepsilon_F}; \qquad (9.13)$$

equation (9.9) provides the last step. Basically, the relationship says that $D(\varepsilon_F)$ times the entire energy range ε_F gives a number of order N, as it ought to.

The thermal domain $0 < T \ll T_F$

Now we turn to physical temperatures above absolute zero but far below the Fermi temperature. Once the temperature is above zero, the Fermi function will vary smoothly with energy ε. It will have the value $\tfrac{1}{2}$ when $\varepsilon = \mu(T)$, the new value of the chemical potential. This information provides the middle entry in table 9.2 and one point in figure 9.2. When ε exceeds $\mu(T)$ by a few kT, where "a few" is 3 or so, the exponential in $f(\varepsilon)$ is large, and so the Fermi function has dropped almost to zero. When ε is less than $\mu(T)$ by a few kT, the exponential is small, and $f(\varepsilon)$ has almost the value 1. The curve drops from approximately 1 to approximately zero over an interval of $2\times$(a few kT). Because we specified $T \ll T_F$, this interval is much less than kT_F and hence is much less than the Fermi energy.

Concerning the interval of variation, we know that the width is much less than ε_F. Where, however, is the interval located relative to ε_F? That is equivalent to asking,

Table 9.2 *Assessing the Fermi function when $0 < T \ll T_F$.*

$\varepsilon - \mu(T)$	$f(\varepsilon)$
a few kT	$\cong 0$
0	$\tfrac{1}{2}$
$-$ (a few kT)	$\cong 1$

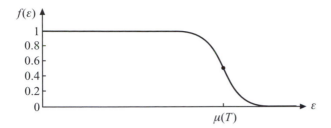

Figure 9.2 The Fermi function in the thermal domain $0 < T \ll T_F$. The dot on the curve represents the point $f|_{\varepsilon=\mu} = \frac{1}{2}$.

where does $\mu(T)$ lie relative to ε_F? Because the physical temperature remains small relative to the system's characteristic temperature T_F, we can expect that the chemical potential will have shifted only a little relative to ε_F, its value at absolute zero. We can confirm this, as follows.

We capture all the physics and can see the results most clearly if we approximate the Fermi function by a trapezoidal shape, as illustrated in figure 9.3. The integral that determines $\mu(T)$ is the continuous version of (9.2), namely equation (9.8). We approximate the integral by three terms:

$$\int_0^{\mu} 1 D(\varepsilon)\, d\varepsilon \; + \; D(\mu + \tfrac{1}{2}\delta\varepsilon)\tfrac{1}{4}\delta\varepsilon \; - \; D(\mu - \tfrac{1}{2}\delta\varepsilon)\tfrac{1}{4}\delta\varepsilon = N. \tag{9.14}$$

The first term corresponds to integrating with a rectangular shape for $f(\varepsilon)$ all the way to $\varepsilon = \mu(T)$. A correction will be applied shortly. The second term incorporates integration over the region beyond μ. The density of states is evaluated at the midpoint, $\mu + \tfrac{1}{2}\delta\varepsilon$, and factored out of the integral. The integral of $f(\varepsilon)$ itself is the area of the little triangle; because the triangle's height is $\frac{1}{2}$, its area is $\frac{1}{2} \times \frac{1}{2} \times \delta\varepsilon$. The third term corrects for the rectangular shape adopted in the first contribution. Using $f(\varepsilon) = 1$ all the way to $\varepsilon = \mu(T)$ over-estimated the integral by the following amount:

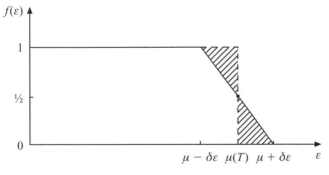

Figure 9.3 The trapezoidal approximation for the Fermi function. The symbol $\delta\varepsilon$ denotes an energy interval of a few kT.

the density of states, evaluated at the mid-point of the upper triangle, $D(\mu - \frac{1}{2}\delta\varepsilon)$, times the triangle's area, $\frac{1}{4}\delta\varepsilon$.

Regardless of whether $\mu(T)$ lies slightly below or above ε_F, we may formally approximate the first contribution in (9.14) as an integral from zero to ε_F (which yields precisely N) plus a product: the remaining integration range, which is $(\mu - \varepsilon_F)$, times the density of states evaluated at ε_F. The second and third terms in (9.14), taken together as a difference, are proportional to the derivative of D at μ, which is adequately represented by $D'(\varepsilon_F)$, the derivative evaluated at ε_F. Thus equation (9.14) becomes

$$N + (\mu - \varepsilon_F)D(\varepsilon_F) + D'(\varepsilon_F)\delta\varepsilon \times \tfrac{1}{4}\delta\varepsilon = N.$$

Canceling N on each side and solving for μ, we find

$$\mu(T) = \varepsilon_F - \frac{1}{4}\frac{D'(\varepsilon_F)}{D(\varepsilon_F)}(\delta\varepsilon)^2.$$

The energy interval $\delta\varepsilon$ is a few kT. If we take "a few" to be 3, then the numerical coefficient is $\frac{9}{4}$. A treatment that uses the exact variation of $f(\varepsilon)$ as displayed in (9.5) yields

$$\mu(T) = \varepsilon_F - \frac{\pi^2}{6}\frac{D'(\varepsilon_F)}{D(\varepsilon_F)}(kT)^2, \tag{9.15}$$

correct through quadratic order in T, and so we did quite well. (The further reading section at the end of the chapter provides a citation for the more elaborate calculation.) Because $D(\varepsilon)$ depends on ε as $\varepsilon^{1/2}$ when the gaseous system exists in three spatial dimensions, the necessary derivative is $D'(\varepsilon_F) = \frac{1}{2}D(\varepsilon_F)/\varepsilon_F$. Thus (9.15) takes the form

$$\mu(T) = \varepsilon_F - \frac{\pi^2}{12}\frac{(kT)^2}{\varepsilon_F}$$

$$= \varepsilon_F - \frac{\pi^2}{12}\left(\frac{T}{T_F}\right)^2 \varepsilon_F. \tag{9.16}$$

The chemical potential shifts to a smaller value, but the fractional shift, $(\mu - \varepsilon_F)/\varepsilon_F$, is of order $(T/T_F)^2$ and hence is quite small.

Looking back, we can see that the only reason that μ shifts at all (to second order in T) is because the density of states varies with ε.

Now we turn to the system's energy and its heat capacity. For the same accuracy—that of the trapezoidal approximation—the calculation is much easier than that for the chemical potential. Look again at figure 9.3 and approximate $\mu(T)$ by ε_F, a step that is justified by now. At finite but low temperature, some electrons have been promoted in energy from $\varepsilon_F - \frac{1}{2}\delta\varepsilon$ to $\varepsilon_F + \frac{1}{2}\delta\varepsilon$, an increase of $\delta\varepsilon$. How many such electrons? Those associated with the little triangle beyond μ, namely, $D(\varepsilon_F + \frac{1}{2}\delta\varepsilon)\frac{1}{4}\delta\varepsilon$ electrons. In the present calculation, we can afford to drop the "$+ \frac{1}{2}\delta\varepsilon$", and so we estimate the system's energy as

$$\langle E \rangle = E_{\text{g.s.}} + \left(\begin{array}{c} \text{number of} \\ \text{shifted electrons} \end{array} \right) \times \left(\begin{array}{c} \text{shift in} \\ \text{their energy} \end{array} \right)$$

$$= E_{\text{g.s.}} + [D(\varepsilon_{\text{F}})\tfrac{1}{4}\delta\varepsilon] \times \delta\varepsilon = E_{\text{g.s.}} + \tfrac{1}{4}D(\varepsilon_{\text{F}})(\delta\varepsilon)^2. \tag{9.17}$$

Because $\delta\varepsilon$ is a few kT, we find that the energy increases quadratically with T and that the numerical coefficient is approximately $\tfrac{9}{4}$. A more precise calculation, carried out to order T^2, yields

$$\langle E \rangle = E_{\text{g.s.}} + \frac{\pi^2}{6} D(\varepsilon_{\text{F}})(kT)^2. \tag{9.18}$$

Succinctly: both the energy shift $\delta\varepsilon$ and the number of electrons promoted are proportional to kT, and so the energy increases quadratically with temperature.

The heat capacity at constant volume, C_V, follows as

$$C_V = \left(\frac{\partial \langle E \rangle}{\partial T} \right)_V = \frac{\pi^2}{3} D(\varepsilon_{\text{F}}) k^2 T$$

$$= \frac{\pi^2}{2} \left(\frac{T}{T_{\text{F}}} \right) Nk; \tag{9.19}$$

equation (9.13) provides the step to the last line. The conduction electrons contribute to the heat capacity a term linear in T. Moreover, because the ratio T/T_{F} is quite small already at room temperature, being of order 0.01, the heat capacity is much smaller than the $\tfrac{3}{2}Nk$ that a classical analysis would predict. Quantum effects dominate.

The Debye model asserts that the vibrations of the crystal lattice—the sound waves—contribute to C_V a term cubic in T at low temperature (low now relative to the Debye temperature θ_{D}). Combining the results from equations (9.19) and (6.57), we have

$$\frac{C_V}{Nk} = \frac{\pi^2}{2} \left(\frac{T}{T_{\text{F}}} \right) + \frac{12\pi^4}{5} \left(\frac{T}{\theta_{\text{D}}} \right)^3 \tag{9.20}$$

as the heat capacity for a metallic solid when $T \leqslant 0.1\theta_{\text{D}}$. A plot of C_V/NkT versus T^2 should yield a straight line. The intercept along the vertical axis as $T^2 \to 0$ should be finite. According to our theory, the intercept will have the value $\pi^2/(2T_{\text{F}})$. In any event, if the electronic contribution to the heat capacity is linear in T, the experimental intercept will give the numerical coefficient of the electronic heat capacity. Figure 9.4 shows a fine example of C_V/NkT plotted against T^2. The data fall nicely along a straight line.

For the elemental metals listed in table 9.1, the theoretical values of T_{F} predict intercepts that agree with the experimental intercepts to within a factor of 2. If we stop to think about the context, the agreement is remarkably good. After all, the conduction electrons interact with the positive ions of the periodic crystal lattice; the force is an attractive Coulomb interaction. Moreover, the electrons interact among themselves

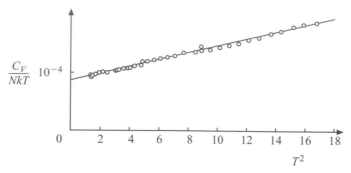

Figure 9.4 The heat capacity of copper at low temperature. The literal temperature range is $1.1 \leqslant T \leqslant 4.2$ K. The upper limit is approximately 1 percent of the Debye temperature: $\theta_D = 343$ K. Experimentally, the intercept as $T^2 \rightarrow 0$ is 8.27×10^{-5} K^{-1}, accurate to within 1 percent. Equation (9.20) and table 9.1 predict an intercept of $\pi^2/(2T_F) = 6.05 \times 10^{-5}$ K^{-1}. [*Source*: William S. Corak *et al.*, *Phys. Rev.* **98**, 1699–707 (1955).]

with a repulsive Coulomb force. This extensive set of interactions has been treated as though it merely confined the electrons to the volume V but otherwise permitted the conduction electrons to move as free particles: a quantum ideal gas.

To understand the reasons for the success, glance back to equation (9.18), which gives the estimated energy $\langle E \rangle$. The temperature-dependent part is proportional to $D(\varepsilon_F)$, the density of single-particle states at the Fermi energy. A pretty good value for the heat capacity can be gotten even if we have only an approximate value for that density of states. When the Coulomb interaction between the conduction electrons and the periodic lattice of positive ions is treated in detail, then the single-particle states are different from those that we used. The density of states will vary differently with energy ε and will have a different value at the Fermi energy. Nonetheless, so long as the better density of states behaves near its Fermi energy approximately as the density of states for an ideal gas, the analysis will go through as it did for us. The heat capacity will have the structure shown in the first line of (9.19): proportional to T and with $D(\varepsilon_F)$ as the sole factor that depends on the specific metal. The linear dependence on temperature arises primarily because of the Pauli principle and the existence of a Fermi energy. Only the numerical value of the coefficient depends on the interaction of the conduction electrons with the ions of the periodic lattice.

But what about the Coulomb interactions of the conduction electrons among themselves? To the extent that wave functions for the conduction electrons extend over the entire metal sample, the negative charge is distributed more or less uniformly. Rather than thinking of one electron as interacting with $N - 1$ point electrons, we may think of it as interacting with a rather smooth smear of negative charge. Moreover, that negative charge is largely canceled by the positive charge of the ions in the crystal lattice, for one can think of their positive charge as a uniform smear plus a periodic variation. As the one electron wanders through the metal sample, it experiences only residual interactions. For the purposes of calculating the heat capacity, those interactions are relatively insignificant, and the quantum ideal gas provides a good approximation. (To be sure, in another context the residual interactions may be absolutely essential. For example, they produce superconductivity in more than 25 different metals at low temperature.)

Synopsis

For a qualitative summary, we compare an ideal fermion gas at low temperature with a classical ideal gas. The latter is to have the same number density N/V and to be at the same physical temperature T. By ascribing to the classical gas a large particle mass m, we can ensure that its thermal de Broglie wavelength is small relative to $(V/N)^{1/3}$ and hence that the comparison gas behaves classically.

The physical temperature T is to be much less than the Fermi temperature T_F. Relative to the classical gas, the fermion gas has the following properties.

1. Large kinetic energy per particle. For the fermions, the energy per particle is $\langle E \rangle / N = \frac{3}{5} \varepsilon_F$ already at absolute zero, a consequence of the Pauli principle. So long as $T \ll T_F$, the T^2 contribution displayed in (9.18) is insignificant relative to the ground-state term. The ratio of energies is therefore

$$\frac{(\langle E \rangle / N)_{\text{fermions}}}{(\langle E \rangle / N)_{\text{classical}}} = \frac{\frac{3}{5}\varepsilon_F}{\frac{3}{2}kT} = \frac{2}{5}\frac{T_F}{T} \gg 1. \tag{9.21}$$

2. Small heat capacity per particle. Only particles near the Fermi energy are promoted in energy when the temperature rises. Hence the ratio of heat capacities is

$$\frac{(C_V)_{\text{fermions}}}{(C_V)_{\text{classical}}} = \frac{\frac{\pi^2}{2}\left(\frac{T}{T_F}\right)k}{\frac{3}{2}k} = \frac{\pi^2}{3}\left(\frac{T}{T_F}\right) \ll 1. \tag{9.22}$$

3. High pressure. In section 1.2, kinetic theory showed that $P = \frac{2}{3}\langle E \rangle / V$ for a classical ideal gas, provided the speeds are non-relativistic. Section 8.5 showed that the relationship holds quantum mechanically for any non-relativistic ideal gas (provided that the energy is solely kinetic energy). Now write the pressure as

$$P = \frac{2}{3}\frac{\langle E \rangle / N}{V/N} \tag{9.23}$$

and recall that V/N is the same for the fermion gas and the comparison classical gas. Thus the ratio of pressures is the same as the ratio of energies per particle [in relationship (9.21)], and so the ratio is large.

4. Little dependence of pressure on temperature. One last feature is worth noting. The energy $\langle E \rangle$ varies little with temperature; after all, that is what a small heat capacity means. Because the pressure is $P = \frac{2}{3}\langle E \rangle / V$ (when computed non-relativistically), the fermion pressure is virtually independent of temperature. In contrast, a classical ideal gas, which is described by the ideal gas law, $P = (N/V)kT$, has a linear dependence of pressure on temperature, which is a substantial dependence. In detail, the ratio of $(\partial P/\partial T)_V$ for the two gases is the same as the ratio $(\partial\langle E\rangle/\partial T)_V$; hence, by equation (9.22), the ratio is very small.

Indeed, the last paragraph is a good prelude to a bit of terminology. At low temperatures, quantum ideal gases behave very differently from the way a classical ideal gas behaves. We have just seen that property for fermions, and section 9.4 will derive, for bosons, another radical departure from the ordinary. The quantum gases are said to be *degenerate* at low temperature. That is not a moral judgment. Rather, the word "degenerate" is used in the sense of departing markedly from the properties of an "ordinary" classical gas.

9.2 Pauli paramagnetism (optional)

This section and the next describe additional phenomena produced by fermions at low temperature.

Section 5.3 provided our first encounter with paramagnetism. The "system" consisted of a single spatially fixed atom whose spin is $\frac{1}{2}\hbar$. We estimated the component of magnetic moment along an external magnetic field $\mathbf{B}$. In a material like cesium titanium alum, there is one such atom per molecule, but the spatial separation of the magnetic moments is large enough so that—for many purposes—the moments act approximately independently of each other. Even vibration about equilibrium sites in the lattice has little effect. The twin properties of (more or less) fixed location in the lattice and substantial separation are sufficient for the moments to act independently, to good approximation. If there are N such atoms in a macroscopic system, the explicit expression in equation (5.13) need only be multiplied by N. Thus

$$\left\langle \begin{array}{c} \text{total magnetic moment along } \mathbf{B} \\ \text{for } N \text{ spatially fixed atoms} \end{array} \right\rangle = Nm_B \tanh(m_B B/kT). \tag{9.24}$$

Radically different is the situation with conduction electrons in a metal. Even if the typical separation $(V/N)^{1/3}$ is the same as for the spatially fixed moments, the conduction electrons remain coupled quantum mechanically by the Pauli principle. We need to consider all of those electrons simultaneously.

For a single electron, the magnetic energy is $-\mathbf{m}_B \cdot \mathbf{B}$. Even in a magnetic field as large as 2 tesla, the energy is only 10^{-4} electron volts and hence is much less than a typical Fermi energy ε_F. Just as room temperature ($\cong 300$ K) leaves the conduction electrons close to their ground state, similarly the magnetic interaction leaves the electron system close to its original ground-state configuration. For our purposes— estimating the total magnetic moment—it suffices to take the context $T = 0$ and to examine the full ground-state, first without the field $\mathbf{B}$ and then in its presence.

In the absence of a magnetic field, single-particle states with a specified amount of kinetic energy but with oppositely oriented moments have the same energy. Part (a) of figure 9.5 indicates that such states will be fully occupied up to the Fermi energy ε_F. The number of up moments equals the number of down moments, and so the total magnetic moment is zero.

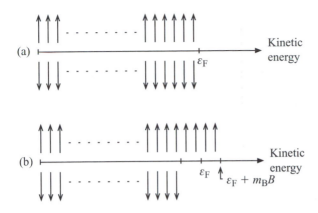

Figure 9.5 Paramagnetism for conduction electrons. The ground state is shown (a) in the absence of a magnetic field and (b) when the field **B** is present (a field that points upward). The arrows represent the magnetic moments. (An electron's spin is anti-parallel to the magnetic moment.) Note that the abscissa is the single-particle *kinetic* energy, not the entire single-particle energy.

When the external field (taken to point upward) is applied, the energy of each single-particle state changes:

$$\Delta\varepsilon = -m_B B \text{ for an up moment;}$$

$$\Delta\varepsilon = +m_B B \text{ for a down moment.} \tag{9.25}$$

To construct the new ground state (the state of lowest total energy), we need to shift some electrons from down-moment states to up-moment states. Each such shift reduces the total energy by $-2m_B B$. The Pauli principle, however, compels the electrons to populate up-moment states with kinetic energy higher than ε_F (which continues to denote the Fermi energy in the absence of the field) because the states of lower kinetic energy are already occupied. How far in kinetic energy can we go and still reduce the total energy (in a net sense)?

Near the zero-field Fermi energy, it is a good approximation (1) to take the density of down-moment states to equal the density of up-moment states and (2) to set each equal to $\frac{1}{2}D(\varepsilon_F)$, where $D(\varepsilon_F)$ is the density function in the absence of the field. Then we can profitably shift electrons until we are taking them from (kinetic energy) $= \varepsilon_F - m_B B$ and promoting them to (kinetic energy) $= \varepsilon_F + m_B B$, a gain in kinetic energy of $+2m_B B$. Altogether, the process yields

$$\left(\begin{array}{c} \text{number of} \\ \text{electrons shifted} \end{array}\right) = \frac{1}{2}D(\varepsilon_F) \times m_B B. \tag{9.26}$$

Each shift increases the total magnetic moment by $+2m_B$. Thus the new ground state has

$$\left(\begin{array}{c}\text{total magnetic moment along } \mathbf{B} \\ \text{for } N \text{ conduction electrons}\end{array}\right) = \tfrac{1}{2}D(\varepsilon_F)m_B B \times 2m_B$$

$$= \frac{3}{2}\frac{m_B B}{\varepsilon_F}Nm_B. \tag{9.27}$$

The step to the second line uses equation (9.13).

To gain an appreciation for this result, let us compare it with the limit $T \to 0$ of equation (9.24), the calculation for spatially fixed magnetic moments. The hyperbolic tangent is a ratio of pairs of exponentials, and its limit is 1. Thus

$$\lim_{T \to 0}\left\langle \begin{array}{c}\text{total magnetic moment along } \mathbf{B} \\ \text{for } N \text{ spatially fixed atoms}\end{array}\right\rangle = Nm_B. \tag{9.28}$$

All the magnetic moments line up along $\mathbf{B}$. In contrast, for conduction electrons, the Pauli principle would make such alignment too expensive (in kinetic energy), and so the value in equation (9.27) is smaller by the factor $\tfrac{3}{2}m_B B/\varepsilon_F$ a small factor indeed.

Wolfgang Pauli worked out the effect of the exclusion principle on electron paramagnetism [*Z. Phys.* **41**, 81–102 (1927)], and so this phenomenon is called *Pauli paramagnetism*.

[Now that the major point has been made, I should note that, as $T \to 0$ for the spatially fixed moments, their mutual magnetic and electrostatic interactions may no longer be neglected. The paramagnetic atoms may develop a ferromagnetic behavior or even an anti-ferromagnetic behavior, but those are separate topics too far from the present mainstream for us to go into them here. Some aspects are developed in chapter 16.

Moreover, in the case of conduction electrons, the external magnetic field affects the translational motion of the electrons, and that induces a (diamagnetic) contribution to the sample's total magnetic moment. Pauli paramagnetism is only part of the electrons' response to an external field.]

9.3   White dwarf stars (optional)

Typical white dwarf stars have a surface temperature of 10,000–30,000 K, substantially higher than that of the sun ($\cong$ 5,800 K), and so they radiate with a bluish white color. Their luminosity (the total radiant energy output), however, is only 0.1 to 1 percent of the solar value, and so the stars must have a small surface area. Indeed, a typical radius R is 1 percent of the solar radius and so is approximately equal to the Earth's radius. The stars are indeed "white" and "dwarf." Yet their mass M ranges over an interval from $\cong 0.2\ M_\odot$ to $1.4\ M_\odot$, where the symbol $M_\odot$ denotes the sun's mass. In order of magnitude, one solar mass of material is compressed into a volume the size of the Earth, producing an average mass density 10^6 times that of water.

At such high density and large mass, gravity must be extremely strong. What supports the material against gravity? The short answer is this: the pressure of a degenerate electron gas. Except for a thin atmosphere, the material in the star is wholly ionized, and so all the electrons are "conduction electrons." Table 9.3 provides parameters from two well-observed white dwarfs and shows that the Fermi energy (calculated non-relativistically) is 10^4 times as large as that for the metals which we studied in section 9.1. The corresponding Fermi temperature is of order 10^9 K. The physical temperature T within the star is of order 10^7 K, which has two consequences, as follows.

1. For the electrons, the physical temperature is low relative to the Fermi temperature, and so our results from section 9.1 are applicable (in part, at least). It is hard to imagine $T = 10^7$ K as being "low" temperature, but $T/T_F \cong 0.01$.
2. The ion cores—that is, the nuclei—are not degenerate (because their individual masses are much larger than that of an electron). The nuclei remain a classical gas and exert a pressure that is smaller than the electron pressure by roughly the ratio T/T_F, and so their pressure is negligible.

To analyze the star, we need consider—for support against gravity—only the electron gas. Moreover, temperature is basically irrelevant to the structure of a white dwarf. Why? Because the pressure of a degenerate electron gas is largely independent of temperature. The electrons are good conductors of microscopic kinetic energy and ensure that the stellar interior has an essentially uniform temperature—and a negligibly low one. For a good approximation, it suffices to represent the pressure by that of an electron gas at $T = 0$. (Because $T/T_F \cong 0.01$, the physical temperature is only 1 percent of the way from absolute zero to T_F. Using $T = 0$

Table 9.3 *Parameters for two white dwarf stars. By charge neutrality, the number density N/V of electrons equals the number density of protons. The stellar mass is dominated by nuclei, such as ^4He, ^{12}C, and ^{16}O, that have as many neutrons as protons. Thus the number density of protons is one-half the number density of nucleons. In short, an adequate approximation for the electrons is $(N/V)_{ave} = \frac{1}{2}(\bar{\rho}/m_{proton})$, where $\bar{\rho} = M/(4\pi R^3/3)$ is the average mass density. The Fermi energy ε_F and the Fermi temperature T_F are computed non-relativistically and are based on $(N/V)_{ave}$. The Fermi momentum p_F is the electrons' maximum momentum magnitude; it is shown divided by the electron rest mass times c. The symbol v_F denotes the corresponding speed (calculated from the relativistic relationship that connects momentum, speed, and rest mass).*

Name	$M/M_\odot$	$R/R_\odot$	$(V/N)_{ave}^{1/3}$ $(10^{-12}$ m)	ε_F $(10^4$ eV)	T_F $(10^9$ K)	p_F/mc	v_F/c
40 Eri B	0.447	0.013	2.3	6.9	0.8	0.52	0.46
Sirius B	1.05	0.0073	0.96	40	4.6	1.2	0.78

Source: Kenneth R. Lang, *Astrophysical Formulae: A Compendium for the Physicist and Astrophysicist*, 2nd edn (Springer, New York, 1980).

is not so outrageous as it might seem at first acquaintance.) This simplification enables us to derive a remarkable relation between the stellar radius R and the total
mass M, as follows.

Non-relativistic domain

Hydrostatic equilibrium relates the central pressure to the stellar mass and radius.
Figure 9.6 shows an imaginary column passing from the center to the stellar surface.
The material in that column is supported against gravity by the pressure at its base, the
central pressure P_{center}. Thus we write

$$P_{\text{center}} A = \left(\begin{array}{c} \text{total mass} \\ \text{in column} \end{array} \right) \times \left(\begin{array}{c} \text{average gravitational force} \\ \text{per unit mass} \end{array} \right)$$

$$\cong (\overline{\rho} R A) \times \left(\frac{1}{2} \frac{GM}{R^2} \right). \tag{9.29}$$

We estimate the total columnar mass by the column volume times the average mass
density $\overline{\rho}$, where $\overline{\rho} = M/(4\pi R^3/3)$. For the average gravitational force per unit mass,
we take half the value at the surface as a reasonable estimate. (Recall that, by
symmetry, the gravitational force per unit mass is zero at the star's center.)

Next, we return to the microscopic theory of section 9.1 According to that section,
the electrons exert a pressure

$$P = \frac{2}{3} \frac{\langle E \rangle}{V} = \frac{2}{5} \varepsilon_F \frac{N}{V}$$

$$= \frac{2}{5} \left(\frac{3}{8\pi} \right)^{2/3} \frac{h^2}{2m} \left(\frac{N}{V} \right)^{5/3}, \tag{9.30}$$

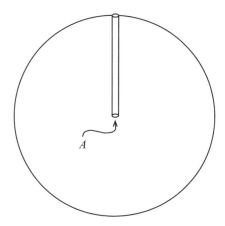

Figure 9.6 A narrow column, of cross-sectional area A, extends from the stellar center to the
surface.

provided their motion is non-relativistic. [For details, see the "synopsis" subsection and equations (9.10) and (9.12).] The electron number density N/V is proportional to the mass density ρ (of electrons and nuclei taken together). The number density at the center will be higher than the average number density, but we can incorporate that effect in a proportionality constant. Thus equation (9.30) implies that $P_{\text{center}} = \text{const} \times (M/R^3)^{5/3}$.

Now substitute the last expression for P_{center} into (9.29):

$$\text{const} \times \left(\frac{M}{R^3}\right)^{5/3} = \text{const}' \times \frac{M}{R^3} \times R \times \frac{M}{R^2}, \tag{9.31}$$

where const and const′ are two constants. This form shows how the central pressure and the gravitational force that the pressure needs to balance depend on M and R. Solving for R, we find

$$R = \left(\frac{\text{const}}{\text{const}'}\right) \times \frac{1}{M^{1/3}}. \tag{9.32}$$

The larger the mass, the smaller the star!

Observations of white dwarfs amply bear out that statement, and table 9.3 gives an example. (Although some things are known about hundreds of white dwarfs, for only two are both mass and radius known to within 10 percent.) A more detailed theory yields an R versus M relationship that is more complicated than a power law, but it can be reasonably approximated by $R \propto M^{-0.57}$ for masses up to one solar mass. The difference in exponent does not mean that we made a computational error. If the non-relativistic expression for the electron pressure were valid throughout the star, then a detailed calculation, taking into account a density that increases monotonically with depth, would produce precisely the power law relation in (9.32). Rather, the difference in exponent arises because the pressure expression in (9.30) must be modified at extremely high densities (because of relativistic effects); we look into that shortly.

If you glance back to equation (9.31), you can see that the left-hand side grows as $1/R^5$ as R decreases, but the right-hand side grows only as $1/R^4$. Thus, contraction enables the electron pressure to grow large enough to support the star against gravity. This conclusion depends, however, on our using a non-relativistic calculation of the electron pressure. Table 9.3 shows that an electron gas at the average separation $(V/N)_{\text{ave}}^{1/3}$ has some electrons moving at half the speed of light. Detailed theory indicates that the central density is roughly 10 times the average density. The maximum electron speeds will be even higher, and so one must consider relativistic effects. We turn to them in the next subsection.

Extreme relativistic regime

Kinetic theory provides a good start. A glance back at equations (1.1) and (1.4) indicates that the pressure of an ideal gas can be expressed as

$$P = \langle p_x v_x \rangle \frac{N}{V}$$

$$= \tfrac{1}{3}\langle \mathbf{p} \cdot \mathbf{v} \rangle \frac{N}{V}. \tag{9.33}$$

The step to the second line reasons that the averages of products of Cartesian components must be equal, for example, $\langle p_y v_y \rangle = \langle p_x v_x \rangle$, and so any one average is one-third of their sum, which is given by the scalar product of the two vectors. Equation (9.33) is relativistically exact. Previously we wrote the momentum $\mathbf{p}$ in the non-relativistic form $m\mathbf{v}$, where m denotes the rest mass, thereby taking the low speed limit. Now, however, we specify an extreme relativistic regime, so that most electrons have a speed near c. (Only this other limit—the high speed limit—lends itself to back-of-the-envelope estimates, which suffice for us.) The Fermi momentum p_F is the maximum momentum that the electrons possess (at absolute zero). Because the tips of the momentum vectors are distributed uniformly within a sphere of radius p_F, 49 percent of the electrons have momentum magnitudes that lie within 20 percent of p_F. [The spherical shell between radii 0.8 p_F and p_F accounts for 49 percent of the spherical volume: $1^3 - (0.8)^3 = 0.49$.] Thus most electrons have a momentum magnitude approximately equal to p_F, and we estimate the pressure as

$$P \cong \frac{1}{3} p_F c \frac{N}{V}. \tag{9.34}$$

The speed v saturates at the speed of light c. (An exact evaluation of $\langle \mathbf{p} \cdot \mathbf{v} \rangle$, outlined in problem 17, replaces the $1/3$ by $(1/4) \times [1 + \text{order of } (mc/p_F)^2]$.)

To determine the Fermi momentum, we return to section 4.1. Equation (4.4) indicates that

$$p_F = n_{\max} \frac{h}{2L}, \tag{9.35}$$

where $n_{\max}$ is the maximum value of the magnitude constructed from the triplet $\{n_x, n_y, n_z\}$. For an electron with spin $\tfrac{1}{2}\hbar$, the generalization of equation (4.6) counts occupied single-particle states as follows:

$$\tfrac{1}{8}(2 \times \tfrac{1}{2} + 1) \frac{4\pi}{3} n_{\max}^3 = N. \tag{9.36}$$

Combining equations (9.35) and (9.36) gives the Fermi momentum:

$$p_F = \left(\frac{3}{8\pi}\right)^{1/3} h \left(\frac{N}{V}\right)^{1/3}. \tag{9.37}$$

(Although we derived this result by considering standing waves, the same result emerges if we represent the electrons with traveling waves, which are closer to the spirit of kinetic theory.)

Now equations (9.34) and (9.37) imply

$$P \cong \tfrac{1}{3}\left(\frac{3}{8\pi}\right)^{1/3} hc \left(\frac{N}{V}\right)^{4/3} \tag{9.38}$$

in the extreme relativistic regime. Because the electron's speed has c as a limit and no longer increases with density, the pressure here depends on number density only to the $4/3$ power (in contrast to the $5/3$ power in the non-relativistic case).

If we use equation (9.38) on the left-hand side of equation (9.29), we find

$$\text{const}'' \times \left(\frac{M}{R^3}\right)^{4/3} = \text{const}' \times \frac{M}{R^3} \times R \times \frac{M}{R^2}. \tag{9.39}$$

The powers of R on the two sides are now the same: $1/R^4$. Once the extreme relativistic regime has been reached, further contraction no longer boosts the pressure faster than the gravitational attraction. That route to stellar equilibrium is blocked. Thus, if the mass M does not satisfy equation (9.39) immediately, the star can *not* adjust its radius to provide equality.

From table 9.3, note that the higher the stellar mass, the higher p_F and the electrons' speeds. In the 1930s, the Indian astrophysicist Subrahmanyan Chandrasekhar calculated that stars with a mass up to the limit $M = 1.4M_\odot$ may evolve to the white dwarf stage. More massive stars can *not* be supported by the pressure of degenerate electrons (subject to theoretical provisos such as no stellar rotation etc.). The relativistic speeds of the electrons in the stellar core produce a high pressure but one that grows too slowly with increasing density.

Today, another effect is known to limit the mass of a white dwarf. At extremely high central density, the electrons have such high energy that inverse beta decay occurs readily. An electron is "squeezed" into a proton: a neutron is formed (and a neutrino is emitted). The reaction removes electrons, causing the electron pressure to drop, which leads to further contraction, more inverse beta decay, and so on. The entire process ceases only when the pressure of a degenerate neutron gas holds gravity at bay. This sequence establishes another limiting mass, which is approximately 10 percent smaller than Chandrasekhar's limiting mass.

Our sun is predicted to shine more or less as it does today for another 5 billion years, then to expand into a red giant phase, next to blow off a substantial fraction of its material via a superwind, and finally to evolve into a white dwarf. We will not be around to see the white dwarf stage, but you know enough thermal physics to understand what will support the sun against further collapse.

9.4 Bose–Einstein condensation: theory

Our attention turns now to bosons, particles with integral spin. Most significant for us is that the Pauli exclusion principle does *not* apply to bosons. Rather, any number of bosons—or any number up to N, if the total number of bosons is fixed—may occupy any given single-particle state. The estimated occupation number $\langle n_a \rangle$ has the structure

$$\langle n_a \rangle = \frac{1}{e^{(\varepsilon_a - \mu)/kT} - 1}. \tag{9.40}$$

In this section, the system is specified to contain N bosons, a fixed number, and so the chemical potential μ is determined by the equation

$$\sum_a \langle n_a \rangle = N. \tag{9.41}$$

Because we will compare our results with experimental properties of the spinless helium isotope ^{4}He, let me specify at the outset that we investigate bosons whose spin is zero. Then the algebraic issue of spin orientations drops out of the problem—and without our losing any significant generality in the results.

As in our study of fermions, we start with the limit $T \to 0$ and then move on to low (but nonzero) temperature, where "low" must be specified by comparison with some characteristic temperature for bosons.

Limit $T \to 0$

As the temperature descends to absolute zero, the system of N bosons settles into its ground state. To construct the full state of lowest energy, we mentally place all bosons in the single-particle state of lowest energy, φ_1. The total energy is then

$$E_{\text{g.s.}} = N\varepsilon_1 \tag{9.42}$$

where

$$\varepsilon_1 = \frac{3h^2}{8m} \frac{1}{V^{2/3}}. \tag{9.43}$$

The estimated occupation number $\langle n_1 \rangle$ must approach N as $T \to 0$:

$$\lim_{T \to 0} \langle n_1 \rangle = \lim_{T \to 0} \frac{1}{e^{(\varepsilon_1 - \mu)/kT} - 1} = N. \tag{9.44}$$

From this relationship, we can assess the behavior of the chemical potential near absolute zero. Because $\langle n_1 \rangle$ is non-negative and finite, the exponential must be greater than 1; that requires $\mu < \varepsilon_1$. Yet the exponential must be extremely close to 1, so that division by the difference between it and 1 yields the large number N. Thus the exponent itself must be small, and we may expand the exponential, writing

$$\frac{1}{1 + (\varepsilon_1 - \mu)/kT + \cdots - 1} \cong N$$

when T is close to zero. Solving this equation for μ yields

$$\mu \cong \varepsilon_1 - \frac{kT}{N}. \tag{9.45}$$

The chemical potential approaches ε_1 from below as $T \to 0$.

The other estimated occupation numbers, $\langle n_\alpha \rangle$ for $\alpha \geq 2$, all vanish as the temperature descends to zero.

Condensation

How does $\langle n_1 \rangle$ vary as the temperature rises from absolute zero? The answer lies hidden in equation (9.41), which we can write as

$$\langle n_1 \rangle + \sum_{\alpha \geq 2} \langle n_\alpha \rangle = N. \tag{9.46}$$

To make progress, we need to replace the summation by an integral with a density of states $D(\varepsilon)$. So we write

$$\langle n_1 \rangle + \int_0^\infty \frac{1}{e^{(\varepsilon - \mu)/kT} - 1} D(\varepsilon)\, d\varepsilon = N. \tag{9.47}$$

Because the standard density of states goes to zero as $\varepsilon^{1/2}$ when ε goes to zero, we may safely let the lower limit of integration be $\varepsilon = 0$. We do not inadvertently double-count the state φ_1 and the term $\langle n_1 \rangle$.

Moreover, that same behavior for the density of states enables us to approximate the value of the chemical potential. For very low temperature, equation (9.45) told us that the temperature dependent part of μ is tiny: $kT/N \cong kT/10^{20}$, say. For the single-particle energy ε_1 that appears in μ, return to (9.43) and note that, even if the volume V is as small as 1 cubic centimeter and if m is the mass of a helium atom, then $\varepsilon_1 = 1.5 \times 10^{-18}$ electron volts. So small an energy is usually insignificant—provided it does not make the difference between finiteness and infinity. But the integral in (9.47) remains finite if we approximate μ by zero, which we proceed to do, as follows:

$$\int_0^\infty \frac{1}{e^{\varepsilon/kT} - 1} D(\varepsilon)\, d\varepsilon = \int_0^\infty \frac{C\varepsilon^{1/2}}{e^{\varepsilon/kT} - 1}\, d\varepsilon = C(kT)^{3/2} \int_0^\infty \frac{x^{1/2}}{e^x - 1}\, dx$$

$$= C(kT)^{3/2} \times \frac{\pi^{1/2}}{2} 2.612 = \text{const} \times T^{3/2}. \tag{9.48}$$

The substitution $x = \varepsilon/kT$ extracts the temperature dependence from the integral. The definite integral that remains is tabulated in appendix A. Altogether, the integral on the left-hand side equals a known constant times $T^{3/2}$.

To tidy things up, we define a characteristic temperature T_B for bosons as the temperature for which the integral in (9.48) equals N. Because $C \times (kT_B)^{1/2}$ equals $D(kT_B)$, the density of states evaluated at kT_B, the definition of T_B is equivalent to the relationship

$$D(kT_B)kT_B \times \frac{\pi^{1/2}}{2} 2.612 = N. \tag{9.49}$$

That equation is analogous to equation (9.13) for fermions, which one can write as

$$D(kT_F)kT_F \times \tfrac{2}{3} = N. \tag{9.50}$$

Upon referring to equation (9.4) for the detailed factors in $D(\varepsilon)$, one finds that (9.49) implies

$$kT_B = \frac{1}{\pi(2.612)^{2/3}} \frac{h^2}{2m} \left(\frac{N}{V}\right)^{2/3}, \tag{9.51}$$

which provides a more explicit expression for T_B.

The characteristic temperatures for bosons and fermions have the same dependence on number density N/V and particle mass m; the numerical coefficients are comparable; and so the two temperatures are very similar, structurally. What the temperatures signify physically, however, are radically different, as we shall see shortly.

To avoid some algebraic substitutions, let us reason as follows. The characteristic temperature T_B was defined so that the integral in (9.48) equals N when $T = T_B$. Therefore the constant at the end of equation (9.48) must be expressible as $N/T_B^{3/2}$. This observation enables us to write equation (9.47) as

$$\langle n_1 \rangle + \left(\frac{T}{T_B}\right)^{3/2} N = N.$$

Solving for $\langle n_1 \rangle$ yields

$$\langle n_1 \rangle = N \left[1 - \left(\frac{T}{T_B}\right)^{3/2} \right]. \tag{9.52}$$

This relationship indicates that $\langle n_1 \rangle$ remains of order N until $T \cong T_B$. Clearly, the expression must lose its validity when T exceeds T_B, but we did specify "low" temperature in our analysis, and so we find that the stipulation $T \leqslant 0.99 T_B$, say, is a sufficient criterion for "low." (Steps to justify this criterion and also some subsequent numerical assertions are offered in problem 21.)

Figure 9.7 shows $\langle n_1 \rangle$ as a function of temperature over a substantial range. At a temperature of $10 T_B$, the bosons act like a nearly classical ideal gas. The first correction to classical values of pressure or energy is of order 1 percent, and the numerical value of $\langle n_1 \rangle$ is much less than 1. (For confirmation of these claims, recall the general analysis in sections 8.4 and 8.5.) Imagine decreasing the temperature at fixed N and V. The value of $\langle n_1 \rangle$ grows continuously and reaches order $N^{2/3}$ when $T = T_B$. The ratio of $N^{2/3}$ to N is $N^{-1/3}$; if $N = 10^{21}$, say, the ratio is only 10^{-7}. On the graph, even a value of order $N^{2/3}$ appears as virtually zero. When T has dropped to $0.99 T_B$, however, equation (9.52) implies that $\langle n_1 \rangle$ is $0.015 N$. Such a value deserves to be called "of order N" and is barely distinguishable from zero on the graph. The temperature T_B signals the onset of a marked increase in $\langle n_1 \rangle$. In analogy with the

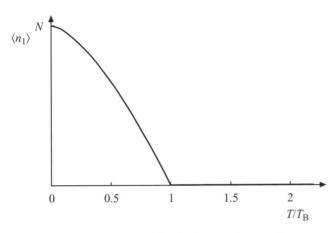

Figure 9.7 The occupancy of the single-particle ground state as a function of temperature.

condensation of water vapor into liquid water, the marked increase is called the *Bose–Einstein condensation*. Note that the bosons "condense" into a specific single-particle state, the state φ_1 of lowest energy. (Because they are particles of an ideal gas, the bosons do *not* cohere and form a droplet.)

The characteristic temperature that signals the onset has several names. The notation T_B is intended to abbreviate the name *Bose* (or *Bose–Einstein*) *temperature*. Sometimes the characteristic temperature is called the "condensation temperature." Other authors call it the "Einstein temperature," for Einstein discovered the "condensation" implicit in Bose's earlier work.

You may wonder, do other occupation numbers, such as $\langle n_2 \rangle$, also become of order N below T_B or below some other characteristic temperature? The answer is no. As the temperature descends from $10T_B$ to zero, the size of $\langle n_2 \rangle$ grows smoothly to a maximum of order $N^{2/3}$ and then declines to zero. Only for φ_1, the single-particle ground state, does the estimated occupation number ever become of order N.

Energy and heat capacity

The formal expression for the system's energy is

$$\langle E \rangle = \sum_\alpha \varepsilon_\alpha \langle n_\alpha \rangle. \tag{9.53}$$

If the temperature is less than T_B, we separate out the term $\varepsilon_1 \langle n_1 \rangle$ and write the remaining sum as an integral:

$$\langle E \rangle = \varepsilon_1 \langle n_1 \rangle + \int_0^\infty \frac{\varepsilon}{e^{(\varepsilon - \mu)/kT} - 1} D(\varepsilon)\, d\varepsilon. \tag{9.54}$$

For the first term, equation (9.52) gives $\langle n_1 \rangle$. For the second term, it suffices to set $\mu = 0$, to extract the temperature dependence with the substitution $x = \varepsilon / kT$, and to look up the remaining definite integral in appendix A. The outcome is

$$\langle E \rangle = \varepsilon_1 N \left[1 - \left(\frac{T}{T_B} \right)^{3/2} \right] + 0.770 \left(\frac{T}{T_B} \right)^{3/2} NkT. \tag{9.55}$$

Because ε_1 is so small, the contribution from the integral dominates (except at extremely low temperature). Thus the energy varies with temperature as $T^{5/2}$, and the heat capacity C_V grows as $T^{3/2}$. Figure 9.8 shows this rise as $T \to T_B$ from below.

If the temperature is greater than T_B, the term $\varepsilon_1 \langle n_1 \rangle$ in (9.53) may be ignored as negligibly small. The value of $\langle E \rangle$ is dominated by an integral over the other single-particle states, but the chemical potential may no longer be approximated by zero. Rather, the chemical potential must be calculated independently from equation (9.41). Let us skip the details and content ourselves with two observations, as follows.

1. If you read section 8.5 on the nearly classical ideal gas, then you can return to equation (8.38), differentiate $\langle E \rangle$ with respect to temperature, and find

$$C_V = \tfrac{3}{2} Nk \left[1 + 0.231 \left(\frac{T_B}{T} \right)^{3/2} + \cdots \right], \tag{9.56}$$

valid for a nearly classical boson gas. While this two-term expression will not suffice all the way down to T_B, it suggests that the heat capacity rises as the temperature

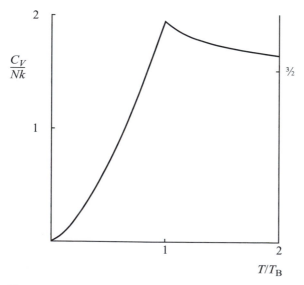

Figure 9.8 The heat capacity C_V of an ideal boson gas as a function of temperature. [*Source*: Fritz London, *Superfluids, Vol. 2: Macroscopic Theory of Superfluid Helium* (Wiley, New York, 1954).]

descends to T_B and that C_V exceeds the classical value of $\frac{3}{2}Nk$. That is, in fact, the case.

2. For the second observation, simply look at figure 9.8, which was computed numerically. The graph of C_V is continuous, has a maximum at T_B, but—for all practical purposes—experiences a discontinuity in slope at the Bose temperature.

This suffices for the theory; now we turn to experiment.

9.5 Bose–Einstein condensation: experiments

Einstein's prediction of a "condensation" appeared in print in 1924. For the next 71 years, only indirect evidence could be found in nature or could be produced in the lab. Even the most notable example was merely suggestive and certainly not definitive.

The prime example consisted of ^{4}He, the spinless isotope of helium and hence a boson. Under atmospheric pressure, gaseous ^{4}He forms a liquid when the temperature is reduced to 4.2 K or below. The liquid and the residual vapor coexist. This condensation is unquestionably caused by real attractive forces between the atoms and has nothing to do with a Bose–Einstein condensation. The two electrons in helium are tightly bound and form a spherical atom. In consequence, the force between atoms is weak. Only when the temperature is quite low and hence the typical kinetic energy is quite small can the weak attractive forces produce a liquid phase. Helium was the last of the common gases to be liquefied—by the Dutch physicist Heike Kamerlingh Onnes in 1908—and for good reason.

Our attention focuses now on liquid ^{4}He. Maintain coexistence of liquid and vapor (at whatever pressure that requires), but slowly reduce the temperature and measure the liquid's heat capacity as a function of temperature. Figure 9.9 shows a modern version of what physicists found in the 1920s. At a temperature of 2.17 K, the heat capacity rises in a sharp spike and then falls again. Below 2.17 K, the liquid behaves in entirely novel ways. For example, the bulk liquid can flow without viscosity, even through tiny capillaries. Circular flow through an annular region, once started and left by itself, will persist indefinitely. Thin films of the liquid can flow up over the lip of a beaker, defying gravity and emptying the beaker as though a siphon were present. As the temperature 2.17 K was passed, the liquid became a *superfluid*.

The shape of the heat capacity curve, reminiscent of the lower case Greek letter lambda, λ, gives the name the *lambda transition* to the change in behavior (which is a "phase transition"). The corresponding temperature is the *lambda point*: $T_\lambda = 2.17$ K.

In 1938, right after the superfluid behavior was discovered, Fritz London suggested that the lambda transition in liquid ^{4}He may be a manifestation of a Bose–Einstein condensation. The heat capacities of helium and of an ideal Bose gas, said London, have similar shape, at least in a rough qualitative way, as figures 9.8 and 9.9 display. Moreover, the peaks occur at similar temperatures, as we can readily show. At the lambda point, the experimental mass density of liquid helium is

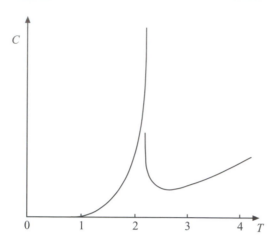

Figure 9.9 The experimental heat capacity C of liquid ^{4}He (in coexistence with its vapor), presented as a function of temperature (in kelvin). [*Source*: K. R. Atkins, *Liquid Helium* (Cambridge University Press, New York, 1959).]

$$\left(\frac{mN}{V}\right)_{\lambda \text{ point}} = 0.146 \times 10^3 \text{ kg/m}^3, \tag{9.57}$$

approximately one-seventh the density of water. The mass density and the mass of a single ^{4}He atom, $m = 6.649 \times 10^{-27}$ kg, provide all the data that one needs in order to compute the Bose temperature T_B of an ideal gas at the same number density N/V. Upon evaluating the factors in equation (9.51), one finds $T_B = 3.15$ K. This temperature is sufficiently close to T_λ to make London's suggestion plausible. Of course, no one has been under any illusions about a perfect match. The Bose–Einstein condensation, as derived by Einstein and by us, occurs in an ideal gas—no inter-particle forces whatsoever—but the atoms in liquid helium certainly exert substantial forces on one another. Nonetheless, there may be some sense in which a macroscopic number of helium atoms are "in the same quantum state" and hence act coherently.

A critical test of London's suggestion became possible after the Second World War, when the isotope ^{3}He became available in quantities sufficient for liquefaction. This helium isotope has a net spin of $\frac{1}{2}\hbar$ (arising from the single, unpaired neutron in the nucleus) and is a fermion. Under atmospheric pressure, ^{3}He condenses at 3.2 K, close to the value of 4.2 K for ^{4}He. The lower temperature for ^{3}He is almost certainly a consequence of the isotope's smaller mass. The crucial question is this: does liquid ^{3}He show a lambda transition? Experiments over the temperature range from 3.2 K down to 3×10^{-3} K say "no." Despite interatomic forces virtually identical to those in ^{4}He and despite a comparable mass and a comparable number density in the liquid phase, ^{3}He shows no evidence of a lambda transition in that great temperature range. The inference is inescapable: the lambda transition in ^{4}He is a consequence of the isotope's boson character.

To be sure, the isotope ^{3}He provides surprises of its own, and they are discussed in section 12.6. Right now, however, we shift our focus from liquids to gases.

BEC in a dilute gas

The acronym "BEC" denotes "Bose–Einstein condensation"—the phenomenon—or "Bose–Einstein condensate"—the atoms in the single-particle ground state, whose number is estimated by $\langle n_1 \rangle$. In the physics literature, you will find both meanings in use. In this subsection, BEC denotes the phenomenon of condensation. The important point is that, in 1995, Eric Cornell, Carl Wieman, and their colleagues in Colorado produced BEC in a dilute gas of rubidium atoms. Rubidium is an alkali atom, following after lithium, sodium, and potassium in the first column of the periodic table. The element is named after a prominent red line in its emission spectrum. The specific isotope used by the Colorado group was ^{87}Rb, which has an even number of neutrons and, of course, equal numbers of electrons and protons. Consequently, the net intrinsic angular momentum must be an integer (or zero) times $\hbar$, and so ^{87}Rb is a boson.

For the moment, suppose that the rubidium atoms had zero spin and were confined in a box, the context that we analyzed in section 9.4. The requirement that must be met for BEC to occur can be expressed in three equivalent ways, as follows.

First, the requirement can be stated as a temperature inequality:

$$T < T_{\mathrm{B}} = \frac{1}{k\pi(2.612)^{2/3}} \frac{h^2}{2m} \left(\frac{N}{V}\right)^{2/3}, \tag{9.58}$$

where the Bose temperature T_{B} was given explicitly in (9.51).

Second, the inequality can be rearranged (and a square root taken) so that it asserts the requirement

$$\lambda_{\mathrm{th}} > (2.612)^{1/3} \left(\frac{V}{N}\right)^{1/3}. \tag{9.59}$$

This form says that the thermal de Broglie wavelength must exceed 1.38 times the average interatomic separation. In short, the thermal wave packets must overlap substantially.

Third, raising both sides of (9.59) to the third power and rearranging yields the form

$$\frac{N}{V}\lambda_{\mathrm{th}}^3 > 2.612. \tag{9.60}$$

Here one sees most clearly that increasing the number density N/V (at fixed temperature) is another way to meet the requirement for BEC. (The physics is really this: increasing the number density crowds the particles together, so that their thermal wave packets overlap and quantum effects become essential.)

Once the atomic species has been chosen and hence the particle mass m has been fixed, an experimenter can try to achieve BEC by lowering the temperature or increasing the number density, or both. The first successful production required both.

The Colorado group first cooled rubidium atoms from room temperature to 20 microkelvin by laser cooling. The frequency of the laser beams that passed through the rubidium vapor was set slightly too low for resonant absorption by atoms at rest. An atom moving "upstream" in the laser beam (that is, moving toward the laser) sees the frequency Doppler-shifted upward, toward resonance. Such a moving atom will absorb a photon in a "head-on collision" and will be slowed down. Later, the atom will emit a photon. The angular distribution of such photons is quite symmetric, and so—on average—emission produces no net vectorial recoil. Thus absorption slows the atoms, but emission has no effect on the motion; the atom's momentum suffers a net reduction, on average. The process, repeated many times, slows down the atoms and cools the gas.

In practice, an atom moves relative to six laser beams, one directed inward along each of six Cartesian axes: $\pm\hat{x}$, $\pm\hat{y}$, and $\pm\hat{z}$. The beams along which the atom moves upstream slow the atom down. For the beams along which the atom moves downstream, the Doppler shift (now away from resonance) reduces the absorption probability sufficiently that—in a qualitative assessment—we may just ignore those beams. The composite result is a slowing force no matter which direction the atom moves. (The context is sometimes called "optical molasses.")

Magnetic fields are required to confine the atoms while they are being cooled—and to keep the cool atoms from falling in the Earth's gravitational field. After preliminary laser cooling, the experimenters increased the field gradients, compressing the cloud of atoms by a factor of 10 (in number density). Further laser cooling took the atoms' temperature down to 20 microkelvin.

Finally, the experimenters removed all laser beams, confined the cool atoms in a different magnetic trap, and began a process of evaporative cooling. The more energetic atoms oscillate farther from the center of the trap. Selectively, the atoms near the periphery were allowed to escape. The atoms that remain have less energy than the original average energy. Continually allowing the faster atoms to escape progressively reduces the average energy of the atoms that remain (because only slow atoms remain) and hence lowers the temperature. The remaining atoms are confined near the center of the trap.

In the 1995 Colorado experiment, the condensate first appeared when the temperature dropped to approximately 170 nanokelvin. Some 2×10^4 atoms remained in the trap. Further evaporative cooling sent almost all surviving atoms into the single-particle ground state: some 2,000 atoms in a volume approximately 10^{-5} meter in diameter, all occupying the same quantum state.

We return now to the rubidium atoms themselves. The nuclear spin of ^{87}Rb is $\frac{3}{2}\hbar$. The electrons, collectively, have zero orbital angular momentum and a net spin of $\frac{1}{2}\hbar$ (contributed by the outermost electron). Thus, in its ground state, ^{87}Rb can have a total intrinsic angular momentum of $1\hbar$ or $2\hbar$. Before turning off their cooling laser beams, the Colorado group used a burst of circularly polarized light to pump all atoms into the $2\hbar$ state and to align all the angular momenta. Therefore only one state of intrinsic angular momentum was populated. Moreover, if a collision, say, changed the angular momentum state, then the magnetic force on the atom would change, the atom would no longer be confined, and it would escape from the trap. Thus only one state of intrinsic angular momentum was ever relevant, and so the analysis for BEC was the same as for a spinless boson.

In one significant respect, however, the Colorado experiment differed from our context in section 9.4. The magnetic fields confined the atoms with forces akin to those of a harmonic oscillator (acting in three dimensions). The density of single-particle states $D(\varepsilon)$ for the three-dimensional harmonic oscillator differs from the density for a box with confining walls, and that change alters the details of the calculation. In short, the Colorado criteria for condensation differ somewhat from those displayed in relations (9.58) to (9.60).

You may wonder, how did the experimenters observe the condensed atoms? Basically, by letting the atoms cast a shadow. The next two paragraphs provide some detail.

Recall that all laser beams were turned off before evaporative cooling began. The atoms were confined solely by a magnetic trap and were in the dark. To "see" the cold atoms, the experimenters (1) turned off the electric currents that produced the trap's magnetic fields, (2) allowed the now-free atoms to fly apart for 0.06 second (to provide a wider cloud of atoms), (3) illuminated the expanded cloud with a pulse of laser light (whose frequency was set—for maximum scattering—at the atoms' resonant frequency), and (4) measured the intensity of the transmitted light as a function of position on a two-dimensional planar detecting surface. Where the atomic cloud was dense, much light was scattered from the laser pulse, and thus the high-density regions cast a strong shadow.

At a temperature of approximately 100 nanokelvin, the shadow has a "darkness" profile that consists of two distinct regions: a wide, gently rounded hill and a prominent, narrow central spire. The spire is produced by the condensed atoms, which have small momenta and hence cannot fly far from the center of the former trap in the 0.06 second expansion time. The atoms that were in single-particle states above the ground state generate the gently rounded hill.

9.6 A graphical comparison

The chapter concludes with a graphical comparison of how fermions and bosons behave. We examine the fashion in which the particles are distributed in energy at various temperatures. The essential quantity is this:

$$\frac{\left(\begin{array}{c}\text{estimated number of particles} \\ \text{per unit energy interval near } \varepsilon\end{array}\right)}{N}, \tag{9.61}$$

where N is the total number of particles. Usually, the numerator is the product of $\langle n_\alpha \rangle$, the estimated number of particles occupying the single-particle state φ_α and having energy ε_α, times $D(\varepsilon)$, the number of single-particle states per unit energy interval. Thus, for fermions,

$$\frac{\left(\begin{array}{c}\text{estimated number of fermions} \\ \text{per unit energy interval near } \varepsilon\end{array}\right)}{N} = \frac{f(\varepsilon)D(\varepsilon)}{N} = \frac{D(\varepsilon)/N}{e^{(\varepsilon-\mu)/kT}+1}. \tag{9.62}$$

For bosons, the corresponding expression is

$$\frac{\left(\begin{array}{c}\text{estimated number of bosons}\\ \text{per unit energy interval near } \varepsilon\end{array}\right)}{N} = \frac{\langle n \rangle_B D(\varepsilon)}{N} = \frac{D(\varepsilon)/N}{e^{(\varepsilon - \mu)/kT} - 1}. \tag{9.63}$$

For bosons, the single-particle ground state, φ_1, and its occupation number $\langle n_1 \rangle$ must be treated separately. The reason is the following. At low temperature, the boson function $\langle n_1 \rangle$ differs numerically by a huge amount from $\langle n_2 \rangle$ and from other $\langle n_a \rangle$, where α is a small integer. A plot of $\langle n_a \rangle$ as a function of the integer α would show a spike at $\alpha = 1$. Using a smooth density of states, $D(\varepsilon) = C \times \varepsilon^{1/2}$, to group adjacent states is permissible if the states all exhibit nearly the same value for $\langle n_a \rangle$. That step may be taken for states with $\alpha \geq 2$, but the state φ_1 shows such different behavior that it must be treated separately.

Figure 9.10 compares the distributions in the limit of absolute zero. The integral of the expression in (9.61) over the range $0 \leq \varepsilon \leq \infty$ must yield 1, and so the area under the fermion curve is 1. In a pure continuum picture, the condensed bosons would be represented by a Dirac delta-function at $\varepsilon = \varepsilon_1$. Using a dot at the finite height that corresponds to the fraction of atoms in the condensate is more informative.

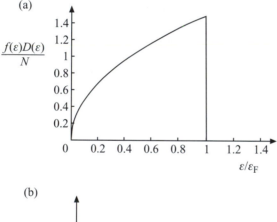

(a)

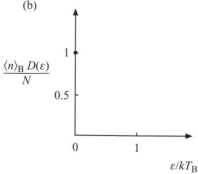

(b)

Figure 9.10 How fermions and bosons are distributed in energy at $T = 0$. (a) Fermions; (b) bosons. The dot on the boson graph represents the condensed particles; all the bosons are in the single-particle ground state.

Next, figure 9.11 shows how the fermion distribution spreads out as the temperature is raised. Already at $T = 3T_F$, the distribution is nearly classical, that is, the limiting expressions in sections 8.4 and 8.5 are becoming valid.

Finally, figure 9.12 displays the boson distribution at temperatures above and below T_B. When $T > T_B$, the distribution has a maximum at some energy substantially greater than ε_1, the lowest single-particle energy. The area under those curves equals 1, for only a negligible fraction of the bosons are in the single-particle ground state φ_1. Already when $T = 3T_B$, the distribution is approaching the classical shape.

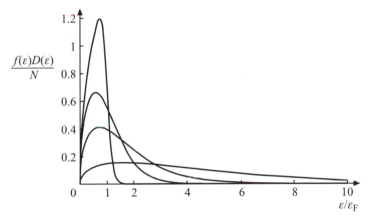

Figure 9.11 The fermion distribution at four temperatures: $T/T_F = 0.1, 0.5, 1,$ and 3. The higher the temperature, the more spread out the distribution.

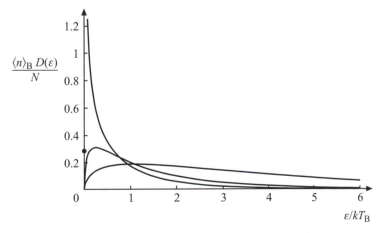

Figure 9.12 The boson distribution at three temperatures: $T/T_B = 0.8, 1.65,$ and 3. The dot represents the fraction of bosons in the single-particle ground state when $T = 0.8T_B$.

When the temperature is below T_B, the peak in the continuous distribution lies extremely close to the vertical axis; for all practical purposes, the distribution declines monotonically with increasing energy. Moreover, the area under the curve is significantly less than 1 because a finite fraction of the bosons are in the single-particle ground state. (Recall that the smooth curve represents the distribution of only those bosons in states φ_α with $\alpha \geqslant 2$.) When $T = 0.8T_B$, that fraction is already 0.28 and is represented by the dot at 0.28 on the vertical axis.

The gamut of behavior: low temperature to classical

At various points in previous sections—and again in this section—the terms "low temperature," "nearly classical," "semi-classical," and "classical" have been used. A review of what they mean—at least in this book—is in order.

Three items lie at the heart of the distinctions:

1. the presence of Planck's constant h in theoretical expressions;
2. the indistinguishability of identical particles (acknowledged at least in some measure);
3. a difference in behavior between fermions and bosons.

In a "classical" expression, none of the three items pertains. Examples are the relations $\langle E \rangle = \frac{3}{2}NkT$ and $P = (N/V)kT$.

Nonetheless, under the same physical circumstances (of temperature and number density), other expressions may include h or recognize indistinguishability (but show no difference between fermions and bosons). Such expressions are called "semi-classical." Examples are the semi-classical partition function, $Z_{\text{semi-classical}} = (Z_1)^N/N!$, the entropy expression in section 5.6, and the chemical potentials in section 7.4.

As the temperature is reduced or the number density is increased, a difference in behavior between fermions and bosons arises. When that difference is still small relative to the behavior that they have in common, the behavior is called "nearly classical." The prime example (so far in this book) is provided by the calculations of energy and pressure in section 8.5.

Finally, at "low temperature"—that is, at temperature below the characteristic temperatures such as T_F and T_B—all three items pertain and strongly so.

9.7 Essentials

1. The *Fermi function*,

$$f(\varepsilon) \equiv \frac{1}{e^{(\varepsilon - \mu)/kT} + 1},$$

is the form of $\langle n_a \rangle_F$ appropriate for a continuous energy spectrum.

The quantitative statements about fermions that follow pertain to fermions whose spin is $\frac{1}{2}\hbar$ and which are free to move in a three-dimensional box.

2. At absolute zero, the Fermi function equals 1 out to the *Fermi energy* ε_F, where

$$\varepsilon_F = \left(\frac{3}{8\pi}\right)^{2/3} \frac{h^2}{2m} \left(\frac{N}{V}\right)^{2/3},$$

and is zero beyond that energy. The chemical potential has the limiting behavior

$$\lim_{T\to 0} \mu(T) = \varepsilon_F.$$

The ground-state energy of the entire system of N fermions is proportional to $N \times \varepsilon_F$:

$$E_{\text{g.s.}} = \tfrac{3}{5} N \varepsilon_F.$$

3. The Fermi energy defines a characteristic temperature, the *Fermi temperature* T_F:

$$kT_F \equiv \varepsilon_F.$$

4. In the thermal domain $0 < T \ll T_F$, the chemical potential differs little from its value at absolute zero:

$$\mu(T) = \varepsilon_F - \frac{\pi^2}{12} \left(\frac{T}{T_F}\right)^2 \varepsilon_F.$$

Some electrons near the Fermi energy are shifted upward in energy by an amount of order kT. The number of such electrons is proportional to $D(\varepsilon_F)kT$. Thus the estimated total energy $\langle E \rangle$ grows by an amount proportional to $(kT)^2$. In consequence, the heat capacity varies linearly with temperature:

$$C_V = \frac{\pi^2}{2} \left(\frac{T}{T_F}\right) Nk.$$

In comparison with a classical ideal gas at the same temperature and number density, the fermion gas (1) has a large kinetic energy per particle, (2) has a small heat capacity per particle, (3) exerts a high pressure, and (4) experiences little dependence of pressure on temperature.

5. At absolute zero, all bosons are in the single-particle ground state: $\langle n_1 \rangle_B = N$. (Here and hereafter the bosons are taken to be spinless and free to move in a three-dimensional box.)

6. For bosons, the characteristic temperature T_B is given by

$$kT_B = \frac{1}{\pi(2.612)^{2/3}} \frac{h^2}{2m} \left(\frac{N}{V}\right)^{2/3}$$

and is called the *Bose* or *Bose–Einstein temperature*. The dependence on h, m, and N/V is the same as for the Fermi temperature; only the numerical coefficient differs (slightly).

7. As the temperature is lowered below T_B, a significant fraction of the bosons condense into the single-particle ground state. The quantitative relationship is

$$\langle n_1 \rangle = N \left[1 - \left(\frac{T}{T_B} \right)^{3/2} \right],$$

provided $T < T_B$.

8. In qualitative terms, condensation into the single-particle ground state commences when the thermal de Broglie wavelength λ_{th} becomes larger than the average inter-particle separation.

Further reading

For a more nearly exact evaluation of integrals containing the Fermi function $f(\varepsilon)$, a good reference is appendix 5, "Integrals for the Fermi gas," in David S. Betts and Roy E. Turner, *Introductory Statistical Mechanics* (Addison-Wesley, Reading, Massachusetts, 1992), pp. 278–81.

White dwarf stars are well-described in an elementary way by Frank H. Shu, *The Physical Universe*: *An Introduction to Astronomy* (University Science Books, Mill Valley, CA, 1982). In general, Shu's book has a ratio of insight to words that is much higher than in the usual textbook. A more detailed development of white dwarfs is provided by A. C. Phillips, *The Physics of Stars* (Wiley, New York, 1994). Phillips's book is a fine place to see thermal physics at work in astronomy.

J. Wilks and D. S. Betts provide a splendid survey of both ^{4}He and ^{3}He in *An Introduction to Liquid Helium*, 2nd edition (Oxford University Press, New York, 1987).

The discovery paper for BEC in a dilute gas is "Observation of Bose–Einstein condensation in a dilute atomic vapor," by M. H. Anderson, J. R. Ensher, M. R. Matthews, C. E. Wieman, and E. A. Cornell, *Science* **269**, 198–201 (1995). Commentary was provided by Gary Taubes, "Physicists Create New State of Matter," *Science* **269**, 152–3 (1995), and by Graham P. Collins, "Gaseous Bose–Einstein condensate finally observed," *Physics Today* **48**, 17–20 (1995).

Carl E. Wieman described the discovery and his role in it in the superb paper, "The Richtmyer memorial lecture: Bose–Einstein condensation in an ultracold gas," *Am. J. Phys.* **64**, 847–55 (1996).

A fine review article by Keith Burnett, "Bose–Einstein condensation with evaporatively cooled atoms," was published in *Contemporary Physics* **37**, 1–14 (1996).

Quantitative differences arise between BEC in a box and in a trap that is described by a power-law potential. The issue is explored by V. Bagnato, D. E. Pritchard, and D. Kleppner, "Bose–Einstein condensation in an external potential," *Phys. Rev. A*, **35**, 4354–8 (1987).

Bose–Einstein Condensation, edited by A. Griffin, D. W. Snoke, and S. Stringari (Cambridge University Press, New York, 1995) collects the papers and insights of a conference on BEC held in 1993.

Problems

1. The metal zinc provides two conduction electrons per atom. The average separation of the conduction electrons is $(V/N)^{1/3} = 1.97 \times 10^{-10}$ meters. Give numerical values at $T = 300$ K for the following quantities.

(a) Fermi energy (in eV),
(b) Fermi temperature,
(c) fractional shift in the chemical potential (relative to its value at absolute zero),
(d) $C_V/(\frac{3}{2}Nk)$ for the conduction electrons, and
(e) the ratio of the heat capacities for the conduction electrons and the lattice vibrations, $C_{electrons}/C_{lattice}$.

2. Use a computer to graph the Fermi function $f(\varepsilon)$ as a function of energy at five values of temperature: $T/T_F = 0.01, 0.1, 0.5, 1,$ and 3. You will need to determine the associated values of the chemical potential. Plot your values of $\mu(T)$ as $\mu(T)/\varepsilon_F$ versus T/T_F, for these are the natural dimensionless variables.

3. At fixed volume and zero temperature, the total energy of an ideal fermion gas is *not* proportional to the number N of fermions present. Since, by the specification "ideal," there are no forces between the fermions, how can this be? Doesn't it indicate that the entire analysis is rotten at the core?

4. A hypothetical system of N fermions has a single-particle density of states given by the linear relation $D(\varepsilon) = \varepsilon/\varepsilon_0^2$, where ε_0 is a positive constant with the dimensions of energy. The fermions do not interact among themselves. Calculate

(a) the system's Fermi energy,
(b) the chemical potential as a function of T under the conditions $0 \leqslant T \ll T_F$, and
(c) the total energy $\langle E \rangle$ and heat capacity under the same conditions.

In parts (b) and (c), it suffices to work through the lowest non-vanishing order in powers of the temperature, that is, through the first temperature-dependent term. Express all final answers in terms of N, ε_0, k, and T.

5. Consider the same system of fermions as in the preceding question, but examine the semi-classical limit. For $T \gg T_F$, calculate

(a) the single-particle partition function,
(b) the partition function for all N fermions,
(c) the total energy $\langle E \rangle$,
(d) the chemical potential, and
(e) the total entropy.

The first non-vanishing term will suffice. Express all final answers in terms of N, ε_0, k, and T.

6. In a hypothetical system, the single-particle energy eigenstates are nondegenerate and have energies given by $\varepsilon_\alpha = \alpha \varepsilon_0$, where $\varepsilon_0 = 10^{-38}$ J. The index α runs over the positive integers. The system contains $N = 10^{20}$ fermions and is at temperature $T = 300$ K.

(a) What is the Fermi temperature?
(b) Provide good estimates of the total energy $\langle E \rangle$ and the heat capacity, expressing your answers in terms of N, ε_0, k, and T.

7. *Heating by adiabatic expansion.* Initially, an ideal gas of N fermions (of spin $\frac{1}{2}\hbar$) is confined to a volume V_i and has zero temperature. Then a valve is opened, and the gas expands adiabatically into an evacuated region; the combined spaces have a volume V_f. When the gas settles to a new thermal equilibrium, its behavior is found to be that of a classical ideal gas. (Note. The walls are rigid and insulating throughout.)

(a) What is the final temperature of the gas?
(b) Derive an inequality for the ratio V_f/V_i and explain your reasoning.
(c) The title of this problem is surely an oxymoron. Explain the sense in which the title is a contradiction in terms and also the sense in which the title is a legitimate use of words.

8. *Fermions in two dimensions.* A total of N fermions (of spin $\frac{1}{2}\hbar$ and mass m) are restricted to motion in two dimensions on a plane of area A. There are no mutual interactions.

(a) For a temperature that satisfies the inequalities $0 < T \ll T_F$, calculate the average energy $\langle E \rangle/N$, the heat capacity C (at fixed area), and the total entropy S.
(b) Calculate the same three quantities when the temperature is sufficiently high that the fermions behave like a semi-classical two-dimensional gas.

(Note. All answers are to be given solely in terms of k, T, m, h, N, and A plus numerical constants.)

9. Consider $N = 10^{22}$ electrons at a number density N/V like that in copper. Place them in a magnetic field of 2 tesla. Calculate the magnetization of the sample at $T = 0$, that is, the net magnetic moment per unit volume. Ignore any contribution from the motion of the electrons.

10. *Pauli paramagnetism revisited.* Split the conduction electrons into two sets: magnetic moment up ($\uparrow$) and moment down ($\downarrow$). At $T = 0$, the number $N_\uparrow$ of up-moment electrons is

$$N_\uparrow(\mu) = \int_{-m_B B}^{\mu} D_\uparrow(\varepsilon)\, d\varepsilon,$$

where the density of states for up-moment electrons is $D_\uparrow(\varepsilon) = \frac{1}{2}C \times (\varepsilon + m_B B)^{1/2}$ and where the constant C was defined in equation (9.4).

(a) Explain why the offset by $-m_B B$ arises in the limit of integration and why the offset by $+m_B B$ appears in the function that gives the density of states. Then write out analogous expressions for the down-moment electrons.
(b) Calculate the difference $N_\uparrow - N_\downarrow$ to first order in the field B. Make the provisional assumption that $\mu \cong \varepsilon_F$, where ε_F is the zero-field value of the Fermi energy, and set $\mu = \varepsilon_F$ temporarily. Develop an expression for the total magnetic moment along **B** and compare it with the result in section 9.2.
(c) Use the requirement $N_\uparrow + N_\downarrow = N$ to calculate $(\mu - \varepsilon_F)/\varepsilon_F$ through quadratic order in the field B. Estimate the relative shift numerically for a field of 2 tesla. In what way does your result justify the approximations in part (b)?

11. Why is the magnetic behavior of lithium-7 nuclei in a crystal lattice so different from the magnetic behavior of conduction electrons? After all, both are fermions with spin $\frac{1}{2}\hbar$.

12. Treat a gas of neutrons as an ideal Fermi gas at $T = 0$ and with number density N/V. A neutron has spin $\frac{1}{2}\hbar$, and its magnetic moment has magnitude $1.04 \times 10^{-3} m_B$.

(a) Will the neutrons form a degenerate gas?
(b) What is the largest value of the number density for which the neutrons can be completely polarized by an external magnetic field of 10 tesla? (Ignore the mutual magnetic interactions of the neutrons.)

13. Suppose N paramagnetic atoms are confined to one-dimensional motion along the direction of a uniform external magnetic field **B**. The length of the region is L, and each atom (of spin $\frac{1}{2}\hbar$) has a magnetic moment of magnitude m_B. Pretend that the atoms do not interact with one another.

Calculate the density of single-particle states as a function of single-particle energy. Provide a graph of your result (with axes labeled). Specify any auxiliary assumptions that you make. Be sure to check your answer by seeing whether it reduces to sensible results in special circumstances. For example, does it yield the correct single-particle partition function if . . . ?

14. *Stars with degenerate but non-relativistic electrons throughout.* Determine the proportionality constant in our theoretical relation, $R/R_\odot = \text{const} \times (M/M_\odot)^{-1/3}$, first algebraically and then numerically. How does the relationship fare when compared with the radius and mass of the white dwarf star 40 Eri B? (For a sense of scale, note that observed stellar masses span a wide range: from approximately 0.1 $M_\odot$, the least mass that can initiate fusion, to approximately 30 $M_\odot$, for extremely bright but short-lived stars.)

15. *Stars with degenerate and extremely relativistic electrons throughout.* Equation (9.39) defines a characteristic mass: the stellar mass that would be in equilibrium (admittedly, in unstable equilibrium) if the electron gas were extremely relativistic throughout the entire star.

(a) Determine that mass in terms of fundamental constants (such as h, c, G, and m_{proton}). You will need to return to equation (9.29) for a starting point.
(b) Why does the electron mass not enter?
(c) Compare the characteristic mass numerically with Chandrasekhar's limiting mass: 1.4 $M_\odot$.

16. *Relativistic classical ideal gas.* This problem uses equations (9.33), (8.18), and (13.6) to calculate the pressure exerted by a classical ideal gas when the typical particle speed may have any value between zero and the speed of light c.

(a) Confirm explicitly that the speed v may be written as

$$v = \frac{pc^2}{\varepsilon} = \frac{d\varepsilon}{dp},$$

where $\varepsilon = (p^2c^2 + m^2c^4)^{1/2}$ is the full relativistic energy of a free particle.
(b) Calculate $\langle pv \rangle$ as an integral over phase space, using the classical limit of the quantum occupation numbers. The relation $v = d\varepsilon/dp$ and one integration by parts will enable you to evaluate $\langle pv \rangle$ by comparison with the integral that determines the chemical potential in terms of N and other parameters.
(c) Finally, determine the pressure P. Is the result familiar? Is its range of validity surprising?

17. *Relativistic electron pressure at T = 0.* At absolute zero, the Fermi function is 1 out to the relativistic energy that corresponds to the Fermi momentum p_F and is zero beyond. Thus

$$\langle \mathbf{p} \cdot \mathbf{v} \rangle = \frac{1}{N} \int_0^{p_F} pv \, \frac{2V4\pi p^2}{h^3} \, dp.$$

The integral employs an expression from equation (13.6) in place of our usual density of states.

(a) Why is division by N required?

(b) If you express v in terms of p and the rest mass m, you can cast the integral into a form that is tabulated, for example, in H. B. Dwight, *Tables of Integrals and Other Mathematical Data*, 4th edition (Macmillan, New York, 1961).

(c) What do you find for the leading term in the pressure P when the strong inequality $p_F/mc \gg 1$ holds?

(d) Compare your result in (c) with the pressure exerted by a photon gas.

18. The metal beryllium has an especially large Fermi energy. Is special relativity theory required if one wants to calculate ε_F correct to 1 percent? As part of your response, compute the ratio v_F/c, where v_F denotes the speed of an electron whose momentum equals the Fermi momentum p_F. Beryllium provides two conduction electrons per atom, and the average separation of the conduction electrons is $(V/N)^{1/3} = 1.60 \times 10^{-10}$ meter.

19. An electron and a proton can react to form a neutron (and a neutrino) provided sufficient kinetic energy is available (in the center-of-mass coordinate system). The neutron–proton mass difference is $m_n - m_p = 2.31 \times 10^{-30}$ kg.

(a) What is the numerical value of the minimum kinetic energy?

(b) Take the proton to be at rest in the lab frame, and ignore the small difference between the lab and CM reference frames. Specify thermal equilibrium at a temperature of $T = 0$. What is the minimum number density N/V of free electrons for which the neutron-forming reaction can proceed?

(c) Under the specification that the number of nucleons (protons plus neutrons) equals the number of electrons, estimate the minimum mass density of a neutron star. Compare that density with the average mass density of the Earth, which is 5.5×10^3 kg/m³. A supernova explosion often produces a neutron star, a stellar object of approximately solar mass but composed solely (almost) of neutrons. Pulsars are believed to be rapidly rotating neutron stars.

20. BEC with sodium. The second experiment to produce BEC in a dilute gas used sodium atoms. The number density was $N/V = 10^{20}$ atoms/m³. The mass of a sodium atom is $m = 3.82 \times 10^{-26}$ kg. As with the rubidium experiment, only one state of intrinsic angular momentum was populated.

(a) If the trap that confined the atoms were adequately approximated by a box with rigid walls, at what temperature would you expect BEC to set in (as one lowered the temperature)?

(b) How low a temperature would be required for 90 percent of the atoms to be in the single-particle ground state?

(c) The common, stable isotope of sodium has 12 neutrons and is the isotope referred to above: ^{23}Na. The unstable isotope ^{21}Na has ten neutrons, the same nuclear spin, and a half-life of 23 seconds. In the following, suppress the possibility of radio-active decay.

A box of volume 1 cm^3 contains 10^{14} sodium atoms at a temperature $T = 1.3 \times 10^{-6}$ K. The atoms form a dilute gas, and only one state of intrinsic angular momentum is populated. Determine whether the heat capacity C_V is an increasing or decreasing function of temperature if

(i) all atoms are ^{23}Na atoms;

(ii) half are ^{23}Na and half are ^{21}Na.

21. *Some details for the Bose–Einstein condensation.*

(a) Chemical potential when $T \leqslant 0.99T_B$. Solve equation (9.52) for the chemical potential μ when $T < T_B$. Show that the size of μ is consistent with our treatment of the integral in (9.47) provided $T \leqslant 0.99T_B$, to give a conservative upper limit. This calculation provides a self-consistency check.

(b) $\langle n_2 \rangle$ relative to $\langle n_1 \rangle$. Estimate $\langle n_2 \rangle$, in order of magnitude, when $T \leqslant 0.99T_B$. For one route, show first that $\varepsilon_2 - \mu = \varepsilon_1 + (\varepsilon_1 - \mu)$, that $\varepsilon_1 = kT_B/O(N^{2/3})$, and that $\varepsilon_1 - \mu = kT/O(N)$. The symbol "$O(\ldots)$" denotes "a number of order $(\ldots)$."

(c) $\langle n_1 \rangle$ when $T = T_B$. Computing $\langle n_1 \rangle$ when $T = T_B$ is a delicate business. Start by introducing shifted energies: $\varepsilon'_\alpha = \varepsilon_\alpha - \varepsilon_1$ and $\mu' = \mu - \varepsilon_1$. On the shifted scale, the energy of the lowest single-particle state is zero, but the estimated occupation numbers remain as before. Equation (9.47) becomes

$$\frac{1}{e^{-\mu'/kT} - 1} + C(kT)^{3/2} \int_0^\infty \frac{x^{1/2}}{e^{x - \mu'/kT} - 1}\, dx = N.$$

The integral is $(\pi^{1/2}/2)\,2.612$ when $-\mu'/kT = 0$. Its value *decreases* when $-\mu'/kT > 0$ because the divisor is larger. The integral has an expansion as

$$\frac{\pi^{1/2}}{2}[2.612 - 3.54(-\mu'/kT)^{1/2} + \ldots],$$

as derived by John E. Robinson, *Phys. Rev.* **83**, 678–9 (1951). Use this information to solve for $-\mu'/kT$ and then $\langle n_1 \rangle$, in order of magnitude, when $T = T_B$.

22. *Energies qualitatively.* On one graph, sketch the average energy per particle (divided by k) versus temperature for three monatomic ideal gases: classical, fermion, and boson. That is, sketch $\langle E \rangle / Nk$ versus T over the range from absolute zero to "high temperature," so high that the quantum gases are in their semi-classical domain. Where you can, specify the functional form of the curves, that is, their dependence on T.

10 The Free Energies

10.1 Generalities about an open system

10.2 Helmholtz free energy

10.3 More on understanding the chemical potential

10.4 Gibbs free energy

10.5 The minimum property

10.6 Why the phrase "free energy"?

10.7 Miscellany

10.8 Essentials

This chapter has several goals. One is to increase your understanding of the chemical potential. Another is to describe the changes that arise in basic thermodynamic laws when particles may enter or leave "the system." A third goal is to study additional properties of the Helmholtz free energy, which first appeared in chapter 7. And a fourth goal is to introduce another free energy, the Gibbs free energy, and to explore its properties. Clearly, a lot is going on, but the section structure should enable you to maintain your bearings.

10.1 Generalities about an open system

In this section, we examine changes that arise in basic thermodynamic laws when particles may enter (or leave) what one calls "the system." What might be examples of such entering or leaving? The migration of ions in an aqueous solution provides an example. So does the motion of electrons from one metal to another at a thermocouple junction. When one studies the equilibrium between argon in the gaseous phase and argon adsorbed on glass (as we did in section 7.4), one may take the adsorbed atoms to constitute "the system." Then atoms may enter the system from the vapor phase and also leave it. The coexistence of two phases—solid, liquid, or vapor—provides a final example in which one may take a specific phase as "the system." Atoms or molecules may enter that phase or leave it.

When particles may enter or leave "the system," one says that the system is *open* (to particle transfer). Otherwise, the system is *closed*.

Next, we recapitulate the First and Second Laws as we have known them. In section 1.3, energy conservation led us to the First Law in the form

$$q = \Delta E + w, \tag{10.1}$$

where w denotes a small (or infinitesimal) amount of work done by the system on its surroundings. The system was presumed closed to transfer of particles across its boundary.

Sections 2.4 and 2.6 gave us the succinct form of the Second Law:

$$\Delta S \geqslant \frac{q}{T}, \tag{10.2}$$

with equality if the process is "slow" and hence reversible. When the equality holds, equation (10.2) tells us precisely how energy input by heating changes the system's entropy.

In both equations (10.1) and (10.2), the symbol q denotes "energy input by heating" and does *not* represent a state function. To derive a relationship among strictly state functions, we begin by eliminating q. Multiply (10.2) on both sides by T and then substitute for q from (10.1), finding

$$\boxed{T\Delta S \geqslant \Delta E + w, \qquad\qquad\qquad (10.3)}$$

with equality if the process is slow. This relationship is quite general, although it does presuppose a system closed to particle transfer. To be sure, because the work term w is not a state function, we must certainly go on.

Now we make two stipulations.

1. We specify slow and hence reversible changes, so that the equality sign holds.
2. We specify that only pressure–volume work is done, so that $w = P\Delta V$.

The upshot is the relationship

$$T\Delta S = \Delta E + P\Delta V. \tag{10.4}$$

This equation connects changes in the variables S, E, and V when no particles enter or leave the system.

Now let us allow particles of one species to enter (or leave) the system; for example, the particles might be electrons or argon atoms. The number of such particles in the system is denoted by N, and so ΔN is positive whenever particles *enter* the system. How should we generalize equation (10.4) to accommodate variations in N? That is, how should we fill in the blank in the form

$$T\Delta S = \Delta E + P\Delta V + (\text{something}) \times \Delta N? \tag{10.5}$$

A discussion at the end of section 4.2 noted that the entropy is a function of E, V, and N: $S = S(E, V, N)$. Think of solving this equation for the energy E as a function of S, V, and N:

$$E = E(S, V, N). \tag{10.6}$$

When the independent variables are altered in such a functional relationship, the change in E takes the form

$$\Delta E = E(S + \Delta S, \, V + \Delta V, \, N + \Delta N) - E(S, \, V, \, N)$$

$$= \left(\frac{\partial E}{\partial S}\right)_{V,N} \Delta S + \left(\frac{\partial E}{\partial V}\right)_{S,N} \Delta V + \left(\frac{\partial E}{\partial N}\right)_{S,V} \Delta N. \tag{10.7}$$

(If such a Taylor's series is not familiar, consult appendix A, which provides a discussion of the procedure.)

Both (10.5) and (10.7) are true statements about how ΔE is related to other differentials. To learn more, we compare coefficients of corresponding differentials in the two equations.

When we compare the coefficients of ΔS in equations (10.5) and (10.7), we infer that

$$T = \left(\frac{\partial E}{\partial S}\right)_{V,N}. \tag{10.8}$$

This equality is just the reciprocal of equation (4.21), namely, the general symbolic definition of absolute temperature. The proviso that the number N of particles be kept fixed while we differentiate appears explicitly here.

Comparing the coefficients of ΔV implies

$$P = -\left(\frac{\partial E}{\partial V}\right)_{S,N}. \tag{10.9}$$

This relationship is comfortably analogous to what we found in section 5.4: the pressure in energy eigenstate Ψ_j is given by $-\partial E_j / \partial V$.

The remaining term in (10.7) needs a counterpart in equation (10.4), and we provided a blank for it: the "something" in equation (10.5). Thus we must generalize (10.4) to read

$$T\Delta S = \Delta E + P\Delta V - \left(\frac{\partial E}{\partial N}\right)_{S,V} \Delta N. \tag{10.10}$$

The derivative $(\partial E / \partial N)_{S,V}$ gives the change in the system's energy when one particle is added under the conditions of constant entropy and constant volume. These are stringent conditions and probably are not at all familiar. Shortly, however, we shall find that this derivative provides another way to express the chemical potential μ. Then we shall have full justification for writing the generalization (10.10) as

$$T\Delta S = \Delta E + P\Delta V - \mu\Delta N. \tag{10.11}$$

We turn now to the necessary derivation.

10.2 Helmholtz free energy

Equation (10.6) told us that the internal energy is a function of the variables $\{S, V, N\}$. The volume and number of particles are usually easy for an experimenter to set and to control, but entropy is not. Temperature, however, is relatively easy to set and to maintain. To devise an energy-like expression that is a function of the variables $\{T, V, N\}$, subtract $S \times (\partial E/\partial S)_{V,N}$ from E, that is, subtract the product ST:

$$F = E - S \times \left(\frac{\partial E}{\partial S}\right)_{V,N} = E - TS. \tag{10.12}$$

The subtraction will cancel the term in ΔS that arises when one forms the differential of E, and therefore F will not be a function of S. The algebraic expression on the far right-hand side, $E - TS$, is precisely the Helmholtz free energy that we met in section 7.1, and so the abbreviation F is appropriate.

To confirm that F indeed depends on the set $\{T, V, N\}$, form the differential of (10.12),

$$\Delta F = \Delta E - T\Delta S - S\Delta T, \tag{10.13}$$

and then use (10.10) to eliminate ΔE on the right-hand side:

$$\Delta F = -S\Delta T - P\Delta V + \left(\frac{\partial E}{\partial N}\right)_{S,V}\Delta N. \tag{10.14}$$

The differences on the right-hand side are solely ΔT, ΔV, and ΔN; thus F is a function of the variables $\{T, V, N\}$, as desired. (In case you wondered, the coefficients S, P and μ can indeed be expressed in terms of the variables $\{T, V, N\}$.) The technique that we used in (10.12), namely, subtracting a judiciously chosen term, is called a *Legendre transformation*. Problem 3 explores the Legendre transformation in a more geometric fashion.

From equation (10.14), we can read off expressions for S, P, and $(\partial E/\partial N)_{S,V}$ in terms of partial derivatives of F:

$$S = -\left(\frac{\partial F}{\partial T}\right)_{V,N}, \tag{10.15}$$

$$P = -\left(\frac{\partial F}{\partial V}\right)_{T,N}, \tag{10.16}$$

$$\left(\frac{\partial E}{\partial N}\right)_{S,V} = \left(\frac{\partial F}{\partial N}\right)_{T,V}. \tag{10.17}$$

The first two relations can be filed away for now. We concentrate on the last and note that we defined the chemical potential μ in (7.13) by $(\partial F/\partial N)_{T,V}$, precisely the derivative that appears here also. Thus $(\partial E/\partial N)_{S,V}$ is indeed the chemical potential, for which we have two numerically equivalent but superficially distinct expressions:

$$\left(\begin{array}{c}\text{chemical}\\\text{potential } \mu\end{array}\right) = \begin{cases}\left(\dfrac{\partial E}{\partial N}\right)_{S,V} & \tag{10.18a}\\[2ex]\left(\dfrac{\partial F}{\partial N}\right)_{T,V} & \tag{10.18b}\end{cases}$$

The succinct form of the differential relation for F becomes

$$\Delta F = -S\Delta T - P\Delta V + \mu\Delta N. \tag{10.19}$$

10.3   More on understanding the chemical potential

Section 7.1 introduced the chemical potential and described it as follows: the chemical potential measures the tendency of particles to diffuse. This characterization focuses on the chemical potential as a function of spatial location. Particles diffuse from regions of high chemical potential to those of low chemical potential.

In the present section we focus on the numerical value taken by the chemical potential. Equation (10.18a), which expresses the chemical potential as a derivative of the internal energy, will enable us to understand why μ takes on certain numerical values.

Figure 10.1 displays the chemical potential, separately for fermions and for bosons, as a function of temperature. In previous chapters, we evaluated μ explicitly in three domains: absolute zero, nonzero but low temperature (in a sense made more precise in the context), and the semi-classical domain. Here we examine the first and last of those domains.

Absolute zero

In the limit as the temperature is reduced to absolute zero, the canonical probability distribution implies that the system settles into its ground state. (Any degeneracy of the ground state, if present, would be insignificant, and so the description assumes none.) Only one quantum state (of the entire system) is associated with the macrostate of zero absolute temperature. Thus the multiplicity is merely 1. Adding a particle at constant entropy requires that the multiplicity remain 1. Moreover, after the addition, the system must again be in thermal equilibrium. Thus the system must be in the ground state of the new system of $(N+1)$ particles. (One could preserve "multiplicity

(a)

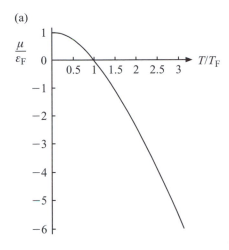

(b)

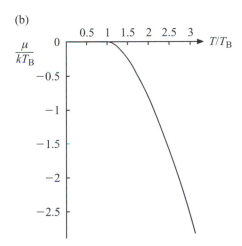

Figure 10.1 The chemical potential μ as a function of temperature for (a) fermions and (b) conserved bosons. When $T = T_F$, the chemical potential for fermions is close to zero: $\mu/\varepsilon_F = -0.0215$. As $T \to 0$, the curve for bosons approaches ε_1/kT_B, where ε_1 is the energy of the single-particle ground state. The numerical value of ε_1 is so small that, on the graph, ε_1/kT_B is indistinguishable from zero.

$= 1$" by using a single state somewhat above the ground state, but that procedure would not meet the requirement of thermal equilibrium.)

For a fermion, we construct the new ground state from the old by adding a new single-particle state at the Fermi level ε_F and filling it. Thus the additional particle must bring energy ε_F with it, and that must be the value of the chemical potential, as indeed figure 10.1 confirms.

The bosons that we studied in chapters 8 and 9, such as helium atoms, obey a conservation law: the number of bosons is set initially and remains constant in time (unless we explicitly add or subtract particles). For such bosons, we construct the new ground state by adding a single-particle state at the lowest single-particle energy and filling it. The chemical potential should equal the lowest single-particle energy, and that is what we found in chapter 9.

Semi-classical ideal gas

For an ideal gas in the semi-classical domain, the probability that any given single-particle state is occupied is quite small. Thus an additional atom could go into any one of a great many different single-particle states. Consequently, it is more productive to think of the energy change for the entire gas (of N or $N + 1$ atoms). Moreover, we may use classical reasoning about multiplicity and entropy. Adding an atom, which may be placed virtually anywhere, surely increases the spatial part of the multiplicity and hence tends to increase the entropy. To maintain the entropy constant, as stipulated by equation (10.18a), requires that the momentum part of the multiplicity decrease. In turn, that means less kinetic energy, and so the inequality $\Delta E < 0$ holds, which implies that the chemical potential is negative, in agreement with the display in figure 10.1 and with section 7.4.

Here we find that adding one atom to an ideal gas reduces the system's energy. Because the atoms of a structureless monatomic ideal gas possess energy only in the form of kinetic energy, which is inherently positive, the addition must occur under special circumstances if the system's energy is to decrease. Indeed, the addition does: conditions of constant volume and constant entropy.

This semi-classical context points out that equation (10.10) is *not* a mere generalization of the First Law of Thermodynamics to encompass transfer of particles to or from the system. A proper generalization of the First Law would read as follows:

$$q = \Delta E + w - \left(\begin{array}{c} \text{energy per particle when} \\ \text{particles are outside the system} \end{array} \right) \Delta N. \tag{10.20}$$

An accelerator could shoot particles into the system with any energy one might wish, and so the coefficient of ΔN here can have any value one chooses. In contrast, equation (10.10) requires that the system be in thermal equilibrium both before and after the small changes in the variables S, E, V, and N. The restriction to maintaining thermal equilibrium is what gives the coefficient of ΔN in equations (10.10) and (10.11) a unique value.

Photon gas

We calculated the properties of a photon gas in chapter 6, well before we developed the Bose–Einstein occupation number analysis in chapter 8. Nonetheless, the idea of an estimated occupation number played a role in both chapters. The estimated number of photons in a mode with frequency ν emerged as

$$\bar{n}(\nu) = \frac{1}{e^{h\nu/kT} - 1}. \tag{10.21}$$

Photons are bosons, for their intrinsic angular momentum—their spin—is $\hbar$ in magnitude and hence is integral. Thus $\bar{n}(\nu)$ should be an instance of the general Bose–Einstein expression

$$\langle n_a \rangle_B = \frac{1}{e^{(\varepsilon_a - \mu)/kT} - 1}. \tag{10.22}$$

A comparison indicates that the chemical potential μ for a photon gas is zero, indeed is zero at all temperatures. How can we best understand this?

The key insight is that photons are *not* subject to a conservation law for number. The atoms in a container wall readily absorb and emit photons. Even if one invoked energy conservation, an atom could absorb one photon of high energy and emit two photons, each of lower energy.

When we calculated in chapter 6, we did not fix the number of photons initially or even specify what the estimate $\langle N \rangle$ for the total number should become. Rather—although only implicitly—we considered all conceivable numbers of photons in each possible mode and weighted the corresponding states by the probability of occurrence according to the canonical probability distribution. The calculation presumed that we knew the volume V of the container and the temperature T, a temperature common to both walls and photon gas. At thermal equilibrium, everything about a photon gas is determined by T and V.

Given this information, we can readily compute the chemical potential for a photon gas. In general, for a gas consisting of a single species of particle, the Helmholtz free energy is a function of the set $\{T, V, N\}$. Because everything about a photon gas is determined by T and V, its Helmholtz free energy must be a function of T and V only:

$$F_{\text{photon gas}} = F(T, V). \tag{10.23}$$

(If you want more evidence for this conclusion, problem 6 offers a route.) The chemical potential μ now follows from (10.18b):

$$\mu_{\text{photon gas}} = \left(\frac{\partial F_{\text{photon gas}}}{\partial N}\right)_{T,V} = 0 \tag{10.24}$$

because $F_{\text{photon gas}}$ does not depend on N.

Summary

For the most part, section 10.3 can be summarized as follows. The expression

$$\mu = \left(\frac{\partial E}{\partial N}\right)_{S,V} \tag{10.25}$$

provides a way to understand the numerical value of the chemical potential: μ is the energy increment when one particle is added at constant entropy and fixed external parameters. Constant entropy is a stringent requirement and will—in the semi-classical domain—force the chemical potential to be negative.

10.4   Gibbs free energy

The way in which thermodynamic quantities scale with the system's size becomes vital now. To recapitulate from section 5.6, imagine two identical macroscopic systems, each in thermal equilibrium. Two identical blocks of ice provide an example. Put the two systems together and in contact. Some quantities, such as temperature and density, will remain the same. Quantities that remain the same when one scales up the system in this fashion are called *intensive*. Some other quantities, such as internal energy and volume, will double (provided that surface effects are negligible). Quantities that double when one doubles the system are called *extensive*.

Table 10.1 lists many of the quantities that have appeared so far and their behavior under scaling.

A glance back at equation (10.6) shows that the internal energy E is a function of three extensive variables: S, V, and N. (This is true under our prior stipulations: namely, only pressure–volume work arises, and particles of only one species are present.) The Legendre transformation to the Helmholtz free energy replaced the extensive variable S by the intensive variable T. An energy-like expression that depends solely on intensive variables plus particle numbers has nice features (as we shall see shortly). How can we construct it?

Another Legendre transformation will replace the extensive variable V by the

Table 10.1 *Scaling properties of some common thermodynamic quantities. In section 5.6, devoted to the semi-classical ideal gas, we noted the linear scaling of entropy, which makes S an extensive variable. As the ratio of changes in two extensive quantities, the chemical potential is an intensive quantity. The presence of the Gibbs free energy anticipates a bit.*

Intensive	Extensive
Temperature: T	Energy: E
Pressure: P	Volume: V
Chemical potential: μ	Entropy: S
Number density: N/V	Number of particles: N
Specific heat	Heat capacity: C_V or C_P
External magnetic field: $\mathbf{B}_{ext}$	Helmholtz free energy: F
	Gibbs free energy: G
	Log of partition function: $\ln Z$

intensive variable P. From F, subtract the product $V \times (\partial F / \partial V)_{T,N}$, that is, subtract $V \times (-P)$:

$$G \equiv F - V \times \left(\frac{\partial F}{\partial V}\right)_{T,N} = F - V \times (-P)$$

$$= E - TS + PV. \tag{10.26}$$

This transformation defines the *Gibbs free energy*, denoted by G.

To confirm that G depends on the set $\{T, P, N\}$, form the differential of equation (10.26),

$$\Delta G = \Delta E - T\Delta S - S\Delta T + P\Delta V + V\Delta P, \tag{10.27}$$

and then use (10.11) to eliminate ΔE on the right-hand side:

$$\Delta G = -S\Delta T + V\Delta P + \mu \Delta N. \tag{10.28}$$

Differentials of only T, P, and N appear on the right-hand side, and so G is indeed a function of the set $\{T, P, N\}$. Moreover, we can read off expressions for the coefficients in terms of partial derivatives of G:

$$S = -\left(\frac{\partial G}{\partial T}\right)_{P,N}, \tag{10.29}$$

$$V = \left(\frac{\partial G}{\partial P}\right)_{T,N}, \tag{10.30}$$

$$\mu = \left(\frac{\partial G}{\partial N}\right)_{T,P}. \tag{10.31}$$

At the moment, the noteworthy result is the last: when all intensive variables are held fixed, the derivative of G with respect to N gives the chemical potential.

In the second line of equation (10.26), each term has one extensive variable, and so the Gibbs free energy is an extensive quantity. Moreover, G depends on only one extensive variable, namely, the number N of particles, and so G must be proportional to N. A glance back at (10.31) tells us that the proportionality constant is the chemical potential. Thus the structure must be

$$G = \mu N, \tag{10.32}$$

a charmingly simple outcome.

Generalizations

The relations that we have just worked out, in particular the definition of G in (10.26), the differential relation in (10.28), and the structure in (10.32), are the most common ones. At times, however, they may need to be generalized. The need can arise in two ways, as follows.

1. If magnetic work, say, is possible in addition to pressure–volume work, then the very definition of G may need to be augmented with magnetic terms. Why? Because the over-riding goal of the Legendre transformation is to remove all extensive variables other than particle numbers. The Gibbs free energy is to be a function of intensive variables plus particle numbers.
2. If more than one species of particle may change in number, then the Gibbs free energy becomes a sum of terms like that in (10.32):

$$G = \sum_i \mu_i N_i, \tag{10.33}$$

where the sum goes over all particle species and each species has its own chemical potential. [To derive this result, one need only go back to equation (10.6) and generalize it to read $E = E(S, V, N_1, N_2, \ldots)$, where $N_1, N_2, \ldots$ denote particle numbers for all species. All subsequent steps in the derivation are merely algebraic extensions of the steps we took.]

Functional dependence of the chemical potential

A last point concerns the functional dependence of the chemical potential. Equation (10.31) tells us how the chemical potential is to be computed when the independent variables are the set $\{T, P, N\}$. As the derivative of an extensive variable with respect to a particle number, the chemical potential itself is an intensive quantity. The variables on which an intensive quantity depend must be intensive or the ratio of extensive quantities. Only in this way will the intensive quantity remain unchanged if two identical systems are combined. (The ratio of two extensive quantities is effectively intensive and hence is admissible.) When the independent variables are the set $\{T, P, N\}$, the chemical potential must be a function of the intensive variables T and P only:

$$\mu = \mu(T, P). \tag{10.34}$$

When several particle species are present, the Gibbs free energy takes the form displayed in (10.33), where each species has its own chemical potential μ_i. The corresponding variables for G are the set $\{T, P, N_1, N_2, \ldots, N_{\#sp}\}$, where #sp denotes the number of species. Temperature and pressure are, of course, intensive variables, but further effectively intensive variables can be formed from the ratios of the particle numbers. Thus, for example, a full set of intensive variables would be

$$\left\{ T, P, \frac{N_2}{N_1}, \frac{N_3}{N_1}, \ldots, \frac{N_{\#sp}}{N_1} \right\}. \tag{10.35}$$

Altogether, there are $2 + (\#sp - 1)$ independent intensive variables, and each chemical potential may depend on all of them. Knowing about this dependence will be vital in chapter 12, where we follow Willard Gibbs and determine the number of phases that may coexist at thermal equilibrium.

10.5 The minimum property

In section 7.2, we used the canonical probability distribution to show that the Helmholtz free energy has a minimum at what thermodynamics calls the equilibrium state. The great merit of that derivation lay in our being able to see the details. Now we derive the same conclusion by a more general but also more abstract line of reasoning.

Specify that the system

(a) is in thermal contact with its environment, whose temperature T remains constant throughout,

(b) does no work on its surroundings (for example, the volume V remains constant),

(c) is closed to entry or exit of particles.

Then equation (10.3) becomes

$$T\Delta S \geqslant \Delta E. \tag{10.36}$$

Subtracting $T\Delta S$ from both sides and invoking the isothermal nature of our context, we may write the relationship as

$$0 \geqslant \Delta E - T\Delta S = \Delta(E - TS) = \Delta F,$$

that is,

$$0 \geqslant \Delta F. \tag{10.37}$$

Read from right to left, (10.37) says that the change in F is negative or zero.

In a small spontaneous change, the inequality sign holds in equation (10.2): $\Delta S > q/T$; our analysis in section 2.6 led to that conclusion. The inequality sign carries along to (10.37), and so the Helmholtz free energy decreases in a small spontaneous change. One can follow such small decreases in F until the system settles down to equilibrium. Then no further spontaneous change in F is possible, and so F has evolved to a minimum.

[For an example, take the context of figure 7.1, insert a valve in the vertical tube, shut the valve, and specify equal number densities in the upper and lower volumes to start with. Keep both volumes in thermal contact with some heat reservoir at a fixed temperature. Now open the valve for a short time. Because of gravity, more molecules will diffuse from the upper volume to the lower volume than in the opposite direction; N_l will increase. Shut the valve, and let the molecules in each volume re-equilibrate among themselves and with the heat reservoir. As figure 7.4 showed, the composite Helmholtz free energy will have decreased. Next, repeat the steps until no further (macroscopic) change occurs. Then you may toss away the valve and say that F has evolved to a minimum.]

If one imagines a small reversible step away from equilibrium, then the equality sign holds in equation (10.2): $\Delta S = q/T$. Again, the sign carries along to (10.37), and so the equation $\Delta F = 0$ holds for a small, imagined step away from equilibrium.

An analogous minimum property holds for the Gibbs free energy. The conditions differ from those for the Helmholtz free energy in only one respect. Item (b) is replaced by the following stipulation.

(b′) The (external) pressure P remains constant, and only pressure–volume work may be done on the environment.

Then equation (10.3) becomes

$$T\Delta S \geqslant \Delta E + P\Delta V. \tag{10.38}$$

Under conditions of constant T and P, we may rearrange the terms so that they assert

$$0 \geqslant \Delta G. \tag{10.39}$$

The reasoning for a minimum at equilibrium proceeds as before.

We stipulated no transfer of particles to or from the system. That does *not*, however, preclude changes in how the particles are aggregated *within* the system as the system evolves to equilibrium. Argon atoms, say, could move from the vapor phase to adsorption sites. Molecules of H_2 and O_2 could react to form molecules of H_2O, or water molecules could dissociate. These possibilities are vital when one studies phase equilibrium or chemical reactions. We will work out the consequences later. The point of this paragraph is merely to alert you to an easily over-looked aspect of the conditions for a minimum.

10.6   Why the phrase "free energy"?

You may wonder, why is the function $F \equiv E - TS$ called a "free energy"? To answer this question, we return to equation (10.3) and subtract ΔE from both sides, thus isolating the work terms on the right-hand side:

$$-\Delta E + T\Delta S \geqslant w. \tag{10.40}$$

If the system and its environment are at the same temperature T and if that temperature remains constant during whatever small changes occur, then we may write the left-hand side in terms of the Helmholtz free energy:

$$-\Delta F \geqslant w \tag{10.41}$$

for an isothermal process. Note the minus sign. Reading (10.41) from right to left, we find that the total work done is less than or equal to the decrease in the Helmholtz free energy. In a slow and hence reversible process, the equal sign applies, and thus the work done is precisely equal to the decrease in F. Thus the change in F determines the amount of energy "free" for work under the given conditions.

The German physicist (and physiologist) Hermann von Helmholtz had in mind the property described above when he coined the phrase "free energy" for $E - TS$ in 1882.

An example

The isothermal expansion that was illustrated in figure 2.6 provides a ready-made example. Suppose the gas starts at an initial pressure P_i that is well above atmospheric pressure. The gas expands slowly from initial volume V_i to final volume V_f. What is the decrease in the Helmholtz free energy? For a semi-classical ideal gas, the internal energy E depends on temperature but not on volume; hence $\Delta E = 0$ holds here. For the entropy change, we can refer to equation (5.41) and find

$$\Delta S = kN[\ln V_f - \ln V_i] = kN \ln(V_f/V_i).$$

Thus, for this reversible isothermal process,

$$\Delta F = \Delta E - T\Delta S$$

$$= -TkN \ln(V_f/V_i).$$

The decrease in the Helmholtz free energy is $NkT \ln(V_f/V_i)$, and that expression should give the work done by the system.

The work done can also be calculated directly: evaluate the integral $\int_{V_i}^{V_f} P \, dV$, using the ideal gas law to express P in terms of V and T. The integral route confirms the conclusion derived by the free energy route.

Note that all the energy expended as work came from the "warm brick" that maintained the gas at constant temperature. This highlights the importance of the context: an *isothermal* process, guaranteed by some environmental source capable of delivering energy by heating (or absorbing it by cooling). To be sure, if we acknowledged intermolecular forces in the gas, then the internal energy E would change also. In the generic situation, both ΔE and $T\Delta S$ contribute to the isothermal change in Helmholtz free energy and hence to the work done.

10.7   Miscellany

A few points about this chapter remain to be made or should be emphasized.

Thermodynamic potentials and natural variables

We used Legendre transformations to move our focus from the energy E to the Helmholtz free energy F and thence to the Gibbs free energy G. Table 10.2 displays the variables upon which we conceived those functions to depend. Also shown are the functions that one can generate by partial differentiation with respect to the independent variables. Looking across the first line, we see that the relevant state functions for the system are seven in number: E, S, V, N, T, P, and μ. The quantities in every other line enable one to express those seven state functions in terms of the line's three independent variables. To be sure, some of the seven quantities may be only implicit. For example, although E does not appear in the second line, it can be expressed as $E = F + TS$.

In mechanics, the potential energy serves as a function from which the force can be computed by differentiation with respect to position. In an analogous language, one says that the functions E, F, and G serve as *thermodynamic potentials* from which the remaining state functions can be computed by partial differentiation. For each thermodynamic potential, the variables that enable one to do this are called the *natural variables* for the potential.

Focus for a moment on the internal energy E as a thermodynamic potential. Its natural variables are S, V, and N. For a monatomic classical ideal gas, one could easily write E as a function of T and N: $E = \frac{3}{2}NkT$. The variables T and N, however, would not enable one to compute the pressure, entropy, or chemical potential by partial differentiation. Thus, although T and N are permissible as variables to use in expressing E, they are not the most appropriate when the goal is to generate other functions from E by differentiation. The same judgment holds for the more general set $\{T, V, N\}$: permissible, but not most appropriate for E.

In short, a state function like E may be thought of as a function of several different sets of independent variables, but only one such set is appropriate—and hence

Table 10.2 *The natural variables for the thermo-dynamic potentials, followed by the state functions that partial differentiation produces.*

Potential	Variables	Partial derivatives
E	S, V, N	$T, -P, \mu$
F	T, V, N	$-S, -P, \mu$
G	T, P, N	$-S, V, \mu$

"natural"—when the goal is to compute other state functions by partial differentiation.

Extremum principles

There is another sense in which the free energies, F and G, are desirable functions. Recall our formulations of the Second Law of Thermodynamics in chapter 2. In one version, the Second Law asserts that the entropy S evolves to a maximum under conditions of fixed E, V, and N. That is an extremum principle.

The free energies provide extremum principles under different but analogous conditions. For example, section 10.5 established that the Helmholtz free energy F achieves a minimum under conditions of fixed T (for the environment), V, and N. In an essential sense, the functions F and G carry over the basic extremum principle of the Second Law to other sets of independent variables. [Because F and G contain the term $-TS$, where one should note the minus sign, an entropy maximum is converted to a free energy minimum (provided that the temperature is positive).]

In summary, because the free energies F and G are functions of convenient variables (such as N, T, V, or P) and because they provide extremum principles, they are sure to be useful theoretically.

Text and context

The notation and conventions in this chapter are the most common ones. In other texts and other contexts (especially magnetic), differences may arise. The key to keeping your bearings is this: for each free energy, look to see what its *independent* (and natural) variables are. If they are temperature, external parameters, and particle numbers, then the free energy has the properties of our Helmholtz free energy F. If they are temperature, other intensive variables, and particle numbers, then the free energy has the properties of our Gibbs free energy G.

When is $\mu\Delta N$ needed?

Lastly, we consider the question, when must one include a term like $\mu\Delta N$ in an equation? If particles enter or leave the system, either literally or as an imagined process, then one must use an equation like (10.11),

$$T\Delta S = \Delta E + P\Delta V - \mu\Delta N, \tag{10.42}$$

or its analog for one of the other thermodynamic potentials. Equation (10.42) might need to be generalized for multiple species of particles or for other work done. In any event, it presupposes that the initial and final situations are states of thermal equilibrium.

If no particles enter or leave the system, then one may remain with equations (10.1) and (10.2), statements of the First and Second Laws:

$$q = \Delta E + w,$$

$$\Delta S \geqslant \frac{q}{T},$$

where, obviously, no $\mu \Delta N$ terms appear. Both of these equations hold even if particles move from one phase to another *within* the system or engage in chemical reactions that change the number of particles of various species, increasing some while decreasing others. Moreover, these two equations suffice for proving that the two free energies, F and G, attain a minimum at equilibrium. (To be sure, certain environmental conditions are required for the proof also, as listed in section 10.5. For example, an isothermal context is required.) Thus, for a small step away from equilibrium in which one imagines particles to move from one phase to another within the system or to engage in a chemical reaction, one may indeed invoke the minimum property and assert that the free energy does not change:

$$\Delta F = 0 \quad \text{or} \quad \Delta G = 0. \tag{10.43}$$

For us, the minima are easily the most important results of the present chapter. They lead to relations among chemical potentials at equilibrium, but the way in which those relations arise is best seen in chapters 11 and 12, to which we now turn.

10.8 Essentials

1. When particles may enter or leave "the system," one says that the system is *open* (to particle transfer). Otherwise, the system is *closed*.

2. Eliminating the energy-input-by-heating term q between the First and Second Laws yields the relationship

$$T\Delta S \geqslant \Delta E + w,$$

with equality if the process is slow. This relationship is quite general, although it does presuppose a system closed to particle transfer.

3. If particles enter or leave the system, either literally or as an imagined process, then the equation

$$T\Delta S = \Delta E + P\Delta V - \mu \Delta N$$

holds, provided that the initial and final situations are states of thermal equilibrium.

4. The energy E has the set $\{S, V, N\}$ as its natural independent variables. The Helmholtz and Gibbs free energies are introduced because they are energy-like expressions with different independent variables, variables that correspond to different—and more easily attained—physical contexts.

5. The *Gibbs free energy* is defined by

$$G \equiv E - TS + PV$$

in a context where only pressure–volume work may be done. In general, the Gibbs free energy is to be a function of intensive variables plus particle numbers.

6. When only a single species is present, G depends on the set $\{T, P, N\}$. Moreover, G is then proportional to the chemical potential:

$$G = \mu N.$$

7. Equivalent expressions for the chemical potential are the following:

$$\mu = \left(\frac{\partial F}{\partial N} \right)_{T,V} = \left(\frac{\partial E}{\partial N} \right)_{S,V} = \left(\frac{\partial G}{\partial N} \right)_{T,P}.$$

8. The chemical potential for a photon gas is zero: $\mu_{\text{photon gas}} = 0$.

9. The free energies F and G attain a minimum at thermal equilibrium under the following conditions: the system

(a) is in thermal contact with its environment, whose temperature T remains constant throughout,
(b) is closed to entry or exit of particles,
(c) for F, does no work on its surroundings (for example, the volume V remains constant) or, for G, the (external) pressure P remains constant, and only pressure–volume work may be done on the environment.

For an imagined step away from the minima (under the stipulated conditions), the free energies do not change:

$$\Delta F = 0 \text{ or } \Delta G = 0.$$

Further reading

The Legendre transformation is developed in detail by Herbert B. Callen, *Thermodynamics and an Introduction to Thermostatistics*, 2nd edition (Wiley, New York, 1985), pages 137–51.

Problems

1. *The chemical potential from the relation* $\mu = (\partial E/\partial N)_{S,V}$. Equation (5.41) gives the entropy S of a monatomic semi-classical ideal gas in terms of V, N, and λ_{th}.

(a) Display the entropy as a function of V, N, and the total energy E.
(b) Compute the chemical potential from the relation $\mu = (\partial E/\partial N)_{S,V}$ and compare your result with that in equation (7.20).

2. *On F and* $\ln Z$. In section 5.4, we derived expressions for P and $\langle E \rangle$ in terms of derivatives of $\ln Z$. Section 7.1 established a connection between the Helmholtz free energy and $\ln Z$. Use the connection and the relations in sections 5.4 and 5.5 to express P, $\langle E \rangle$, and S in terms of operations on F. Then compare your results with the expressions in section 10.2.

3. *Legendre transformation, geometrically conceived.* Given a functional relation,

$$y = y(x), \tag{1}$$

with slope or tangent $t \equiv dy/dx$, how can one replace the original independent variable x by the tangent t as the new independent variable while retaining all the information contained in equation (1)? Conceive of the curve $y = y(x)$ as generated by the envelope of its tangent lines, as illustrated in parts (a) and (b) of figure 10.2. If you do not have the curve in front of you, then, to construct the set of tangent lines, you need to know the y-intercept of each tangent line as a function of the slope. That is, you need $I = I(t)$, where I denotes the y-intercept. The letter I is a mnemonic for *I*ntercept.

(a) Use part (c) of the figure to determine the intercept I in terms of y, x, and t. How is I related to the function $y - x \times (dy/dx)$, which is the Legendre transform of y with respect to x?
(b) Next, show that I can indeed be considered a function of t alone. (Forming the differential dI is a good first step.)
(c) What is the derivative dI/dt in terms of the original variables?
(d) Does there exist a second Legendre transformation that would take one from the pair I and t back to the pair y and x?

4. *Maxwell relations.* If a function $f(x, y)$ is sufficiently smooth, then the order of partial differentiation is irrelevant: $\partial(\partial f/\partial x)/\partial y = \partial(\partial f/\partial y)/\partial x$. James Clerk Maxwell put this property to use with functions derivable from a thermodynamic potential.

(a) Use the Helmholtz free energy as an intermediary to express the derivative $(\partial S/\partial V)_T$ in terms of a derivative of the pressure P. Then specify that the ideal gas law holds, evaluate your derivative of P, and finally integrate to determine the volume dependence of the entropy of a classical ideal gas.

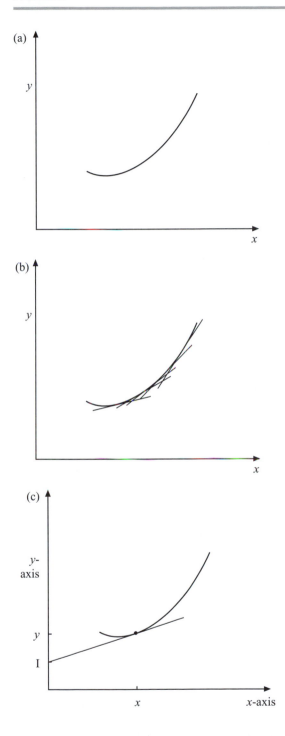

(a)

(b)

(c)

Figure 10.2 Diagrams for the Legendre transformation, conceived geometrically. (a) The curve $y = y(x)$. (b) The curve defined by the envelope of its tangent lines. (c) One tangent line and its relation to x, y, t, and $I(t)$.

(b) Use equation (10.19) and Maxwell's idea to relate the derivative $(\partial\mu/\partial V)_{N,T}$ to another derivative and then determine the volume dependence of the chemical potential for a classical ideal gas.

5. *Enthalpy.*

(a) Use a Legendre transformation to construct a new energy-like function $H \equiv E + $ (product of two variables) such that H is naturally a function of S, P, and N. The function is called the *enthalpy.*
(b) Determine the thermodynamic functions that can be gotten by differentiating H with respect to S, P, and N.

6. *Helmholtz free energy for a photon gas.* The Helmholtz free energy for a single electromagnetic mode must have the form $-kT \ln Z_{\text{that single mode}}$. For many independent modes, the system's full partition function is formed multiplicatively, and so the free energy is formed additively.

(a) Use the reasoning outlined above to derive the result

$$F_{\text{photon gas}} = \int_0^\infty D_{\text{EM}}(\nu) \times \left(-kT \ln \frac{1}{1 - e^{-h\nu/kT}} \right) d\nu.$$

(b) Extract the dependence of $F_{\text{photon gas}}$ on T and V. An integration by parts will enable you to cast the remaining dimensionless integral into a form that you have met. Provide a detailed treatment of the terms at $\nu = 0$ and $\nu = \infty$ that arise in the integration by parts.
(c) Use $F_{\text{photon gas}}$ to compute P, S, E, and μ. Then compare with previous expressions in chapter 6 and in this chapter.

7. Because the chemical potential for a photon gas is zero, the Gibbs free energy should be zero also. Are the results for $\langle E \rangle$, S, and P in section 6.2 consistent with this expectation?

8. *Information from the Gibbs free energy.* You are given the following Gibbs free energy:

$$G = -kTN \ln\left(\frac{aT^{5/2}}{P} \right),$$

where a is a constant (whose dimensions make the argument of the logarithm dimensionless). Compute (a) the entropy, (b) the heat capacity at constant pressure C_P, (c) the connection among V, P, N, and T, which is called the "equation of state," and (d) the estimated energy $\langle E \rangle$.

9. *Chemical potential from S.* The chemical potential can be expressed as a partial derivative in many different ways, each useful in some context.

(a) Determine the proportionality constant (really, a proportionality function) in the relation

μ is proportional to $(\partial S/\partial N)_{E,V}$.

Rearranging equation (10.11) provides a good route.

(b) Test your relationship in some context where you can work out all factors explicitly.

11 Chemical Equilibrium

11.1 The kinetic view

11.2 A consequence of minimum free energy

11.3 The diatomic molecule

11.4 Thermal ionization

11.5 Another facet of chemical equilibrium

11.6 Creation and annihilation

11.7 Essentials

Chemical equilibrium is certainly a topic for chemists. Why does it appear in a book on thermal physics for physicists and astronomers? For two reasons, at least. First, chemical reactions play vital roles in astrophysics, in atmospheric studies, and in batteries, to mention just three topics that physicists and astronomers may find themselves investigating. Second, chemical equilibrium provides more practice with the chemical potential, which is truly an indispensable tool for any practitioner of thermal physics.

The chapter starts with a kinetic view of how chemical equilibrium arises. That will show us how chemical concentrations are related at equilibrium. Then we will use a free energy argument to derive the same structure from the Second Law of Thermodynamics. Applications follow.

11.1 The kinetic view

A glass flask is filled, let us suppose, with a dilute gaseous mixture of diatomic hydrogen (H_2) and diatomic chlorine (Cl_2). Through the electric forces acting during collisions, a chemical reaction of the form

$$H_2 + Cl_2 \rightleftharpoons 2\,HCl \tag{11.1}$$

is possible. The reaction forms hydrogen chloride (HCl) and, run backward, leads to the dissociation of HCl. For simplicity's sake, we specify that the reaction (11.1) is the only reaction that H_2, Cl_2, and HCl participate in. The number density of HCl is written as $[HCl] \equiv N_{HCl}/V$, where N_{HCl} denotes the total number of HCl molecules and V is the flask's volume. For brevity's sake, I shall occasionally refer to the number density as the "concentration."

The concentration of HCl will be governed by a rate equation with the structure

$$\frac{d[\text{HCl}]}{dt} = \text{formation rate} - \text{dissociation rate.} \tag{11.2}$$

How do these rates depend on the number densities of the various molecular species? If the concentration of Cl_2 is held fixed but the concentration of H_2 is doubled, we can expect twice as many collisions per second that form HCl. Thus the formation rate ought to be proportional to the hydrogen concentration $[\text{H}_2]$. By a similar line of reasoning, the rate ought to be proportional to the concentration of $[\text{Cl}_2]$. Thus

$$\text{formation rate} = f_{\text{formation}}(T)[\text{H}_2][\text{Cl}_2], \tag{11.3}$$

where the proportionality constant $f_{\text{formation}}(T)$ depends on the electric forces and the temperature T, for the temperature sets the scale of the kinetic energy with which a typical collision occurs and also governs the distribution of internal states of the molecules, that is, their rotation or vibration.

The reasoning for the dissociation rate is only a little more difficult. Two HCl molecules must come close together in order to react and produce H_2 and Cl_2. The total number of HCl molecules is proportional to the concentration $[\text{HCl}]$. The probability that another HCl molecule is close enough to react with a specific HCl molecule is also proportional to the concentration. Thus the dissociation rate ought to be proportional to the square of the HCl concentration:

$$\text{dissociation rate} = f_{\text{dissociation}}(T)[\text{HCl}]^2, \tag{11.4}$$

where, again, the proportionality constant depends on electric forces and the temperature.

Inserting the rate expressions into equation (11.2), we find

$$\frac{d[\text{HCl}]}{dt} = f_{\text{formation}}(T)[\text{H}_2][\text{Cl}_2] - f_{\text{dissociation}}(T)[\text{HCl}]^2. \tag{11.5}$$

At thermal equilibrium, the concentration of HCl does not change with time (when viewed macroscopically), and so the rates of formation and dissociation are equal. Setting the derivative in (11.5) to zero and isolating all concentrations on the left-hand side, we find

$$\frac{[\text{HCl}]^2}{[\text{H}_2][\text{Cl}_2]} = \frac{f_{\text{formation}}(T)}{f_{\text{dissociation}}(T)}. \tag{11.6}$$

No matter how much H_2, Cl_2, and HCl one starts with, the reaction—running either forward or backward—will adjust concentrations so that equation (11.6) is satisfied. The combination of concentrations on the left-hand side will settle to a value that depends on only the temperature (and, of course, the molecular species). The right-hand side, regarded as a single function of temperature, is called the *equilibrium constant* for the reaction.

Equation (11.6), taken as a whole, is an instance of the *law of mass action*. In the usage here, the word "mass" means "the amounts of chemicals" or the magnitudes of the concentrations. The word "action" means "the effect on the reaction rate or the equilibrium." How the current amounts of chemicals affect the reaction is seen most clearly in a rate equation like (11.5). After equilibrium has been reached, algebraic steps give equation (11.6). Now the effect of concentrations is not so apparent, but that equilibrium structure is what today carries the name "the law of mass action."

Two Norwegians, the mathematician Cato Maximilian Guldberg and the chemist Peter Waage, developed the law of mass action in the 1860s by rate arguments similar to ours. In the next section, a line of reasoning based on a free energy will lead to a structure identical to (11.6) but with two salient advantages. (1) A method for calculating the equilibrium constant from molecular parameters will emerge. (2) The relationship will be shown to hold, at equilibrium, regardless of what other reactions the molecular species may engage in. The pathways of chemical reactions are often extremely complex and depend on species—often ions—that never appear in an equation of the form "reactants" $\rightarrow$ "products," such as (11.1). Our reasoning with rates of formation and dissociation, with its tentativeness of "ought to," is best regarded as suggestive, not definitive. The basic idea of balancing rates is valid, but the details may be more complex than is readily apparent.

11.2   A consequence of minimum free energy

As a preliminary step, we generalize the chemical reaction under study. We can write the HCl reaction (11.1) in the algebraic form

$$-H_2 - Cl_2 + 2\,HCl = 0, \tag{11.7}$$

which expresses—among other things—the conservation of each *atomic* species (H and Cl) during the reaction. Adopting this pattern, we write the generic form for a chemical reaction as

$$b_1\,B_1 + b_2\,B_2 + \cdots + b_n\,B_n = 0, \tag{11.8}$$

where each molecular species is represented by a symbol B_i and the corresponding numerical coefficient in the reaction equation is represented by the symbol b_i. Table

Table 11.1 *The stoichiometric coefficients for the reaction* $H_2 + Cl_2 \rightarrow 2\,HCl$, *which can be written* $-H_2 - Cl_2 + 2\,HCl = 0$.

B_1: H_2	$b_1 = -1$
B_2: Cl_2	$b_2 = -1$
B_3: HCl	$b_3 = +2$

11.1 shows the correspondences for the HCl reaction. For the products of a reaction, the coefficients b_i are positive; for the reactants, they are negative. Altogether, the set $\{b_i\}$ gives the number change in each molecular species when the reaction occurs once. The coefficients $\{b_i\}$ are called *stoichiometric coefficients* (from the Greek roots, *stoikheion*, meaning "element," and *metron*, meaning "to measure"). The pronunciation is "stoi' key a met' rik."

We ask now, what consequence for chemical equilibrium does the minimum property of the free energies entail? Take first the situation where the chemicals are in a glass flask of fixed volume V. The flask itself is immersed in a water bath at temperature T, so that the chemical reaction comes to equilibrium under isothermal conditions. The Helmholtz free energy will attain a minimum. Imagine taking one step away from equilibrium: the number N_i of molecular species B_i changes by ΔN_i, which equals the stoichiometric coefficient b_i. Then the change in the Helmholtz free energy is

$$\Delta F = \sum_i \left(\frac{\partial F}{\partial N_i} \right)_{T,V,\text{other } Ns} \Delta N_i$$

$$= \sum_i \mu_i b_i = 0. \tag{11.9}$$

The partial derivatives are precisely the chemical potentials, and the zero follows because the imagined step is away from the minimum. Equilibrium for the chemical reaction implies a connection among the various chemical potentials:

$$\sum_i \mu_i b_i = 0. \tag{11.10}$$

If the chemical reaction had come to equilibrium under conditions of fixed temperature and pressure, we would have invoked a minimum in the Gibbs free energy. The analog of equation (11.9) would have had G in place of F, but its second line would have been the same, and so the relationship (11.10) would have emerged again.

At this point, a remark by the Austrian physicist Ludwig Boltzmann comes to my mind. In his *Lectures on Gas Theory*, Boltzmann wrote,

> I have once previously treated the problem of the dissociation of gases, on the basis of the most general possible assumptions, which of course I had to specialize at the end.

Equation (11.10) is extremely general: it holds for gas phase reactions in a dilute gas or a dense gas, and it holds for reactions occurring in solution, whether the solvent is water or an organic liquid. But we shall make rapid progress only if we specialize to semi-classical ideal gases, which we do now.

Chemical equilibrium in semi-classical ideal gases

Both our derivation in chapter 7 and our confirmation in section 8.4 tell us that the chemical potential for species B_i will have the form

$$\mu_i = -kT \ln (Z_{1i}/N_i). \tag{11.11}$$

The symbol Z_{1i} denotes the single-molecule partition function for species B_i.

Next, we ask, what is the structure of the single-molecule partition function? Dropping the index i for a moment, we write

$$Z_1 = \sum_{\substack{\text{states } \varphi_\alpha \text{ of} \\ \text{one molecule}}} e^{-\varepsilon_\alpha/kT}. \tag{11.12}$$

The single-molecule energy ε_α splits naturally into a term representing the translational motion of the molecular center of mass (CM) and another portion that reflects the internal state: rotation, vibration, etc. Thus

$$\varepsilon_\alpha = \varepsilon_\alpha(\text{CM}) + \varepsilon_\alpha(\text{int}). \tag{11.13}$$

Insert this decomposition into (11.12) and factor the exponential:

$$Z_1 = \sum_{\substack{\text{states } \varphi_\alpha \text{ of} \\ \text{one molecule}}} e^{-\varepsilon_\alpha(\text{CM})/kT} \times e^{-\varepsilon_\alpha(\text{int})/kT}$$

$$= \frac{V}{\lambda_{\text{th}}^3} \times \sum_{\substack{\text{internal states} \\ \text{of one molecule}}} e^{-\varepsilon_\alpha(\text{int})/kT} = \frac{V}{\lambda_{\text{th}}^3} \times Z(\text{int}), \tag{11.14}$$

where

$$Z(\text{int}) \equiv \sum_{\substack{\text{internal states} \\ \text{of one molecule}}} e^{-\varepsilon_\alpha(\text{int})/kT}. \tag{11.15}$$

For each specific set of internal quantum numbers, we are to sum over all possible quantum numbers for the translational motion. That sum we worked out in section 5.6 and found to be V/λ_{th}^3. The mass m that enters into the definition $\lambda_{\text{th}} \equiv h/\sqrt{2\pi mkT}$ is the mass of the entire molecule. The remaining sum goes over all the internal states, and the symbol $Z(\text{int})$ suffices as an abbreviation. The sum $Z(\text{int})$ is called the *internal partition function*.

Now insert the structure (11.14) into the expression for μ_i but isolate the concentration dependence as much as possible:

$$\mu_i = +kT \ln (N_i/V) - kT \ln \{\lambda_i^{-3} Z_i(\text{int})\}. \tag{11.16}$$

The ratio N_i/V is the number density $[B_i]$, and λ_i denotes the thermal de Broglie wavelength for species B_i.

We return to the connection among chemical potentials at equilibrium, equation (11.10). Inserting the form (11.16) and splitting the sum into two natural pieces, we find

$$\sum_i b_i \ln [B_i] = \sum_i b_i \ln \{\lambda_i^{-3} Z_i(\text{int})\}. \tag{11.17}$$

To tidy this up, move the stoichiometric coefficients b_i inside the logarithm operation as exponents. For example, $b_i \ln [B_i] = \ln ([B_i]^{b_i})$. Then recall that a sum of logarithms equals the logarithm of the product of the arguments. Thus (11.17) becomes

$$\ln \prod_i [B_i]^{b_i} = \ln \prod_i \{\lambda_i^{-3} Z_i(\text{int})\}^{b_i}.$$

The arguments of these two logarithms must be equal, and so

$$\prod_i [B_i]^{b_i} = \prod_i \{\lambda_i^{-3} Z_i(\text{int})\}^{b_i}. \tag{11.18}$$

The minimum in the free energy, either F or G, has this relation among concentrations as its consequence. Equation (11.18) is the generic law of mass action whenever the participating molecules may be treated as semi-classical ideal gases.

An example will make the result more familiar. Reference to table 11.1 gives us the stoichiometric coefficients for the HCl reaction, and thus (11.18) implies

$$\frac{[HCl]^2}{[H_2][Cl_2]} = \frac{\lambda_{HCl}^{-6} Z_{HCl}(\text{int})^2}{\lambda_{H_2}^{-3} Z_{H_2}(\text{int}) \times \lambda_{Cl_2}^{-3} Z_{Cl_2}(\text{int})}. \tag{11.19}$$

The combination of concentrations on the left-hand side is precisely what the kinetic reasoning in section 11.1 generated. The combination here on the right-hand side is a function of temperature and molecular parameters; thus it corresponds well to the right-hand side of (11.6) and is an equilibrium constant.

Indeed, we define the general equilibrium constant $K(T)$ by the right-hand side of (11.18):

$$K(T) \equiv \prod_i \{\lambda_i^{-3} Z_i(\text{int})\}^{b_i}. \tag{11.20}$$

Then equation (11.18) takes the succinct form

$$\prod_i [B_i]^{b_i} = K(T). \tag{11.21}$$

You will sometimes find the law of mass action, (11.18) or (11.21), expressed in terms of different variables. Problem 2 explores two of these alternatives.

If any of the species B_i engages in another chemical reaction, then we can carry through the same line of reasoning for the other reaction. Each reaction in which the species engages leads to an equation like (11.10). Moreover, if the molecules behave like semi-classical ideal gases, then each reaction leads also to an equation like (11.18). The algebraic complexity grows, of course, and we will not let ourselves be dragged into it. I mention these possibilities merely to point out that equations (11.10) and (11.18) hold regardless of whether the molecules engage in other reactions or not.

Before we can calculate the equilibrium constant for the HCl reaction, we need to know more about diatomic molecules, and so we turn to that topic now.

11.3   The diatomic molecule

Although diatomic molecules are the simplest molecules, their thermal physics is surprisingly complex. Rest assured that this section has a summary figure—figure 11.2—that displays $Z(\text{int})$ for the most common situation. Understanding how the factors in that figure arise provides a good foundation for understanding diatomic molecules in general.

Energies

The atoms in a diatomic molecule are bound together by electric forces. Those forces are responsible for the attractive portion of the interatomic potential energy that figure 11.1 displays. The strong repulsion at short distance arises from two effects: (1) the electric repulsion between the positively charged nuclei and (2) the Pauli exclusion principle as it affects the interaction among the atomic electrons. The potential energy has a minimum, but one must expect the separation of the nuclei to oscillate about that minimum. Physically, this means that the two nuclei periodically move toward each other and then away—although not by much in either direction. The minimum in the

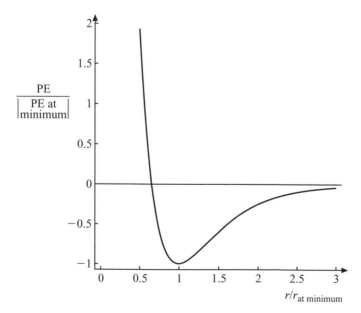

Figure 11.1 The interatomic potential energy (PE) for a typical diatomic molecule. The relative separation r is the distance between the two nuclei.

potential well can be approximated as parabolic (for small amplitudes of oscillation), and so the vibrational motion is approximately that of a harmonic oscillator.

In addition to vibrating, the molecule can rotate in space about its center of mass. The rotational energy is determined by the molecule's rotational angular momentum $J\hbar$ and by its moment of inertia I, computed about the molecule's center of mass. The molecular rotation is about an axis perpendicular to the inter-nuclear axis (at least in a classical view of things), and so the relevant moment of inertia is calculated by adding two terms of the form "(nuclear mass) × (square of distance from CM)."

The translational motion of the center of mass contributes kinetic energy, but we focus in this section on energy relative to the center of mass.

The view outlined above enables us to write down expressions for the energy *internal* to the molecule:

$$\varepsilon(\text{int}) = \begin{pmatrix} \text{energy when nuclei are} \\ \text{at rest at minimum in} \\ \text{interatomic potential energy} \end{pmatrix} + \begin{pmatrix} \text{vibrational} \\ \text{energy} \end{pmatrix} + \begin{pmatrix} \text{rotational} \\ \text{energy} \end{pmatrix}$$

$$= \Delta\varepsilon + (n + \tfrac{1}{2})h\nu + J(J+1)\frac{\hbar^2}{2I}. \tag{11.22}$$

The symbol $\Delta\varepsilon$ denotes the change in energy, relative to free atoms, because of the binding into a diatomic molecule. It consists of the changes in electrostatic potential energies (among electrons and nuclei) and the change in electronic kinetic energy. The zero of energy corresponds to free atoms at rest at infinite separation from each other. Thus the very existence of a bound diatomic molecule means that $\Delta\varepsilon$ is negative. Also, note that our focus here is on the state of lowest (electronic) energy. Binding into excited electronic states is often possible, but such states are largely irrelevant at modest temperatures, and so we ignore them.

The vibrational energy has the familiar harmonic oscillator form, where n is zero or a positive integer and ν is the frequency of vibration.

The rotational energy is governed by the square of the rotational angular momentum, $J(J+1)\hbar^2$. The possible values of J are restricted to zero or a positive integer. (But, as we will discuss later, even some of these values may not be permitted for molecules whose nuclei are identical. For example, odd integral values are excluded if diatomic oxygen is formed with two atoms of the isotope ^{16}O.) The moment of inertia I is taken as a constant, equal to its value when n and J are zero. (For a good first approximation, we exclude any increase in I due to vibrational motion or due to a centrifugal effect of rotation.)

The energy of the molecular ground state, $\varepsilon_{\text{g.s.}}$, is given by summing the lowest value of each term in (11.22). This means

$$\varepsilon_{\text{g.s.}} = \Delta\varepsilon + \tfrac{1}{2}h\nu, \tag{11.23}$$

and $\varepsilon_{\text{g.s.}}$ will be negative if the molecule exists stably, a condition that we specify.

Table 11.2 displays the ground state energies of some common diatomic molecules. In order of magnitude, the energies range from -2 to -10 electron volts.

Partition function

Equations (11.22) and (11.23) enable us to write the internal partition function as

$$Z(\text{int}) = \sum_{\substack{\text{internal states} \\ \text{of one molecule}}} \exp\left[-\left(\varepsilon_{\text{g.s.}} + nh\nu + J(J+1)\frac{\hbar^2}{2I}\right)\bigg/ kT\right]. \tag{11.24}$$

The exponential can be factored into the product of three factors, and so the sums over n and J can be split off as separate, independent sums. This step leads to the form

$$Z(\text{int}) = e^{-\varepsilon_{\text{g.s.}}/kT} \times \sum_{n} e^{-nh\nu/kT}$$

$$\times \sum_{J} \left(\begin{array}{c}\text{spin degeneracy}\\ \text{factor}\end{array}\right)(2J+1)e^{-J(J+1)\hbar^2/2IkT}. \tag{11.25}$$

Of course, this expression needs additional explanation. For each permitted value of J, the rotational angular momentum may take on various orientations in space. If one chooses a direction as the "quantization axis," then the projection of the rotational angular momentum along that axis may take on $2J + 1$ different values. Each value corresponds to a different quantum state for the molecule. Because the partition function is a sum over distinct *states* (not merely over distinct energies), we must include the factor $(2J + 1)$ when summing over permitted values of J.

Even after one has chosen the vibrational and rotational quantum numbers, the quantum state may not be fully specified because the nuclear spins (if nonzero) may point in various directions. If the nuclear spins are not correlated or otherwise constrained, then each spin may take on $(2 \times \text{spin} + 1)$ distinct orientations and hence contributes a factor $(2 \times \text{spin} + 1)$ to the counting of states. The hydrogen nucleus, being a single proton, has a spin of $\frac{1}{2}$ (in units of $\hbar$) and contributes the factor $(2 \times \frac{1}{2} + 1)$. The predominant isotopes of chlorine have spin $\frac{3}{2}\hbar$ and hence contribute a factor $(2 \times \frac{3}{2} + 1)$. The electron spins, taken all together, may also contribute to the degeneracy factor. In the case of HCl, they do not. Thus the spin degeneracy factor for common HCl is the product of the hydrogen and chlorine nuclear factors: (2×4), a constant independent of J. The spin degeneracy factor will be constant whenever the two nuclei differ. The case of two identical nuclei is deferred a bit.

Vibrational portion of the partition function

Now we turn to evaluating the sums in detail. The vibrational energy changes in increments of $h\nu$. The size of $h\nu$ relative to kT determines which vibrational states will have substantial probability of being occupied; therefore the size determines which states contribute significantly to the sum. To facilitate comparison, define a characteristic vibrational temperature θ_v by the equation

$$k\theta_v \equiv h\nu. \tag{11.26}$$

(This definition is analogous to the definition of the Debye temperature in section 6.5.) Table 11.2 shows that θ_v is often several thousand kelvin. If both nuclei are especially massive (or if the chemical bond is particularly weak), then the vibrational frequency will drop. In turn, that will cause θ_v to drop below 10^3 K, as it does for the three relatively massive molecules, chlorine, bromine, and iodine.

In the most common situation, the strong inequality $T \ll \theta_v$ holds, and so the sum over vibrational states cuts off exponentially fast. In fact, often it suffices to take just the first term, which is 1. The physical temperature T is so low (relative to θ_v) that the vibrational motion is "frozen out."

Rotational portion of the partition function

The situation with rotational motion is quite different. To be sure, again there is a characteristic temperature, defined now by the equation

$$k\theta_r \equiv \hbar^2/2I. \tag{11.27}$$

Table 11.2 indicates that the characteristic rotational temperature θ_r varies over a great range, from 88 K for H_2 to 0.05 K for relatively massive nuclei like those in diatomic

Table 11.2 *Parameters for some common diatomic molecules. "Separation" denotes the inter-nuclear separation in the gound state. The zero of energy corresponds to free atoms at rest and at infinite separation from each other. The values for all columns come primarily from spectroscopic data.*

	θ_v (K)	θ_r (K)	Separation (10^{-10} m)	Ground-state energy (eV)
H_2	6,320	87.5	0.742	-4.476
N_2	3,390	2.89	1.094	-7.37
O_2	2,270	2.08	1.207	-5.08
CO	3,120	2.78	1.128	-9.14
NO	2,740	2.45	1.151	-5.29
HCl	4,300	15.2	1.275	-4.43
HBr	3,810	12.2	1.414	-3.75
HI	3,320	9.42	1.604	-3.06
Cl_2	813	0.351	1.988	-2.48
Br_2	465	0.116	2.284	-1.97
I_2	309	0.054	2.667	-1.54

Source: Gerhard Herzberg, *Molecular Spectra and Molecular Structure, Vol. 1: Spectra of Diatomic Molecules* (Van Nostrand, New York, 1950).

iodine, a range of three orders of magnitude. The decrease in θ_r reflects an increase in the moment of inertia I. In turn, the moment of inertia increases for two readily assignable reasons: as one moves through the periodic table, (1) the atoms get larger, which increases the relative separation in the diatomic molecule, and (2) the nuclei become more massive. The increase in mass dominates, contributing more than two orders of magnitude for the molecules in the table, and the increase in size makes up the rest of the readily assignable variation in θ_r. (Although atomic sizes increase by only a factor of 3 or so, the moment of inertia depends on the square of the inter-nuclear separation, and so a factor of 10 is attainable. To be sure, atomic sizes typically decrease as one advances along a row in the periodic table, say from lithium to fluorine, but the row-average increases as one moves through the periodic table. Alternatively, if one progresses down a column at either end of the table, atomic size grows monotonically. Variations in the strength of the chemical bond will also affect the relative separation and hence θ_r. In short, the factors that determine the relative separation can become quite complex.)

For a molecule like HCl and for room temperature, the characteristic rotational temperature θ_r is much less than the physical temperature. The exponential drops off slowly with increasing J (at least while its magnitude is still significant). Moreover, the spin degeneracy factor is a constant. Consequently, one may approximate the sum by an integral:

$$Z_{\text{rot}} \equiv \sum_j (2J + 1)e^{-J(J+1)\hbar^2/2IkT}$$

$$\cong \int_0^\infty e^{-J(J+1)\theta_r/T}(2J + 1)\, dJ \tag{11.28}$$

$$= T/\theta_r,$$

where the last line follows because $(2J + 1)\, dJ$ is the differential of the $J(J + 1)$ in the exponent. Before being used in equation (11.25), the approximation T/θ_r is to be multiplied by the spin degeneracy factor.

A bit of vocabulary will be useful. Molecules whose nuclei differ are called *heteronuclear*. Those whose nuclei are identical are called *homonuclear*. For hetero-nuclear diatomic molecules in the context $\theta_r \ll T \ll \theta_v$, we have determined the internal partition function $Z(\text{int})$, and figure 11.2 summarizes the result. Now we go on to consider a homonuclear molecule.

If the two atoms are identical, as would be true for Cl_2 formed from two atoms of the isotope ^{35}Cl, then the indistinguishability of identical particles causes the spin degeneracy factor to depend on J. If the spin of the nuclei is integral, the nuclei are bosons, and the wave function for the diatomic molecule must be symmetric with respect to the nuclei. For half-integral spin, the nuclei are fermions, and antisymmetry is required. The nuclear spins can be combined into quantum states that have definite symmetry: either symmetric or antisymmetric. Each of these two kinds of spin states must be paired with certain J values, either even J or odd J. Figure 11.3 displays the

$$Z(\text{int}) = e^{-\varepsilon_{\text{g.s.}}/kT} \times 1 \times \left(\begin{array}{c} \text{electronic spin} \\ \text{degeneracy factor} \end{array} \right)$$

$$\times (2 \times \text{spin}_1 + 1)(2 \times \text{spin}_2 + 1) \times \left(\frac{T}{\theta_r} \right) \times \left\{ \begin{array}{l} 1 \text{ if heteronuclear} \\ \frac{1}{2} \text{ if homonuclear} \end{array} \right\}.$$

Figure 11.2 Summary for $Z(\text{int})$ when the temperature satisfies the strong inequalities $\theta_r \ll T \ll \theta_v$. The ground-state energy $\varepsilon_{\text{g.s.}}$ must be calculated from the same zero for all atoms or molecules in the reaction. Usually the zero corresponds to free atoms at rest at infinite separation. The numbers "spin$_1$" and "spin$_2$" refer to the nuclear spins of nuclei 1 and 2 (which may be identical).

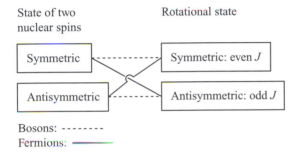

Bosons: --------
Fermions: ————

Figure 11.3 When the two nuclei are identical, a specific rotational state is paired with only certain states of the nuclear spins. Bosons and fermions require different pairings.

Note. The total number of distinct states in the left-hand column is $[2 \times (\text{spin of one nucleus}) + 1]^2$.

connections. Thus there are two distinct values for the spin degeneracy factor: one value equal to the number of symmetric nuclear spin states, the other equal to the number of antisymmetric spin states. The sum over J in (11.25) splits, in effect, into two sums, one over even J, the other over odd J, each with its separate (and constant) spin degeneracy factor. At temperatures of order θ_r and lower, those two sums must be done in numerical detail. The situation is simpler at temperatures well above θ_r. Because the exponential varies slowly, a sum over all even J is approximately one-half of the sum over all J, namely $\frac{1}{2}(T/\theta_r)$. The same is true for a sum over just the odd values of J. Thus the pair of sums, even J and odd J, amounts to $\frac{1}{2}(T/\theta_r)$ times the sum of the two nuclear spin degeneracy factors. The latter sum is the total nuclear spin degeneracy, $[2 \times (\text{spin of one nucleus}) + 1]^2$. Therefore, in the context $T \gg \theta_r$, we find

$$\left(\begin{array}{c} \text{rotational and nuclear spin} \\ \text{part of } Z(\text{int}) \end{array} \right) = [2 \times (\text{spin of one nucleus}) + 1]^2 \times \frac{1}{2}\left(\frac{T}{\theta_r} \right)$$

$$(11.29)$$

when the nuclei are identical. This result completes the homonuclear version of $Z(\text{int})$ in figure 11.2.

Application to the HCl reaction

Now we pull everything together for the heteronuclear molecule HCl in the thermal context $\theta_r \ll T \ll \theta_v$. Invoking equations (11.25) and (11.28), we emerge with the expression

$$Z(\text{int})_{\text{HCl}} = e^{-\varepsilon_{\text{g.s.}}/kT} \times 1 \times (2 \times 4) \times \frac{T}{\theta_r}. \tag{11.30}$$

The calculation for the other diatomic molecules in the HCl reaction is similar. For the homonuclear molecule H_2, equations (11.25) and (11.29) give

$$Z(\text{int})_{\text{H}_2} = e^{-\varepsilon_{\text{g.s.}}/kT} \times 1 \times (2)^2 \times \frac{1}{2}\left(\frac{T}{\theta_r}\right). \tag{11.31}$$

The result for the homonuclear molecule Cl_2 will be similar except that, because the nuclear spin is $\frac{3}{2}\hbar$, the factor $(2)^2$ is replaced by $(4)^2$. Of course, the values of $\varepsilon_{\text{g.s.}}$ and θ_r differ among all three molecules, and the notation will soon have to display that.

Finding a single temperature T that satisfies the strong inequalities $\theta_r \ll T \ll \theta_v$ for all three species is not easy, but $T = 260$ K gives inequalities of a factor of 3 even in the worst cases, namely $\theta_r = 85$ K for H_2 and $\theta_v = 810$ K for Cl_2. One would like a factor of 10, but—in the spirit of maintaining maximum simplicity—let us suppose that a factor of 3 is good enough and insert our expressions into the right-hand side of (11.19):

$$K(T) = \frac{m_{\text{HCl}}^3}{m_{\text{H}_2}^{3/2} m_{\text{Cl}_2}^{3/2}} \times 2^2 \times \frac{\theta_{r,\text{H}_2} \theta_{r,\text{Cl}_2}}{\theta_{r,\text{HCl}}^2}$$

$$\times \exp\left[\left(-2\varepsilon_{\text{g.s.,HCl}} + \varepsilon_{\text{g.s.,H}_2} + \varepsilon_{\text{g.s.,Cl}_2}\right)/kT\right]. \tag{11.32}$$

The mass factors come from the thermal de Broglie wavelengths. The factor 2^2 arises because H_2 and Cl_2 are homonuclear molecules. The nuclear spin factors cancel between numerator and denominator. A dependence on temperature survives only in the exponential, but that would not be true in other temperature regimes for the HCl reaction. Moreover, if the number of product molecules (here two) differs from the number of reactant molecules (also two here) for a single step of the reaction, then the thermal wavelength factors will leave a residual dependence on T at every temperature. Nonetheless, the exponential usually carries the dominant temperature dependence. If the ground-state energies of the products (weighted by the stoichiometric coefficients) are lower than those of the reactants (again weighted), then the exponent is positive. One can think of a drop in energy as favoring the products exponentially, but one should not focus on such a detailed interpretation at the expense of other factors that influence the equilibrium concentrations. We explore some of the other factors in the following sections and in the homework problems.

Some practicalities

Figure 11.2 provided a summary for $Z(\text{int})$ in the most common situation. Given a specific molecule, you may wonder, how do I determine the numerical values of the nuclear spins? And also the "electronic spin degeneracy factor"? The short answer is this: by looking them up. There is no easy way to calculate a nuclear spin (when given merely the number of protons and neutrons). Likewise, there is no easy way to compute the electronic spin degeneracy factor. For example, in diatomic nitrogen, the electrons pair up to give a net electronic spin of zero, and so the degeneracy factor is just 1. In diatomic oxygen, however, the electrons combine to produce a net electronic spin of $1\hbar$, and so the factor is $(2 \times 1 + 1) = 3$. The molecules, N_2 and O_2, are otherwise quite similar. Only experiment or the details of a quantum mechanical calculation can reliably determine the electronic spin degeneracy factor.

Another practical item is the zero of energy. Energy *differences* are what matter in chemical equilibria, and so any convenient situation may be assigned a value of zero energy. That assignment, of course, must be used consistently throughout a calculation. If the chemical reaction consists of rearranging the way that atoms are aggregated into molecules, the convention is to assign zero energy to free atoms (of all species) when at rest and at infinite separation from one another. Then the molecules, as bound aggregates of atoms, have negative potential energy. (This is the scheme that we used for the HCl reaction.) If, however, the reaction is the ionization of a free atom, then the electron and the positive ion are (usually) assigned zero energy when at rest and infinitely separated. Now the atom, as a bound aggregate of electron and ion, has negative potential energy. You get to decide which situation is assigned zero energy, but be explicit and consistent.

Now we leave the complexity of molecules and return to single atoms.

11.4 Thermal ionization

Hot gases at low density occur frequently in astrophysics. Here we consider atomic hydrogen (H) in thermal equilibrium with its dissociation products, ionized hydrogen (H^+, which is just a free proton) and a free electron. The "chemical reaction" is written

$$H \rightleftharpoons H^+ + e. \tag{11.33}$$

According to equations (11.20) and (11.21), the law of mass action takes the form

$$\frac{[H^+][e]}{[H]} = K(T), \tag{11.34}$$

where

$$K(T) = \frac{\lambda_{H^+}^{-3} \lambda_e^{-3}}{\lambda_H^{-3}} \frac{Z_{H^+}(\text{int}) Z_e(\text{int})}{Z_H(\text{int})}. \tag{11.35}$$

In treating H^+ and the electrons as ideal gases, we are supposing that electrical interactions are negligible (except during recombination or dissociation). The masses of H^+ and H are so nearly identical that, for all practical purposes, the corresponding thermal wavelengths cancel. The "internal" partition function for a free electron is merely its spin degeneracy factor, $(2 \times \frac{1}{2} + 1)$, which is 2. The same is true for H^+.

The partition function for the hydrogen atom is a sum over all electronic and nuclear states. For now, we take only the electronic ground state and justify that severe truncation later. Thus we make the approximation

$$Z_H(\text{int}) = 2 \times 2 \times e^{-\varepsilon_{\text{g.s.}}/kT}, \tag{11.36}$$

where the factors 2×2 represent the nuclear and electron spin degeneracy factors and where $\varepsilon_{\text{g.s.}} = -13.6$ electron volts, the ground-state energy of a hydrogen atom, a negative quantity. The upshot of the numbers and approximation is that the equilibrium constant becomes

$$K(T) = \lambda_e^{-3} e^{\varepsilon_{\text{g.s.}}/kT}. \tag{11.37}$$

The characteristic temperature θ_H for hydrogen is

$$\theta_H = |\varepsilon_{\text{g.s.}}|/k = 160{,}000 \text{ K}. \tag{11.38}$$

This temperature is so high that the exponential in $K(T)$ is less than 10^{-2} whenever $T < 35{,}000$ K. Nonetheless, the electron's thermal wavelength is small and compensates partially for the small exponential. A surprisingly large degree of ionization can arise at temperatures well below the characteristic temperature θ_H.

If all the free electrons come from the reaction (11.33), then the concentrations of electrons and H^+ are equal:

$$[e] = [H^+]. \tag{11.39}$$

In that case, equation (11.34) may be written as

$$\frac{[H^+]^2}{[H]} = K(T). \tag{11.40}$$

Furthermore, if the sum $[H] + [H^+]$ is known and also the temperature T, then that information plus equations (11.37) and (11.40) form an algebraically complete set from which $[H]$ and $[H^+]$ can be calculated by solving nothing worse than a quadratic equation. In any specific astrophysical context, however, one may be able to judge whether $[H]$ or $[H^+]$ dominates and then simplify accordingly.

For example, suppose $T = 7{,}000$ K and

$$[H] + [H^+] = 10^{-6} \times (2.5 \times 10^{25}) \text{ particles per cubic meter,}$$

that is, 10^{-6} times the number density of air under typical room conditions. One might guess that the gas would be overwhelmingly atomic hydrogen. Then one can solve for the ratio $[H^+]/[H]$ by dividing (11.40) on both sides by $[H]$, taking a square root, and approximating on the right-hand side:

$$\frac{[H^+]}{[H]} = \left(\frac{K(T)}{[H]}\right)^{1/2}$$

$$\cong \left(\frac{2.3 \times 10^{17}}{2.5 \times 10^{19}}\right)^{1/2} = 0.096.$$

Approximately 10 percent of the hydrogen is ionized—despite a temperature that is extremely low relative to θ_H.

The basic reason for the surprisingly large degree of ionization is this: ionization increases the number of free particles (here from 1 to 2). More free particles means more ways to place the particles in space, thus larger multiplicity, and hence larger entropy. The ionization is driven by an increase in entropy. In the next section, we shall see this entropic influence as a generic aspect of chemical equilibrium. Before turning to that, however, we need to consider again the partition function for the hydrogen atom and our approximation.

The sum over electronic states goes over increasingly higher energies: $\varepsilon_n = -13.6/n^2$, where n denotes the principal quantum number. Thus the Boltzmann factor,

$$\exp(-\varepsilon_n/kT) = \exp(+13.6/n^2 kT),$$

becomes successively smaller and approaches 1 as n goes to infinity. Because the terms remain finite and because there are infinitely many of them, the formal sum diverges. How do we extricate ourselves from this embarrassment? The size of the atom grows with n; the atomic radius is essentially n^2 times the Bohr radius. In a dilute gas, no atom can become larger than the typical separation between atoms and yet avoid ionization by its neighbors. Interaction with adjacent ions or atoms would strip off the distant electron, especially because it is bound so weakly. Thus the sum for the partition function extends only until the atomic size is comparable to the typical separation. Such a sum is necessarily finite in value. For the conditions of temperature and density in our example, the realistic sum increases our one-term approximation by less than 0.01 percent. This result implies, also, that almost all of the hydrogen atoms are still in the electronic ground state.

Figure 11.4 displays the run of fractional ionization over the temperature range 5,000 K to 12,000 K.

The Indian physicist Megh Nad Saha was the first to investigate such ionization equilibrium (in 1920), and he applied it to stellar atmospheres, including the solar chromosphere. Eliminating $K(T)$ between equations (11.34) and (11.37) produces an instance of the *Saha equation*:

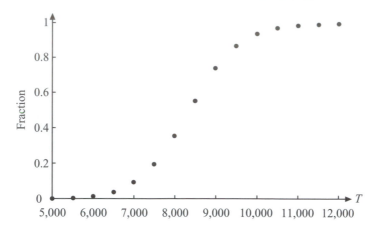

Figure 11.4 Fractional ionization of hydrogen. The fraction plotted is the ratio of ionized hydrogen to "all hydrogen:" $[H^+]/([H] + [H^+])$. The denominator has the numerical value already used in the text: $10^{-6} \times (2.5 \times 10^{25})$ particles per cubic meter. Note that the ionization fraction rises from virtually zero to almost one over a narrow temperature interval: about 4,000 K, relative to the 160,000 K of the characteristic temperature θ_H.

$$\frac{[H^+][e]}{[H]} = \lambda_e^{-3} e^{\varepsilon_{g.s.}/kT}$$

$$= \lambda_e^{-3} \exp(-|\text{ionization energy}|/kT). \tag{11.41}$$

The second line displays explicitly the dependence on the energy that is needed to ionize atomic hydrogen from its ground state. Invariably, a large ionization energy inhibits dissociation.

The general name for the effect that we have studied is *thermal ionization*. Saha focused especially on thermally ionized calcium but considered hydrogen and a host of other elements as well. His first paper, "Ionization in the solar chromosphere" (*Phil. Mag.* **40**, 472–88, 1920), is a landmark in stellar astrophysics.

11.5   Another facet of chemical equilibrium

To understand better what "drives" a chemical reaction, we return to two items: (1) the basic equilibrium condition, equation (11.10), which is

$$\sum_i \mu_i b_i = 0, \tag{11.42}$$

and (2) the generic chemical potential for the semi-classical ideal gas, expression (11.11), which is

$$\mu_i = -kT \ln(Z_{1i}/N_i). \tag{11.43}$$

Equation (11.14) showed that Z_{1i}, the partition function for a single molecule of species B_i, is proportional to the system's volume V. Consequently, we can write

$$Z_{1i} = V \times \left(\begin{array}{c} \text{partition function for a single molecule} \\ \text{of species } B_i \text{ in unit volume} \end{array} \right). \tag{11.44}$$

We use this factorization in (11.43):

$$\mu_i = +kT \ln (N_i/V) - kT \ln \left(\begin{array}{c} \text{partition function for a single molecule} \\ \text{of species } B_i \text{ in unit volume} \end{array} \right)$$

$$= +kT \ln (N_i/V) + \langle \varepsilon_i \rangle - T\tilde{s}_i. \tag{11.45}$$

Reference to equations (7.9) and (7.10) tells us that the second term in the first line is the Helmholtz free energy for one molecule when in unit volume. The Helmholtz free energy is always an energy minus T times an entropy. The symbol $\langle \varepsilon_i \rangle$ denotes the estimated energy of a molecule of species B_i, and $\tilde{s}_i$ is the entropy that a single molecule would have if present by itself in unit volume. Substituting the decomposition (11.45) of the chemical potential into equation (11.42) and rearranging much as we did before, one finds

$$\prod_i [B_i]^{b_i} = \exp \left[\frac{1}{kT} \times \sum_i b_i \left(- \langle \varepsilon_i \rangle + T\tilde{s}_i \right) \right]. \tag{11.46}$$

This equation is, of course, just another form of the law of mass action. The right-hand side is not especially suitable for computation—partition functions are better for that—but it lends itself admirably to interpretation. We proceed in two steps.

(1) The energy sum:

$$\sum_i b_i(- \langle \varepsilon_i \rangle). \tag{11.47}$$

If the products of the reaction, weighted by their stoichiometric coefficients, have a lower estimated energy than the reactants, again weighted, then the energy sum is positive, and the exponential will favor the products. The reaction "rolls down hill" toward the lower energy of the products. The energy sum is akin to the energy terms in equation (11.32).

(2) The entropy sum:

$$\sum_i b_i \tilde{s}_i. \tag{11.48}$$

If the products of the reaction, weighted by their stoichiometric coefficients, have a greater entropy than the reactants, again weighted, then the entropy sum is positive, and the exponential will favor the products. Because entropy is the logarithm of a multiplicity (times k), we find that larger multiplicity for the products favors them relative to the reactants.

For the hydrogen dissociation reaction, the entropy sum is

$$-\tilde{s}_H + \tilde{s}_{H^+} + \tilde{s}_e.$$

Because the masses for H and H^+ are nearly identical, their entropy terms cancel (for all practical purposes). The free electron's entropy term drives the equilibrium toward substantial ionization despite an unfavorable energy sum and the relatively low temperature.

In a loose but useful way of speaking, one may say that the energy sum represents the effect of forces and that the entropy sum represents the effect of multiple possibilities. The balance between these effects sets the chemical equilibrium.

11.6 Creation and annihilation

In the reactions that we have considered so far, the nuclei have not changed, nor has the number of electrons changed. Nuclei and electrons have merely been rearranged or freed. For calculating chemical potentials and partition functions, the zero of energy could be taken as the energy when the essential constituents are infinitely separated and at rest. This is both convenient and conventional. It does not, however, cover all cases.

In the extremely hot interior of stars, gamma rays can form an electron–positron pair,

$$\gamma + \gamma \rightleftharpoons e^- + e^+, \tag{11.49}$$

and the pair can mutually annihilate to form two gamma rays. Moreover, this process occurred ubiquitously in the early evolution of our universe. The general principles of section 11.2 apply here, too. If thermal equilibrium prevails, then the sum of the chemical potentials, weighted by the stoichiometric coefficients, must be zero:

$$-2\mu_\gamma + \mu_{e^-} + \mu_{e^+} = 0. \tag{11.50}$$

The novelty is that some particles are being created and others are being destroyed. How does the chemical potential acknowledge this?

Among the equivalent expressions for the chemical potential is the statement

$$\mu_i = \left(\frac{\partial E}{\partial N_i}\right)_{S,V,\text{other } Ns} \tag{11.51}$$

for species B_i. When particles are being created or destroyed, we need to include the energy associated with their very existence, that is, the energy associated with their rest mass: $m_i c^2$, where m_i denotes the rest mass of species B_i. The system's total energy E will have the form

$$E = \sum_i m_i c^2 N_i + \left(\begin{array}{c}\text{the usual kinetic and}\\ \text{potential energies}\end{array}\right). \tag{11.52}$$

The free energies will be augmented by the same sum over rest energies.

If species B_i acts like a semi-classical ideal gas, then the derivative process for calculating the chemical potential yields

$$\mu_i = m_i c^2 - kT \ln \left(Z_{1i}/N_i \right), \tag{11.53}$$

where the second term was derived in section 7.3 and confirmed in section 8.4. In this context, each Boltzmann factor in the single-particle partition function,

$$Z_{1i} = \sum_{\text{states } \varphi_\alpha} e^{-\varepsilon_\alpha/kT}, \tag{11.54}$$

contains the kinetic and potential energy only, *not* the energy associated with the rest mass. To be sure, the rest energy $m_i c^2$ in equation (11.53) can be tucked inside the logarithm, so that the equation becomes

$$\mu_i = -kT \ln \left\{ \frac{\sum\limits_{\alpha} \exp\left[-(m_i c^2 + \varepsilon_\alpha)/kT\right]}{N_i} \right\}, \tag{11.55}$$

where the exponent now contains the full single-particle energy. This form is good for showing the continuity with our previous calculations, but it is cumbersome, and so we use equation (11.53)

Returning now to the reaction in equations (11.49) and (11.50), we note that the chemical potential for the electron is

$$\mu_{e^-} = m_e c - kT \ln \left(\frac{2\lambda_{\text{th}}^{-3} V}{N_{e^-}} \right) \tag{11.56}$$

provided the electrons form a semi-classical ideal gas. The factor of 2 accounts for the two possible spin orientations. The chemical potential for the positrons will be the same except that the number density N_{e^+}/V replaces that for the electrons. The chemical potential for photons is zero, as we noted in section 10.3. Substitution into (11.50) and rearrangement yield

$$[e^-][e^+] = 4\lambda_{\text{th}}^{-6} \exp\left(-2m_e c^2/kT\right). \tag{11.57}$$

The core of a hot, massive star (of mass equal to 12 solar masses, say) may have a temperature of $T = 10^9$ K. The right-hand side of (11.57), which is a function of temperature only, then has the value 1.65×10^{65} m^{-6}. If $[e^-] = 3 \times 10^{33}$ m^{-3}, a reasonable value for such a star, then $[e^+] = 5.52 \times 10^{31}$ m$^{-3} = 0.018 \times [e^-]$. That is, there is one positron—however fleetingly—for every 50 electrons.

The chemical potential plays a vital role in astrophysics and cosmology. The major point of this section is to alert you: in those contexts, the rest energy $m_i c^2$ may need to appear explicitly in the chemical potential.

11.7 Essentials

1. The generic chemical reaction takes the form

$$b_1 B_1 + b_2 B_2 + \cdots + b_n B_n = 0,$$

where each molecular species is represented by a symbol B_i and where the corresponding numerical coefficient in the reaction equation is represented by the symbol b_i. The coefficients $\{b_i\}$ are called *stoichiometric coefficients*. For the products of a reaction, the coefficients b_i are positive; for the reactants, they are negative.

2. The minimum property of the free energies implies

$$\sum_i \mu_i b_i = 0$$

when the chemical reaction has come to equilibrium.

3. For semi-classical ideal gases, item 2 implies the relations

$$\prod_i [B_i]^{b_i} = K(T)$$

and

$$K(T) \equiv \prod_i \{\lambda_i^{-3} Z_i(\text{int})\}^{b_i},$$

where $K(T)$ is called the *equilibrium constant*. The symbol $[B_i]$ denotes the concentration of species B_i. The *internal partition function* $Z(\text{int})$ is

$$Z(\text{int}) = \sum_{\substack{\text{internal states} \\ \text{of one molecule}}} e^{-\varepsilon_\alpha(\text{int})/kT},$$

where $\varepsilon_\alpha(\text{int})$ is the full single-molecule energy minus the translational kinetic energy of the center of mass.

4. The internal energy of a diatomic molecule may be decomposed as

$$\varepsilon(\text{int}) = \begin{pmatrix} \text{energy when nuclei are} \\ \text{at rest at minimum in} \\ \text{interatomic potential energy} \end{pmatrix} + \begin{pmatrix} \text{vibrational} \\ \text{energy} \end{pmatrix} + \begin{pmatrix} \text{rotational} \\ \text{energy} \end{pmatrix}$$

$$= \Delta\varepsilon + (n + \tfrac{1}{2})h\nu + J(J+1)\frac{\hbar^2}{2I}.$$

5. A diatomic molecule has two characteristic temperatures:

- vibrational temperature θ_v: $k\theta_v \equiv h\nu$,
- rotational temperature θ_r: $k\theta_r \equiv \hbar^2/2I$.

6. When the temperature satisfies the strong inequalities $\theta_r \ll T \ll \theta_v$, the internal partition function takes the form

$$Z(\text{int}) = e^{-\varepsilon_{\text{g.s.}}/kT} \times 1 \times \begin{pmatrix} \text{electronic spin} \\ \text{degeneracy factor} \end{pmatrix}$$

$$\times (2 \times \text{spin}_1 + 1)(2 \times \text{spin}_2 + 1) \times \left(\frac{T}{\theta_r}\right) \times \begin{Bmatrix} 1 \text{ if heteronuclear} \\ \tfrac{1}{2} \text{ if homonuclear} \end{Bmatrix}.$$

7. The ideas of chemical equilibrium can be applied to *thermal ionization*, the equilibrium of electrons, ions, and neutral atoms or molecules. The fractional ionization (at fixed concentration of the nuclei) switches from almost zero to almost one over a relatively short temperature interval.

8. Just where a chemical reaction reaches equilibrium can be viewed as a consequence of a competition between reactants and products, a competition with respect to both energy and entropy.

If the products of the reaction, weighted by their stoichiometric coefficients, have a lower estimated energy than the reactants, again weighted, then the reaction "rolls down hill" toward the lower energy of the products.

If the products of the reaction, weighted by their stoichiometric coefficients, have a greater entropy than the reactants, again weighted, then the larger multiplicity for the products favors them relative to the reactants.

9. When particles are created or annihilated, one must include the rest energy in the chemical potential:

$$\mu = mc^2 - kT \ln(Z_1/N),$$

where the second term on the right is the usual expression (based on the usual kinetic and potential energies).

Further reading

A fine discussion of thermal ionization in stellar physics is provided by Richard L. Bowers and Terry Deeming, *Astrophysics I: Stars* (Jones and Bartlett, Boston, 1984), section 6.2.

G. H. Nickel discusses the reasons why thermal ionization occurs and provides an alternative derivation of the Saha equation in his article, "Elementary derivation of the Saha equation," *Am. J. Phys.* **48**, 448–50 (1980). (In essence, his derivation is based on a probability maximization like the analysis in our section 7.1. Except for the language, Nickel's derivation is equivalent to ours, which explicitly uses the notion of a chemical potential.)

Problems

1. *How the equilibrium constant varies.* Compute the temperature derivative $d \ln K(T)/dT$ of the equilibrium constant $K(T)$, defined in (11.20). Relate your result to the estimated energy per molecule $\langle \varepsilon_i \rangle$ of the various reactants and products in the chemical reaction. Describe the relationship in words, too. If the reaction, taken as a whole, requires an input of energy, does the equilibrium constant increase or decrease when the ambient temperature increases?

2. *Other versions of the law of mass action.*

(a) Eliminate the variables $[B_i]$ in terms of the *partial pressures* $P_i \equiv N_i kT/V$. Determine the new equilibrium constant in terms of $K(T)$ and kT.
(b) Eliminate the variables $[B_i]$ in terms of the *fractional concentrations* c_i, defined by $c_i \equiv [B_i]/\sum_j[B_j]$, and the total gas pressure P. When the left-hand side of equation (11.18) is expressed as a function of the set $\{c_i\}$ only, what is the new equilibrium constant? And on which variables does it depend?

3. *Heat capacity of diatomic molecules.* Specify that the temperature satisfies the strong inequalities $\theta_r \ll T \ll \theta_v$. Use partition functions.

(a) Calculate the estimated energy $\langle E \rangle$, the entropy S, and the heat capacities C_V and C_P for a semi-classical ideal gas of N diatomic molecules. Supply additional data as needed.
(b) Compare the ratio of heat capacities, C_P/C_V, with the corresponding ratio for a monatomic gas.

4. *Rotational energy.* For a diatomic molecule formed from two different nuclei, the rotational contributions to $\langle E \rangle$ and to C_V come from the "rotational partition function" Z_{rot}, defined by equation (11.28).

(a) Compute and plot the dimensionless quantities Z_{rot}, $\langle E_{rot} \rangle / k\theta_r$, and C_{rot}/k as functions of T/θ_r for the range $0 \leqslant T/\theta_r \leqslant 3$.

(b) How large must the ratio T/θ_r become in order that C_{rot}/k be at least 90 percent of its asymptotic value?

5. The oxygen isotope ^{16}O has a nuclear spin of zero. When a diatomic molecule is formed with this isotope, which values of the rotational angular momentum J are permitted? Consequently, in a sum over J like that in (11.25), what are the nuclear spin degeneracy factors for even J and for odd J?

6. For the reaction $H_2 + I_2 \rightleftharpoons 2\,HI$, determine the equilibrium constant when a flask of the gases is immersed in a bath of boiling water. For the common isotope of iodine, the nuclear spin is $\frac{5}{2}\hbar$. Give the equilibrium constant both analytically and numerically. Specify and justify any approximations that you make.

7. *Dissociation of molecular hydrogen.* Consider the reaction $H_2 \rightleftharpoons 2H$ under the stipulation that $[H] + 2[H_2] \equiv A = 10^{-10} \times 2.5 \times 10^{25}$ particles per cubic meter, where 2.5×10^{25} is the number density of air under typical room conditions. The electron spins in molecular hydrogen are anti-parallel, and so the net electronic spin is zero.

(a) Determine the equilibrium constant $K(T)$ and graph its logarithm (using base 10) for the temperature range $1,000 \leqslant T \leqslant 2,000$ K. (A point every 100 K will suffice.)

(b) Compute the dissociation fraction $f = [H]/([H] + [H_2])$ for the same temperature range and graph it also. [Doing some algebra by hand may be useful. You can express f as a simple function of $K(T)/A$.]

(c) Compare the temperature that gives $f = 0.8$ with the characteristic molecular temperature defined by $|\text{dissociation energy}|/k$.

8. *The population distribution in rotational states.* An experimenter needs to tune a laser so that the photons will induce a transition from the most probable rotational state to some fixed state of much higher energy. (Selecting the most probable state gives a large signal.) Take the heteronuclear diatomic molecule to be gaseous carbon monoxide at $T = 10^3$ K. Let

$$P(J) = \left(\begin{array}{c} \text{probability that molecule has} \\ \text{rotational energy } J(J+1)\hbar^2/2I \end{array} \right).$$

(a) Determine the most probable value of the rotational quantum number J.

(b) Plot $P(J)$ for the range $0 \leqslant J \leqslant 40$.

(c) Estimate the width of the probability distribution $P(J)$.

9. *Equal rotational populations.* A gas of the heteronuclear molecule hydrogen bromide (HBr) is in thermal equilibrium. At what temperature will the population of molecules with $J = 3$ equal the population with $J = 2$? Here $J(J + 1)\hbar^2$ specifies the magnitude of the rotational angular momentum, squared, as described in section 11.3.

10. *On the mechanism of thermal ionization.* The actual dissociation (or recombination) process may require the intervention of another atom, as in the reaction

$$2H \rightleftharpoons H^+ + e + H, \tag{a}$$

or the intervention of a photon, as in

$$H + \text{ photon } \rightleftharpoons H^+ + e. \tag{b}$$

In particular, conservation of energy and momentum may require such extra, "silent" partners.

Why would the more explicit reactions, (a) and (b), lead to the same results as our more abbreviated reaction? Provide a *detailed* response for each reaction separately, but they can be succinct.

11. *Saha equation for a generic atom.*

(a) Work out the analog of equation (11.41) for a multi-electron atom and its first stage of ionization:

$$\text{atom} \rightleftharpoons \text{atom}^+ + e.$$

Couch your result in terms of two internal partition functions, $Z_{atom}(\text{int})$ and $Z_{ion}(\text{int})$, and other, familiar quantities. Describe explicitly the situation that you take for the zero of energy, and describe (relative to that zero) the energies that enter the two partition functions.
(b) Display the form to which your equation reduces if you approximate each partition function by its ground-state term.

12. *Thermal ionization of helium.* The ground state of neutral helium is 79.00 eV below the state of *fully* ionized helium: He^{++}. Already the Bohr theory enables you to work out the energy of singly ionized helium.

(a) Determine the equilibrium constant algebraically for the ionization of neutral helium to its one-electron ion: He^+. Assume that the temperature is below 15,000 K.
(b) Evaluate both the exponential and the entire equilibrium constant numerically for $T = 10^4$ K.

13. *Solar chromosphere.* Many elements contribute (by thermal ionization) to the electron density in the solar chromosphere. Where the chromospheric temperature is

7,000 K, an approximate value for the total electron density is $[e] = 1.5 \times 10^{20}$ electrons per cubic meter.

(a) Calculate the ionization ratio $[H^+]/[H]$ for hydrogen.
(b) The ground state of atomic calcium is 6.113 eV below the ground state of singly ionized calcium. The electronic ground state of the atom is not degenerate, but the ion has an electron spin degeneracy factor of 2. Calculate the ionization ratio $[Ca^+]/[Ca]$.

14. *Positrons in white dwarfs*. In thermal equilibrium, the reaction $\gamma + \gamma \rightleftharpoons e^- + e^+$ produces a steady-state concentration of positrons. Calculate the concentration of positrons, $[e^+] \equiv N_{e^+}/V$, in a typical white dwarf (one whose electrons are non-relativistic). Recalling the connection between the electron Fermi energy at $T = 0$ and the chemical potential may help you.

15. *Electron–positron pairs*. If pure blackbody radiation (devoid of any electrons or other particles) is compressed adiabatically (by an implosion, say), then high temperatures can be achieved, and electron–positron pairs will form spontaneously (from the interaction of gamma rays). Specify that a temperature of $T = 10^8$ K is achieved. *Derive* an expression for the number density of electrons; then evaluate it numerically.

12 Phase Equilibrium

12.1 *Phase diagram*

12.2 *Latent heat*

12.3 *Conditions for coexistence*

12.4 *Gibbs–Duhem relation*

12.5 *Clausius–Clapeyron equation*

12.6 *Cooling by adiabatic compression (optional)*

12.7 *Gibbs' phase rule (optional)*

12.8 *Isotherms*

12.9 *Van der Waals equation of state*

12.10 *Essentials*

In a loose sense, this chapter is about the coexistence of solids, liquids, and gases, usually taken two at a time. What physical condition must be met if coexistence is to occur? What other relations follow? The chapter provides some answers. And in its last section, it develops a classic equation—the van der Waals equation of state—that was the first to describe gas and liquid in coexistence.

12.1 Phase diagram

Figure 12.1 displays three phases of water as a function of pressure and temperature. By the word *phase*, one means here a system or portion of a system that is spatially homogenous and has a definite boundary. [In turn, homogeneity means that the chemical composition (including relative amounts), the crystalline structure (if any), and the mass density are uniform in space. A *continuous* variation produced by gravity, however, is allowed, as in the case of a finite column of air in the Earth's gravitational field.] The liquid phase is typified by a glass of water; the solid, by an ice cube; and the vapor, by the "dry" steam (that is, steam without water droplets) in the turbine of a power plant, as was illustrated in figure 3.1.

Of more interest, however, are the curves in the drawing, locations where two phases coexist. When a pond freezes over and develops a 20 centimeter layer of ice, the lower surface of the ice and the top of the remaining liquid water coexist at a temperature and pressure that lie along the leftward (and upward) tilted curve that emanates from the triple point. The locus of points where solid and liquid coexist is called the *melting curve* (or the *fusion curve*).

If one were to place a lot of ice in a container, pump out the air, and then seal the

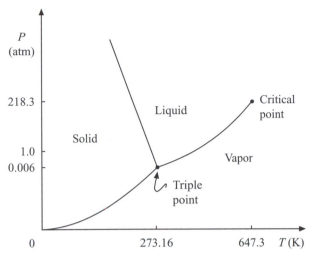

Figure 12.1 The phase diagram for water: the pressure-temperature version. Pressures are cited in atmospheres (where 1 atmosphere = 1.013×10^5 N/m^2, the standard value for atmospheric pressure at sea level); temperatures are given in kelvin. The scales, however, are not linear; rather, they have been compressed or expanded to display the more interesting regions. Nonetheless, the curves are qualitatively faithful.

Water has at least ten distinct solid phases: various kinds of ice that differ in their crystalline structure. Only one solid phase—common freezer ice—exists at pressures below 2,000 atmospheres, and only that ice is shown.

vessel, doing all this at any temperature below the triple point, then some water molecules would escape from the ice's surface and form a gaseous phase in equilibrium with the remaining ice. This evaporative process, in which no liquid phase occurs, is called *sublimation*. The curve running from the origin to the triple point is the *sublimation curve*. [By the way, you may have noticed that ice cubes, dumped out of a tray and lying loose in a bowl in the freezer, slowly diminish in size if left there a long time. The ice is subliming—but never reaching a coexistence with its vapor (because too much vapor leaks out of the freezer compartment). The frost on the backyard lawn can provide another example. If the morning sunshine is intense and if the air temperature remains below freezing, then the frost disappears by sublimation.]

Along the curve that connects the triple point and the critical point, liquid and vapor coexist. That curve is called the *vaporization curve*. The end points are special. At the *triple point*, of course, all three phases coexist simultaneously. Only for temperatures below the *critical temperature* T_c (the temperature of the critical point) can a meaningful distinction be made between a vapor phase and a liquid phase. The volume of a vapor is determined by the size of the container; a gas expands to fill the volume available to it. Not so a liquid, whose volume is determined by temperature, pressure, and number of molecules. If one starts at the triple point on the vaporization curve, the density of the liquid phase (10^3 kg/m^3) greatly exceeds the vapor density (5×10^{-3} kg/m^3). Traveling along the vaporization curve, one finds that the liquid density decreases (as the higher temperature causes the liquid to expand, despite the increasing pressure). Simultaneously, the vapor density increases (as the higher pressure compresses the vapor). As one approaches the critical point, the two densities converge to a single value: 400 kg/m^3. At temperatures higher than the critical temperature, water exists only as a "fluid" that has the volume-filling character of a gas. The microscopic kinetic energy is so large that the attractive forces between molecules cannot form droplets, much less a large-scale cohesive liquid phase.

Indeed, to jump ahead a bit, take a look at figure 12.8, which displays a typical intermolecular potential energy curve and its attractive potential well. An empirical rule of thumb is that $kT_c \cong$ (well depth), a relationship good to 20 percent for helium, hydrogen, and nitrogen. The interpretation is this: if the molecular translational kinetic energy, $\frac{3}{2}kT$, exceeds the well depth, then the molecule will not be bound by the attractive force exerted by another molecule.

To summarize, the *critical point* is a limit point along the vaporization curve: below the temperature of the critical point, liquid and vapor coexist at distinct, unequal densities; at and above the critical temperature, there is only a single "fluid" phase.

Table 12.1 provides experimental data on the critical point and triple point of some common substances.

A line drawn horizontally at a pressure of one atmosphere would intersect the melting curve at 273.15 K, the *normal melting point* of ice, slightly below the temperature of the triple point. The same line would intersect the vaporization curve at 373.12 K, the *normal boiling point* of water. In general, boiling occurs when an originally cool liquid has been heated sufficiently that its vapor pressure equals the ambient pressure (so that tiny bubbles can expand and grow). Thus, whenever boiling occurs, it does so at some point on the vaporization curve. If you have ever camped and cooked at high altitude, at 3,000 meters, say, you know that there the temperature of boiling water is sufficiently lower than 373 K that the time required to cook rice is perceptibly longer than at sea level. An altitude of 3,000 meters corresponds to a

Table 12.1 *Data for the critical point and triple point. The volume per molecule at the critical point, $(V/N)_c$, is cited in units of 10^{-30} m^3 (1 cubic angstrom).*

Substance	T_c (K)	P_c (atm)	$(V/N)_c$ (10^{-30} m^3)	$T_{t.p.}$ (K)	$P_{t.p.}$ (atm)
Water	647	218	91.8	273.16	0.0060
Sulfur dioxide	431	77.8	203	200	0.02
Carbon dioxide	304	72.9	156	217	5.11
Oxygen	155	50.1	130	54.8	0.0026
Argon	151	48.0	125	83.8	0.68
Nitrogen	126	33.5	150	63.4	0.127
Hydrogen	33.2	12.8	108	14.0	0.0712

Sources: *AIP Handbook*, 3rd edn, edited by D. E. Gray (McGraw-Hill, New York, 1972), and others.

pressure of approximately 0.7 atmosphere. A line drawn at that pressure intersects the vaporization curve closer to the triple point, at approximately 363 K, a reduction by 10 K.

12.2 Latent heat

When the tea kettle whistles on the stove, the burner is supplying energy to the water. That energy enables some molecules in the liquid phase to escape the attractive forces of their fellow molecules and to enter the vapor phase. The process occurs at constant temperature ($T = 373$ K) and constant pressure ($P = 1$ atm). In short, the process occurs at a *point* on the vaporization curve of figure 12.1.

For vaporization in general, the amount of energy that must be supplied by heating is called the *latent heat of vaporization* and is denoted by L_{vap}. For one molecule, we write

$$L_{vap} \equiv \left(\begin{array}{l} \text{energy input by heating to promote one molecule} \\ \text{from the liquid to the vapor at constant } T \text{ and } P \end{array} \right). \qquad (12.1)$$

A significant distinction will become evident if we use the First Law of Thermodynamics to express the latent heat in terms of the energies per molecule in the liquid and vapor phases, together with whatever other quantities are required. For each phase separately, let

$$\varepsilon \equiv E/N = \text{average energy per molecule,}$$

$$\qquad (12.2)$$

$$v \equiv V/N = \text{volume per molecule.}$$

By definition, the latent heat of vaporization is the amount of energy *supplied by heating*, and so it corresponds to the variable q in the First Law. Upon referring to equation (1.13) for the First Law, we find that

$$L_{vap} = \Delta\varepsilon + P\Delta v, \qquad (12.3)$$

where

$$\Delta\varepsilon = \varepsilon_{vap} - \varepsilon_{liq} \text{ and } \Delta v = v_{vap} - v_{liq}. \qquad (12.4)$$

Because the volume per molecule is larger in the vapor than in the liquid, the difference Δv is positive. Some of the energy input goes into expanding the entire two-phase system against the external pressure P. Only a certain fraction of the energy supplied by heating goes into $\Delta\varepsilon$, the change in the molecule's energy. Thus L_{vap} and $\Delta\varepsilon$ are distinct quantities (although obviously related).

Table 12.2 *Latent heats for various substances. For the vaporization data, the ambient pressure is atmospheric pressure. Carbon dioxide never exists as a liquid at atmospheric pressure (because its triple point lies at a pressure of 5.11 atmospheres). For CO_2, the latent heat of sublimation at atmospheric pressure is 0.26 eV/molecule (and occurs at $T = 195$ K). The melting temperatures and corresponding latent heats are associated with a medley of different pressures.*

Substance	Melting point (K)	L_{fusion} (eV/molecule)	Boiling point (K)	$L_{vaporization}$ (eV/molecule)
Carbon dioxide	217	0.086		
Chlorine (Cl_2)	172	0.067	239	0.21
Helium (^4He)	1.76	8.7×10^{-5}	4.2	0.00087
Iron	1,810	0.14	3,140	3.6
Mercury	234	0.024	630	0.61
Nitrogen (N_2)	63	0.0075	77	0.058
Silver	1,230	0.12	2,440	2.6
Tungsten	3,650	0.37	5,830	8.5
Water	273	0.062	373	0.42

Source: *AIP Handbook*, 3rd edn, edited by D. E. Gray (McGraw-Hill, New York, 1972).

Precisely because vaporization occurs at constant pressure, equation (12.3) may be rearranged as

$$L_{vap} = \Delta(\varepsilon + Pv). \tag{12.5}$$

For any phase, the quantity $E + PV$ is called the *enthalpy* (from the Greek root, *thalpein*, "to heat"). Thus one may say that L_{vap} equals the change in enthalpy per molecule.

The discussion so far has focused on vaporization. Analogous definitions and relationships hold (1) for melting, the solid to liquid transition, which occurs at a point on the melting (or fusion) curve and (2) for sublimation, the solid to vapor transition, which occurs at a point on the sublimation curve.

Table 12.2 provides values of the latent heat for some common substances.

Latent heat versus heat capacity

Another distinction is worth pointing out: the distinction between latent heat and heat capacity. To begin with, a latent heat characterizes a process *at constant temperature*, but a heat capacity describes a process in which the *temperature changes* (at least infinitesimally). To display the latter feature, we recapitulate equation (1.14):

$$C_X \equiv \left(\begin{array}{c} \text{heat capacity} \\ \text{under conditions } X \end{array} \right) \equiv \frac{\left(\begin{array}{c} \text{energy input by heating} \\ \text{under conditions } X \end{array} \right)}{(\text{ensuing change in temperature})}, \tag{12.6}$$

where X may denote constant volume or constant pressure (or, for a magnetic system, constant external magnetic field). The ratio on the right-hand side presupposes a change in temperature.

Indeed, the historical origin of the adjective "latent" reflects the thermal distinction. Under most circumstances, if one supplied energy by heating, the system's temperature increased. But under certain circumstances, such as melting, the system's temperature remained the same: the effect of supplying energy remained "latent" (so far as temperature was concerned). Today, of course, we know that the energy supplied by heating goes into the subprocesses that we describe with $\Delta\varepsilon$ and $P\Delta v$.

To display the distinction further, we express latent heat and heat capacity in terms of entropy changes. Taking the processes to occur slowly and hence to be reversible, we invoke the Second Law as

$$\Delta S = \frac{q}{T}.$$ (12.7)

The definition of latent heat in equation (12.1) implies that L_{vap} corresponds to q, and so equation (12.7) implies

$$L_{\text{vap}} = T \times (s_{\text{vap}} - s_{\text{liq}}),$$ (12.8)

where $s \equiv S/N$ for each phase.

[Note that the symbol s, as used here and subsequently in this chapter, denotes a ratio: (the total entropy of N molecules)$/N$. Thus the symbol s here differs subtly but significantly from the symbol $\tilde{s}_i$ used in section 11.5 to denote the entropy that a single molecule of species i would have if present *by itself* in unit volume. The lower case letter s has good mnemonic value, and so it has to do double duty. Indeed, because s is used also for "spin" (in units of $\hbar$), the letter does triple duty.]

For heat capacity, equation (12.6) tells us that $C_X\Delta T$ corresponds to q. Then equation (12.7) implies

$$(\Delta S)_X = \frac{C_X\Delta T}{T},$$ (12.9)

where $(\Delta S)_X$ is the entropy change under conditions X. Multiply both sides by $T/\Delta T$ and then take the limit as ΔT approaches zero:

$$C_X = T\left(\frac{\partial S}{\partial T}\right)_X.$$ (12.10)

Here we see that heat capacity is given by a *rate of change with temperature*. In contrast, the latent heat in (12.8) is proportional to a finite change in entropy.

By the way, equation (12.10) does not replace any of our previous expressions for a heat capacity, such as $C_V = (\partial\langle E\rangle/\partial T)_V$. Rather, the expression for C_X in terms of an entropy derivative is an alternative form, particularly useful if one happens to know the system's entropy.

12.3 Conditions for coexistence

The boundaries in the phase diagram, figure 12.1, are intriguing: what determines their location or slope? The chemical potential provides some answers, as this section and section 12.5 demonstrate. For a thorough start, however, we return to the Gibbs free energy and its minimum property at fixed T and P, developed in section 10.5.

Consider a point in the P–T plane that is on the vaporization curve, that is, a point where liquid and vapor coexist. The minimum property of the Gibbs free energy implies that G must not change if we imagine transferring a molecule of water from the liquid phase to the vapor phase. That is, the equation

$$\Delta G = \frac{\partial G}{\partial N_{\text{vap}}} \Delta N_{\text{vap}} + \frac{\partial G}{\partial N_{\text{liq}}} \Delta N_{\text{liq}}$$

$$= \mu_{\text{vap}} \times (1) + \mu_{\text{liq}} \times (-1) = 0 \tag{12.11}$$

must hold. The partial derivatives are taken at constant values of T, P, and the numbers of molecules in other phases. Thus, by equation (10.31), the derivatives are the corresponding chemical potentials. Rearrangement puts equation (12.11) into the succinct form

$$\mu_{\text{vap}}(T, P) = \mu_{\text{liq}}(T, P). \tag{12.12}$$

This relationship determines the vaporization curve, but more detail may be welcome and comes next.

Each phase is formed from water molecules only and hence consists of only one species of particle. Consequently, the chemical potential for each phase depends on the intensive variables T and P only. The functional dependence, however, definitely does change from one phase to the next, being different for a liquid from what it is for a vapor. A simple theoretical model will make this clear, as follows.

A model for the vaporization curve

The semi-classical partition function for a structureless ideal gas is

$$Z_{\text{gas}}(N) = \frac{(V \lambda_{\text{th}}^{-3})^N}{N!}. \tag{12.13}$$

Via $F_{\text{gas}} = -kT \ln Z_{\text{gas}}$, the partition function leads to the chemical potential for the vapor phase:

$$\mu_{\text{vap}} = \frac{\partial(-kT \ln Z_{\text{gas}})}{\partial N} = -kT \ln \left(\frac{V/N}{\lambda_{\text{th}}^3} \right)$$

$$= -kT \ln \left(\frac{kT/P}{\lambda_{\text{th}}^3} \right). \tag{12.14}$$

The last step uses the ideal gas law to eliminate the number density in terms of P and kT. Of course, water molecules are not structureless—the hydrogen and oxygen atoms can rotate and vibrate about the molecular center of mass—but a structureless gas provides a tractable model and captures the essentials. We pursue it.

[Note. The Gibbs free energy G appeared in equation (12.11) because our context is fixed P and T. Therefore we need to start with the free energy that has a minimum, at equilibrium, in the context of fixed P and T, namely, the Gibbs free energy. To calculate the chemical potentials that appear in equation (12.12), however, we may use whichever energy is most convenient. The partition function readily gives us the Helmholtz free energy F, and so we use that energy to compute μ. Knowing several equivalent expressions for the chemical potential is handy; indeed, it is essential.]

Because a liquid is largely incompressible, we model its partition function by replacing a container volume V with the product $N v_0$, where v_0 is a fixed volume of molecular size and N is the number of molecules in the liquid phase. An attractive force (of short range) holds together the molecules of a liquid and establishes a barrier to escape. We model the effect of that force by a potential well of depth $-\varepsilon_0$, saying that a molecule in the liquid has potential energy $-\varepsilon_0$ relative to a molecule in the vapor phase. The parameter ε_0 is positive. Thus the approximate partition function for the liquid is

$$Z_{\text{liq}}(N) = \frac{[(N v_0) \lambda_{\text{th}}^{-3} e^{\varepsilon_0/kT}]^N}{N!}. \tag{12.15}$$

The associated chemical potential follows as

$$\mu_{\text{liq}} = \frac{\partial(-kT \ln Z_{\text{liq}})}{\partial N} = -kT \ln \left(\frac{e v_0 e^{\varepsilon_0/kT}}{\lambda_{\text{th}}^3} \right). \tag{12.16}$$

[The isolated factor of e that multiplies v_0 arises from the differentiation of $N \ln(N v_0)$ and then a return to logarithmic form.] Manifestly, the chemical potentials for the vapor and the liquid have different functional forms.

The condition for coexistence, equation (12.12), becomes

$$-kT \ln \left(\frac{kT/P}{\lambda_{\text{th}}^3} \right) = -kT \ln \left(\frac{e v_0 e^{\varepsilon_0/kT}}{\lambda_{\text{th}}^3} \right).$$

The arguments of the logarithms must be equal, and so the relationship

$$P = \frac{kT}{ev_0} e^{-\varepsilon_0/kT} \tag{12.17}$$

gives the pressure as a function of temperature along the vaporization curve.

To be sure, equation (12.17) is the outcome of a simple model, but an exponential dependence on the reciprocal of temperature is characteristic of experimental vaporization curves. Moreover, we shall see a similar result emerge from a more phenomenological approach later, in section 12.5.

Return to generality

Now we return to general expressions, independent of any model. Equation (12.12) determines the vaporization curve—in the precise sense that it provides one constraint on the two variables, T and P, and thus determines a curve in the $P-T$ plane. Similar reasoning about a minimum in the Gibbs free energy would generate an analogous equation for the melting curve:

$$\mu_{\text{liq}}(T, P) = \mu_{\text{sol}}(T, P). \tag{12.18}$$

Again the functional dependencies will differ, and the equation will determine a curve in the $P-T$ plane.

The coexistence of solid, liquid, and vapor requires a minimum in the Gibbs free energy with respect to all possible imagined transfers of a water molecule: liquid to vapor, solid to liquid, and solid to vapor. Thus equations (12.12) and (12.18) must hold—and their validity already ensures the equality of chemical potentials for solid and vapor. Thus only two independent equations arise, and they constrain two variables: T and P. The solution will be the intersection of the vaporization curve and the melting curve: a point. Thus coexistence of the three phases occurs at a triple *point*.

To be sure, one would be hard pressed to imagine anything else geometrically. Later, in section 12.7, we will consider more than three phases, for example, the many phases of ice as well as the liquid and vapor phases of water. Nonetheless, we will find that no coexistence greater than a triple point can arise (when only one species of particle exists in the entire system). Thermodynamics discloses remarkable constraints on the behavior of real systems.

For a synopsis, let us note that

> *coexistence of two phases requires that their chemical potentials be equal.*

All the rest flows from that succinct statement.

Another way to state the situation is this:

> *at thermal equilibrium, the chemical potential has the same numerical value everywhere in the system.*

Why? Because if it did not, then one could reduce the Gibbs free energy by transferring molecules from regions of high chemical potential to the region of lowest chemical potential. But a characteristic of thermal equilibrium is that the Gibbs free energy has already attained its minimum value (when temperature and pressure are held fixed). Given the statement in italics, the chemical potential of a phase must be numerically the same as that of any other coexisting phase.

Here is yet another way to understand the equality of chemical potentials for coexisting phases, a way that emerges from the very meaning of the chemical potential. Section 7.1 noted that the chemical potential measures the tendency of particles to diffuse. *Coexistence* of phases imples no net diffusion from one phase to the other, and so—of course—the chemical potentials must be equal.

Before we go on to derive a general equation for the slope of a coexistence curve, we need a lemma concerning the chemical potential. The next section develops it.

12.4 Gibbs–Duhem relation

When only a single species of molecule is present, the Gibbs free energy has the structure

$$G = \mu N, \tag{12.19}$$

and the chemical potential is a function of the intensive variables T and P only:

$$\mu = \mu(T, P). \tag{12.20}$$

Section 10.4 developed these expressions. In this context, what physical significance do the derivatives of μ with respect to T and P have?

To answer that question, we compare two expressions for the total differential of G. Differentiating (12.19) gives

$$\Delta G = \left(\frac{\partial \mu}{\partial T}\Delta T + \frac{\partial \mu}{\partial P}\Delta P\right) \times N + \mu \Delta N.$$

Now compare this equation with ΔG as expressed in equation (10.28). The coefficients of corresponding differentials (such as ΔT) must be equal. Thus we find the relations

$$\frac{\partial \mu}{\partial T} = -\frac{S}{N} = -s, \tag{12.21}$$

$$\frac{\partial \mu}{\partial P} = \frac{V}{N} = v. \tag{12.22}$$

The lower case letters, s and v, denote the entropy and volume per molecule, respectively. They provide the physical significance of the derivatives.

Tersely stated, the differential of the chemical potential is

$$\Delta\mu = -s\Delta T + v\Delta P. \tag{12.23}$$

This equation is called the *Gibbs–Duhem relation*. The chemical potential is an intensive variable, and so its derivatives with respect to T and P must also be intensive. The entropy and volume per molecule are intensive variables, as befits their role.

12.5   Clausius–Clapeyron equation

Now we are ready to compute the slope of a coexistence curve. For definiteness, take the vaporization curve, but you will recognize that the method of derivation works for any other coexistence curve in figure 12.1. Figure 12.2 sets the scene. At point A on the vaporization curve, the chemical potential of vapor and liquid are equal:

$$\mu_{\text{vap}}(T_A, P_A) = \mu_{\text{liq}}(T_A, P_A). \tag{12.24}$$

The two potentials are equal at the adjacent point B also:

$$\mu_{\text{vap}}(T_A + \Delta T, P_A + \Delta P) = \mu_{\text{liq}}(T_A + \Delta T, P_A + \Delta P). \tag{12.25}$$

Both μ_{vap} and μ_{liq} may have changed in numerical value, but—if so—then necessarily they changed by the same amount. Subtract equation (12.24) from (12.25) and use the Gibbs–Duhem relation, (12.23), to find the equation

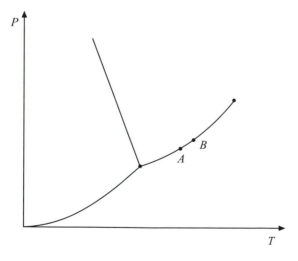

Figure 12.2 The vaporization curve and two adjacent points on it. The differences in the independent variables are defined by the statements $T_B = T_A + \Delta T$ and $P_B = P_A + \Delta P$.

$$-s_{\text{vap}}\Delta T + v_{\text{vap}}\Delta P = -s_{\text{liq}}\Delta T + v_{\text{liq}}\Delta P.$$

Solving for the ratio $\Delta P/\Delta T$ and then passing to the limit (as point B approaches point A), one finds

$$\frac{dP}{dT} = \frac{s_{\text{vap}} - s_{\text{liq}}}{v_{\text{vap}} - v_{\text{liq}}}. \tag{12.26}$$

The ratio on the right-hand side refers to values at the point on the coexistence curve where the slope dP/dT is to be determined.

The entropy difference in the numerator can be re-expressed in terms of the latent heat of vaporization, L_{vap}, and the temperature. Reference to equation (12.8) yields

$$\frac{dP}{dT} = \frac{L_{\text{vap}}}{(v_{\text{vap}} - v_{\text{liq}})T}. \tag{12.27}$$

This equation is called the *Clausius–Clapeyron equation*. (The French mining engineer Émile Clapeyron derived the equation, in a less explicit form, on the erroneous basis of the old caloric theory. In 1850, Rudolf Clausius provided the first wholly legitimate derivation. In the same paper, Clausius gave the earliest statement of the Second Law of Thermodynamics. At the time, he was 28 years old.)

A simple model again

To see what kind of a coexistence curve the Clausius–Clapeyron equation yields, we construct again a simple model. The ideal gas law gives the volume per molecule in the vapor phase as

$$v_{\text{vap}} = \frac{V}{N} = \frac{kT}{P}. \tag{12.28}$$

The volume per molecule in the liquid is much smaller (except near the critical point); so we drop v_{liq} relative to v_{vap}. The latent heat of vaporization (per molecule) should correspond to the well depth parameter ε_0 of section 12.3 and should be approximately constant over modest sections of the vaporization curve. So we simplify equation (12.27) to

$$\frac{dP}{dT} = \frac{L_{\text{vap}}}{(kT/P)T}, \tag{12.29}$$

where L_{vap} is taken as constant. This differential equation integrates readily to

$$P = \frac{P_A}{\exp(-L_{\text{vap}}/kT_A)}\exp(-L_{\text{vap}}/kT), \tag{12.30}$$

where the subscript A indicates values at point A on the vaporization curve. Once

Table 12.3 *The slope dP/dT of coexistence curves in the $P–T$ plane. Note that the experimental melting curve usually tilts rightward and upward from the triple point. Water is one of a few anomalous substances.*

Difference in volumes	Expected sign	Difference in entropies	Expected sign	Expected slope	Experimental slope
$v_{vap} - v_{liq}$	> 0	$s_{vap} - s_{liq}$	> 0	> 0	> 0
$v_{vap} - v_{sol}$	> 0	$s_{vap} - s_{sol}$	> 0	> 0	> 0
$v_{liq} - v_{sol}$	usually^a > 0	$s_{liq} - s_{sol}$	usually^b > 0	> 0 (usually)	usually^c > 0

^{a}But ice floats: $v_{liq} - v_{sol} < 0$ for water.
^{b}But $s_{liq} - s_{sol}$ is negative for ^{3}He below 0.3 K.
^{c}But dP/dT is negative for water and for ^{3}He below 0.3 K.

again, an exponential dependence on the reciprocal of temperature appears, and we see the role played by the experimental latent heat per molecule.

Equation (12.30) suggests that $\ln P$ is a linear function of $1/T$. Real gases, such as argon, xenon, N_2, O_2, CO, and CH_4, exhibit nearly linear behavior over a surprisingly large range of temperatures: from $T \cong 0.98 T_c$ down to the triple point. In part, this is a tribute to a good theory. In part, it reflects a cancellation of potential causes of deviation: (1) the actual variation of L_{vap} with temperature, (2) the neglect of v_{liq}, and (3) corrections to the ideal gas law (as used for v_{vap}) because of high density as the temperature approaches T_c.

Some expectations

For other coexistence curves, the associated precursor to the Clausius–Clapeyron equation has the same structure as in (12.26). The slope dP/dT is given by a ratio: the difference in entropies (per molecule) divided by the difference in volumes (per molecule). Whichever phase comes first in the numerator comes first in the denominator; beyond that, order is irrelevant.

Table 12.3 shows the sign that one can expect for the slope of various coexistence curves in the $P–T$ plane. Ordinary experience suggests the sequence of inequalities $v_{vap} > v_{liq} > v_{sol}$, but ice floats, and so exceptions arise. To vaporize a liquid or a solid, one needs to heat it, and so both $s_{vap} - s_{liq}$ and $s_{vap} - s_{sol}$ should be positive. Melting usually requires heating, and so $s_{liq} - s_{sol}$ is usually positive, but ^{3}He below 0.3 K provides an exception, a topic that we discuss in the next section.

12.6   Cooling by adiabatic compression (optional)

When compressed adiabatically, a classical ideal gas gains energy (because work is done on the gas), and its temperature increases. Section 1.5 provided an explicit relationship:

$$TV^{\gamma-1} = \text{constant},$$

where γ is the ratio of heat capacities, C_P/C_V, and is greater than 1. As the volume V decreases, the temperature T must increase.

Kinetic theory provides another way to understand the increase in energy. When a molecule rebounds from a *moving* piston, its final kinetic energy differs from the initial value. To establish the sign of the effect, think of a slowly moving molecule being approached by a fast, inward-moving, and massive piston. After the collision, the molecule will travel faster than before; it will have gained energy from the piston.

Adiabatic compression always increases the energy of the system that is being compressed—simply because work is being done *on* the system. For most physical systems, the increase in internal energy is accompanied by an increase in temperature, but no law of physics requires an increase in temperature. Indeed, when ^{3}He is on its melting curve and when the initial temperature is 0.3 K, say, then adiabatic compression produces a decrease in temperature. The helium literally becomes colder. Let's see why.

The Clausius–Clapeyron equation will be the central equation:

$$\frac{dP}{dT} = \frac{s_{\text{liq}} - s_{\text{sol}}}{v_{\text{liq}} - v_{\text{sol}}}. \tag{12.31}$$

Thus we need to compare the entropies of solid and liquid ^{3}He. Temperatures of 1 K and below are presumed.

Solid ^{3}He

Picture the solid as a neat crystalline lattice, each helium atom remaining in the neighborhood of its equilibrium site. An atom of ^{3}He has a spin of $\frac{1}{2}\hbar$, which arises from the single neutron in the nucleus. The magnetic interaction between neighboring nuclear magnetic moments is weak; at first, we ignore it. Each nuclear spin has two quantum states available to it, and those states have equal energy—indeed, zero energy—in the absence of magnetic interactions. For a single nucleus, the spin partition function is $Z_{\text{spin}} = 2$ because each Boltzmann factor in the sum for Z_{spin} is exp(zero). The spin contribution to the entropy per atom follows from equation (5.25) as

$$s_{\text{sol,spin}} = \frac{\langle \varepsilon \rangle_{\text{spin}}}{T} + k \ln Z_{\text{spin}}$$

$$= 0 + k \ln 2 = k \ln 2. \tag{12.32}$$

The Debye theory of lattice vibrations enables one to calculate the lattice contribution to the entropy. When $T \leqslant 1$ K, that contribution is negligible in comparison with the nuclear spin contribution. Thus the entropy per atom of the solid is

$$s_{\text{sol}}/k = \ln 2. \tag{12.33}$$

Liquid ^{3}He

When turning to liquid ^{3}He, the first question to ask is this: may we treat the atoms classically, or must we use quantum theory? Table 5.2 showed that liquid ^{4}He at $T = 4$ K has $(V/N)^{1/3}/\lambda_{\mathrm{th}} = (V/N)^{1/3}\sqrt{2\pi mkT}/h = 0.86$. A ratio less than 3 or so indicates that quantum theory is required. An atom of ^{3}He has a smaller mass, and we are interested in temperatures $T \leqslant 1$ K; hence quantum theory is needed *a fortiori*. Indeed, if we treat liquid ^{3}He as an ideal Fermi gas, its Fermi temperature is $T_{\mathrm{F}} = 6$ K. This value is based on $v_{\mathrm{liq}} = V/N = 43 \times 10^{-30}$ m^3/atom at $T = 0.3$ K and $P = 29$ atm, a point on the experimental melting curve. At physical temperatures below the Fermi temperature, quantum theory is mandatory.

The simplest theory regards liquid ^{3}He as a degenerate ideal Fermi gas. Equation (9.19) gives the heat capacity as

$$C_V = \frac{\pi^2}{2} Nk \frac{T}{T_{\mathrm{F}}},$$

and with it we can calculate the entropy S_{liq}:

$$
\begin{aligned}
S_{\mathrm{liq}} &= \int_0^T \frac{C_V(T')}{T'}\, dT' \\
&= \int_0^T \frac{\pi^2}{2} Nk \frac{T'}{T_{\mathrm{F}}} \times \frac{1}{T'}\, dT' = \frac{\pi^2}{2} Nk \frac{T}{T_{\mathrm{F}}}.
\end{aligned}
\tag{12.34}
$$

In section 10.3, we reasoned that the multiplicity of the ground state is 1, and hence its entropy is zero. Thus no additive constant need appear in the first line. For the liquid, the entropy per atom emerges as

$$s_{\mathrm{liq}}/k = \frac{\pi^2}{2} \frac{T}{T_{\mathrm{F}}}. \tag{12.35}$$

Figure 12.3 graphs the two entropies. The solid has the greater entropy,

$$s_{\mathrm{liq}} - s_{\mathrm{sol}} < 0, \tag{12.36}$$

when $T < (2\ln 2/\pi^2)T_{\mathrm{F}}$. In qualitative terms, the situation is the following. The nuclei in the solid phase are located at specific lattice sites, and so, in many respects, the nuclei act independently of each other. Their spins need not be correlated by the Pauli exclusion principle (because the atoms differ already in which lattice site they inhabit). Nuclei in the liquid phase do not enjoy this quasi-independence, and so the Pauli principle causes their spins to be correlated and reduces the spin contribution to the entropy. Because of the correlations, the spin contribution to the entropy per atom is less for the liquid than for the solid. At very low temperatures, the spin contribution dominates over motional and positional contributions. Thus the entropy per atom of the solid is greater than that of the liquid.

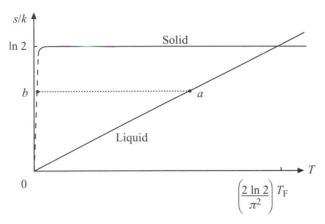

Figure 12.3 Entropy per atom in the coexisting solid and liquid phases of ^{3}He, based on the simplest theory. The dashed extension of the curve for solid helium indicates that, at very low temperatures, the magnetic and electrostatic interactions cease to be negligible, and so s_{sol}/k drops away from ln 2. The dotted line labeled $b \cdots a$ corresponds to slow adiabatic compression from pure liquid to pure solid. [Adapted from I. Pomeranchuk, *Zh. Eksp. i Teor. Fiz.* **20**, 919–24 (1950).]

One last preliminary is needed. When solid ^{3}He melts, the volume per atom increases, that is,

$$v_{liq} - v_{sol} > 0. \tag{12.37}$$

For example, at $T = 0.3$ K, experiment gives $v_{liq} - v_{sol} = 2.1 \times 10^{-30}$ m^3/atom. The inequality (12.37) holds throughout the entire range of temperatures for which the volume change has been measured.

Pomeranchuk's prediction

The Russian physicist Isaak Yakovlevich Pomeranchuk developed a graph like figure 12.3 in 1950. At that time, ^{3}He had only recently been liquefied, and its properties at very low temperature were not yet known. Nonetheless, Pomeranchuk expected the inequality $v_{liq} - v_{sol} > 0$ to hold, as it does for most substances. Using that inequality—as well as the inequality (12.36)—in the Clausius–Clapeyron equation, Pomeranchuk predicted a negative slope for the melting curve at temperatures below the crossover point in figure 12.3. The subsequent experimental findings are displayed in figure 12.4.

Moreover, Pomeranchuk predicted that adiabatic *compression* of coexisting solid and liquid would *cool* the helium. If the initial temperature is below 0.32 K, then an increase in the external pressure moves the helium along the melting curve up and to the left: toward higher pressure and lower temperature. This is the basis of the *Pomeranchuk refrigerator*, a laboratory device that successfully cooled samples from 0.3 K to approximately 1 millikelvin (abbreviated 1 mK).

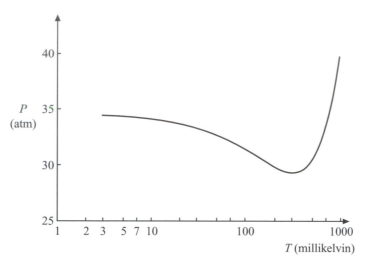

Figure 12.4 The melting curve for ³He between 0.003 K and 1 K. Between 3 millikelvin and the minimum, the melting curve indeed has a negative slope. The minimum occurs at a temperature of 0.32 K and a pressure of 29.3 atmospheres. The extension of the curve to temperatures lower than 3 mK is asymptotically horizontal. [*Source*: J. Wilks and D. S. Betts, *An Introduction to Liquid Helium*, 2nd edn (Oxford University Press, New York, 1987).]

For a more intuitive way to understand the cooling, consider the situation depicted in figure 12.5. Solid and liquid ³He coexist. If the external pressure is increased (adiabatically), one expects the total volume to decrease. The inequality (12.37) suggests that some liquid will turn into solid. Because s_{sol} is greater than s_{liq}, the

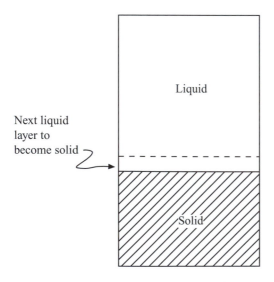

Figure 12.5 Solidification during slow adiabatic compression. (The geometry has been idealized.)

entropy increases in the layer that is becoming solid. The increase in entropy requires energy input by heating: $\Delta S = q/T$. That energy comes from both the existing solid and the remaining liquid, and so they become cooler. While this description may help, it introduces its own paradox: to solidify liquid ^{3}He, heat it!

Energy changes

The issue of energy transfer merits even more attention. By how much does the energy of a helium atom change when it is promoted from the liquid to the solid? On the melting curve, solid and liquid coexist in thermal equilibrium. Therefore their chemical potentials are equal: $\mu_{sol} = \mu_{liq}$. For each phase, the chemical potential may be written as the Gibbs free energy divided by the number of atoms in the phase; that property was expressed by equation (10.32). Thus

$$\frac{G_{sol}}{N_{sol}} = \frac{G_{liq}}{N_{liq}}, \tag{12.38}$$

and so

$$\varepsilon_{sol} - Ts_{sol} + Pv_{sol} = \varepsilon_{liq} - Ts_{liq} + Pv_{liq},$$

where ε denotes E/N, the average energy per atom. Solving for the energy change of the average atom, we find

$$\varepsilon_{sol} - \varepsilon_{liq} = T(s_{sol} - s_{liq}) - P(v_{sol} - v_{liq}). \tag{12.39}$$

At $T = 0.2$ K, the pressure is $P = 29.46$ atmospheres; the volume decrease is

$$v_{sol} - v_{liq} = -2.11 \times 10^{-30} \text{ m}^3/\text{atom};$$

and the entropy increase is

$$s_{sol} - s_{liq} = 2.02 \times 10^{-24} \text{ J}/(\text{K} \cdot \text{atom}).$$

These experimental numbers imply

$$T(s_{sol} - s_{liq}) = 2.52 \times 10^{-6} \text{eV}/\text{atom} \quad \text{and}$$
$$\tag{12.40}$$
$$-P(v_{sol} - v_{liq}) = +39.3 \times 10^{-6} \text{ eV}/\text{atom}.$$

The energy change that we can associate directly with compressive work done *on* the system, $-P\Delta v$, is 16 times as large as the change associated with heating, $T\Delta s$. In this comparison, the heating is relatively insignificant. But the heating must occur— because only it can produce the entropy increase $s_{sol} - s_{liq}$ (for the promotion of an atom from the liquid to the solid) that figure 12.3 displays. And only the heating of some liquid (that becomes solid) cools the remaining liquid and the already existing solid (because, among other reasons, the system's total entropy remains constant). Just knowing that $\varepsilon_{sol} - \varepsilon_{liq} = 41.8 \times 10^{-6}$ eV/atom would not tell us whether slow adiabatic compression lowers the temperature or raises it. The modes of energy change are crucial. As the Clausius–Clapeyron equation shows, to lower the temperature by compression, two conditions must hold: (1) there must be an entropy difference, $s_{sol} - s_{liq} \neq 0$, and (2) the differences $s_{sol} - s_{liq}$ and $v_{sol} - v_{liq}$ must have opposite signs.

As a model, a degenerate ideal Fermi gas captures the essential behavior of liquid ^{3}He. For us, that essence is low entropy per atom. Although interatomic forces influence the behavior significantly, the graph in figure 12.3 is a good first approximation—provided the cross-over point is set at 0.32 K, the temperature of the minimum in the empirical melting curve. Magnetic and electrostatic interactions in the solid cause its entropy to drop away as T heads toward 1 mK. Qualitatively, the interactions correlate the orientations of the nuclei; that reduces the multiplicity and hence the entropy. The entropy of the solid drops sharply to $s_{sol}/k \cong 0.15$ as T decreases from 4 mK to 1 mK, virtually a vertical line on a graph in which temperature is plotted linearly from 0 to 350 mK.

Slow adiabatic compression does not change the system's entropy. The adjective *isentropic* denotes "at constant entropy," and so a slow adiabatic process is an isentropic process. The process corresponds to a horizontal line in figure 12.3. The dotted line labeled with the letters a and b indicates that, in principle, adiabatic compression can cool from 0.2 K to the order of 2 mK. Other horizontal lines would give other ranges. In particular, the lower the initial temperature, the lower the intercept (where s_{sol}/k drops steeply) and hence the lower the final temperature.

Superfluid ^{3}He

In section 9.5, we noted that liquid ^{4}He becomes a superfluid at temperatures below 2.17 K. The conduction electrons in many metals become a superfluid at low temperature; one calls it superconductivity. In lead, for example, conduction electrons form pairs when the temperature drops below 7.2 K; each pair of electrons acts like a boson and gives the characteristic resistance-free flow of electric charge that typifies superconductivity. As soon as the basic electron-pair theory of superconductivity had been developed in the 1950s, physicists wondered whether liquid ^{3}He would form atomic pairs with boson character and the attributes of a superfluid. Predictions for the transition temperature were made, and experimental searches were launched. No luck. By the late 1960s, enthusiasm for the quest had waned, but there was much to study in ^{3}He anyway, and so the isotope remained a focus of active research.

At Cornell University in 1971, Douglas D. Osheroff, Robert C. Richardson, and David M. Lee looked for a phase transition in solid ^{3}He, the transition responsible for the sharp drop in entropy that figure 12.3 displays. To cool the solid, they used a Pomeranchuk refrigerator. The solid in the coexisting mix of solid and liquid was their "sample." What they discovered, serendipitously, were two phase transitions in the liquid. Figure 12.6 shows the pressure–temperature phase diagram for ^{3}He. Cooling by adiabatic compression from 20 mK, the three physicists found a transition to a superfluid phase A at $T = 2.7$ mK and a second transition to a phase B at $T \cong 2$ mK.

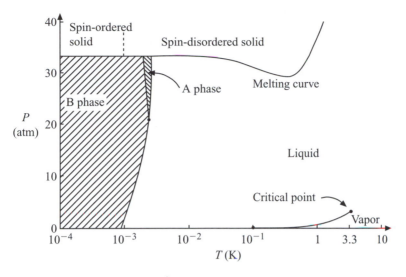

Figure 12.6 Phase diagram for ^{3}He at low temperature (and in the absence of an external magnetic field). The phases A and B are superfluid. Under its own vapor pressure, ^{3}He remains a liquid to the lowest temperatures at which it has been studied, and it is expected to remain a liquid all the way to absolute zero. [*Source*: Dieter Vollhardt and Peter Wölfle, *The Superfluid Phases of Helium 3* (Taylor and Francis, New York, 1990).]

[More recent experiments (in 1994) put the transitions at 2.5 and 1.9 mK, respectively. And in 1996, Osheroff, Richardson, and Lee received the Nobel Prize in Physics for their discovery.]

Superfluid ^{3}He shares with superfluid ^{4}He the property of being able to flow without viscosity. But there is additional richness. When two electrons form a superconducting pair in lead, their spins are oriented oppositely, and the pair—as a boson—has neither net spin nor magnetic moment. Not so with ^{3}He. The two atoms are loosely held into a pair by standard (albeit weak) interatomic forces, and the two nuclear spins are aligned parallel to each other. Thus the pair—as a boson—has a net spin of $1\hbar$ and a net magnetic moment. Moreover, the motion of the atoms relative to their center of mass generates a nonzero orbital angular momentum. The net spin angular momentum and the orbital angular momentum can be coupled in various ways, and such different couplings distinguish the superfluid phases from each other (in too complicated a way for us to pursue here).

The superfluid phases respond surprisingly strongly to magnetic fields. Moreover, when the boson-like pairs are partially aligned by an external magnetic field, the properties of the fluid depend on direction relative to the external field. For example, the attenuation of a sound wave depends on how the wave propagates relative to the magnetic field. The fluid's response is anisotropic. Beyond all this, in the presence of an external magnetic field, yet another superfluid phase makes its appearance, sandwiched between the A phase and the more conventional liquid.

12.7   Gibbs' phase rule (optional)

We return to the theoretical conditions for coexistence of phases. Let the context be set by figure 12.1. If we require that two phases coexist, we may freely choose one intensive variable (T or P), but the other is then fixed (because the point must lie on a coexistence curve). If we require that three phases coexist, no freedom of choice remains. How does this kind of counting generalize when more than one species of molecule is present? For example, if the system consists of a mixture of ^{3}He and ^{4}He atoms, how many phases may coexist? We approach the answers via the following route.

Let

$\#\mathrm{sp} \equiv$ number of molecular species and

$\varphi \equiv$ number of phases that coexist.

Furthermore, stipulate that no chemical reactions occur among the species. After $\#\mathrm{sp}$ and φ have been specified, how many intensive variables may we choose freely?

For each species, the chemical potentials in all φ phases must be equal (because we may imagine transferring molecules from one phase to any other). Table 12.4 illustrates this. In general, there are $\varphi - 1$ independent equations per row, and there is one row for each molecular species. Thus

$$\begin{pmatrix} \text{number of} \\ \text{constraint equations} \end{pmatrix} = \#\mathrm{sp} \times (\varphi - 1). \tag{12.41}$$

How many intensive or effectively intensive variables are there? Equation (10.35) tells us that there are $(\#\mathrm{sp} - 1)$ relative concentrations in each phase. Multiplication by the number of phases φ gives $\varphi \times (\#\mathrm{sp} - 1)$ relative concentrations. To this we add temperature and pressure (which are uniform throughout the system) and find

Table 12.4 *An example. Suppose the species are ^{3}He and ^{4}He, whence $\#\mathrm{sp} = 2$, and suppose that the phases are the familiar vapor, liquid, and solid, whence $\varphi = 3$. Coexistence requires a minimum in the Gibbs free energy with respect to all possible transfers of an atom of each species. Hence the sequences of equalities displayed below must hold. Each chemical potential depends on three intensive variables: T, P, and the ratio of concentrations, $N_{^3\mathrm{He}}/N_{^4\mathrm{He}}$, in the given phase.*

$$\mu_{^3\mathrm{He,vap}} = \mu_{^3\mathrm{He,liq}} = \mu_{^3\mathrm{He,sol}}$$
$$\mu_{^4\mathrm{He,vap}} = \mu_{^4\mathrm{He,liq}} = \mu_{^4\mathrm{He,sol}}$$

$$\begin{pmatrix} \text{number of} \\ \text{intensive variables} \end{pmatrix} = \varphi \times (\#\text{sp} - 1) + 2. \tag{12.42}$$

The number of intensive variables that one may choose freely is the difference of the numbers in equations (12.41) and (12.42):

$$\begin{pmatrix} \text{number of variables that} \\ \text{one may choose freely} \end{pmatrix} = \begin{pmatrix} \text{number of} \\ \text{intensive variables} \end{pmatrix} - \begin{pmatrix} \text{number of} \\ \text{constraint equations} \end{pmatrix}$$

$$= [\varphi \times (\#\text{sp} - 1) + 2] - \#\text{sp} \times (\varphi - 1)$$

$$= \#\text{sp} + 2 - \varphi. \tag{12.43}$$

Freedom to choose grows with the number of molecular species ($\#$sp) and declines with the number of phases (φ) that are required to coexist. Equation (12.43) is *Gibbs' phase rule*.

The maximum number of phases coexist when the freedom of choice has been reduced to zero. Thus

$$\begin{pmatrix} \text{maximum number of} \\ \text{coexisting phases} \end{pmatrix} = \#\text{sp} + 2. \tag{12.44}$$

In the case of pure water, where $\#$sp $= 1$, the maximum is $1 + 2 = 3$, a triple point. At least ten kinds of ice exist—they differ in crystal structure—but it is not possible to have two kinds of ice coexisting with liquid and vapor. To be sure, nothing prevents two or three kinds of ice from coexisting; they just cannot coexist with liquid and vapor as well. Percy Bridgman's classic experiments found several triple points in which three kinds of ice coexist.

If ^{3}He and ^{4}He are mixed, then $\#$sp $= 2$, and the maximum number of coexisting phases is four.

The preceding analysis took as intensive variables the relative concentrations plus temperature and pressure. Additional intensive variables sometimes exist, such as a "spreading pressure" in an adsorbed film. If so, then the counting must be augmented to include them.

Moreover, if chemical reactions are possible, then each reaction imposes another constraint on the chemical potentials (as chapter 11 demonstrated), and the counting must include those constraints.

12.8 Isotherms

The pressure–temperature phase diagram of figure 12.1 displays the phase state of water as a function of two intensive variables, P and T. Precisely because the variables are intensive, there is no need to say how much water is being discussed, and there is no need even to hold constant the amount of water under discussion. (Doubling the amount of water at fixed P and T does not alter the phase in which water exists at that P and T.) Some other aspects of how water behaves are displayed more clearly if pressure and volume are used as the independent variables. Because the volume V is an extensive variable and depends, in part, on how much water is present, one stipulates that the total number of water molecules is held fixed and is the same at all points in the P–V plane. Figure 12.7 shows a portion of the phase diagram that ensues.

The curves are curves of constant temperature: *isotherms*. In the P–T plane of figure 12.1, isotherms are vertical straight lines; in the P–V plane, they are curved (except in special circumstances). To understand the isotherms qualitatively, we begin at temperatures well above the critical temperature, where the ideal gas law—valid for a *dilute* gas—is a good approximation. That law asserts

$$P = \frac{N}{V}kT,$$ (12.45)

and so, as the volume V decreases along an isotherm, the pressure P rises along the arm of a hyperbola. The curve in the upper right portion of figure 12.7 is qualitatively of this shape.

Consider now an isotherm that lies below the critical temperature, at $T = 600$ K, say. Start in figure 12.1 at low pressure, well below the vaporization curve, and then move up along the vertical isotherm. (You will have to draw in the isotherm mentally.) As you increase the pressure, the vapor volume decreases, and you can see that

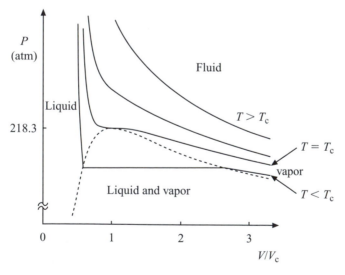

Figure 12.7 The phase diagram for water: the pressure–volume version. The volume is given relative to the volume at the critical point, V_c (because the ratio V/V_c provides an intensive variable). The curves are isotherms. The displayed isotherms all have temperatures higher than the temperature of the triple point, and so no solid phase appears. Within the region outlined by dashes, liquid and vapor coexist.

behavior displayed in figure 12.7. When the system hits the vaporization curve, you can continue to compress it; as you do so, vapor condenses—drop by drop—into liquid. Both temperature and pressure remain constant, and so the process appears as a horizontal line in the $P-V$ plane. (Note, by the way, that the chemical potential is constant along the line segment, for the segment corresponds to a single point on the vaporization curve. Later, that property will be essential.) After compression has squeezed all the vapor into the liquid state, the system moves off the vaporization curve (in the $P-T$ plane) and into the pure liquid phase. Because liquids are relatively incompressible, the pressure (in the $P-V$ plane) shoots up almost vertically as you decrease the volume further.

In the $P-T$ plane, the critical point marks the termination of the vaporization curve. When the number of water molecules has been specified, the volume of water at the critical point is uniquely determined by the critical temperature T_c and the critical pressure P_c. That is to say, there is a unique value for the intensive variable $(V/N)_c$, the volume per molecule at the critical point. Thus the critical *point* of the $P-T$ plane maps into a critical *point* in the $P-V$ plane. The isotherm that passes through the critical point is called the *critical isotherm*. In figure 12.7, the critical point marks the end (in the pressure dimension) of the two-dimensional region where liquid and vapor coexist.

On the horizontal line segments in the region of liquid–vapor coexistence, the derivative dP/dV is obviously zero. One may think of the critical point as the limit as the horizontal segments shrink to zero length. Thus the equation $dP/dV = 0$ holds at the critical point. In different words, the critical isotherm has $dP/dV = 0$ at one, but only one, point.

Moreover, there is good evidence, both experimental and theoretical, that the second derivative, d^2P/dV^2, is also zero at the critical point. That makes the critical point a point of inflection on the critical isotherm.

Having finished this prelude, we turn to a classic equation that seeks to describe behavior in the $P-T$ and $P-V$ planes.

12.9 Van der Waals equation of state

The ideal gas law, displayed most recently in equation (12.45), provides an example of an *equation of state*: a relationship that connects P, T, and the number density N/V, so that knowledge of any two variables enables one to calculate the third. (One could, of course, use mass density in place of number density, and that was the custom in the nineteenth century.) The ideal gas law works well for a dilute gas, but it is not capable of describing a liquid or even a dense gas. More comprehensive equations of state have been proposed, literally dozens of them, ranging from curve fitting of experimental data to elaborate theoretical expressions. In this wealth of proposals, the van der Waals equation of state remains a standard as well as an excellent introduction to the topic. The van der Waals relationship works fairly well in the vapor and liquid regions of figure 12.7 and also in the fluid region somewhat above the critical isotherm. In the region of liquid–vapor coexistence, it predicts nonsense, but that failing can be patched up.

Several distinct routes lead to the van der Waals equation of state, each with individual merits. Our route starts with the partition function for a semi-classical ideal gas and then introduces two modifications: one for the repulsion between molecules at very short range; the other for the attraction between molecules at intermediate range. Thus our starting point is

$$Z_{\text{ideal}} = \frac{(V\lambda_{\text{th}}^{-3})^N}{N!}.$$

(12.46)

(Because pressure is determined primarily by the center-of-mass motion of molecules, we may suppress aspects of molecular structure such as rotation and vibration about the molecular center of mass.)

Figure 12.8 shows a typical intermolecular potential energy curve. With this curve in mind, we introduce the two modifications, as follows.

1. Repulsion. When two molecules come into "contact," a strong repulsive force arises. Each molecule excludes the other from the volume of space that it occupies. In the partition function, we represent this effect by replacing the volume V with $V - Nb$, where the constant b denotes a volume of molecular size, in order of

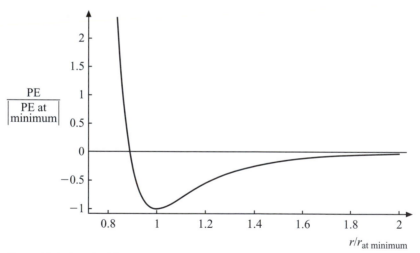

Figure 12.8 A typical intermolecular potential energy (PE) curve. The mutual potential energy of two molecules is shown as a function of the distance r between their centers. Where the slope is positive, the force is attractive; there the potential energy varies as $1/r^6$ (except near the bottom of the potential well). Where the slope is negative, the force is repulsive. There the slope is so steep that virtually any value of r in the repulsive region may be regarded as corresponding to two molecules "in contact" and hence as a separation of "two molecular radii." Strictly speaking, a mutual potential energy that depends on only separation presupposes spherical molecules (before they interact). The form displayed here is qualitatively correct for monatomic gases like argon and helium and is quite adequate for small diatomic molecules such as H_2 and N_2.

magnitude. For example, if the molecules exert repulsive forces as though they were hard spheres of radius r_0, then the minimum center-to-center separation would be $2r_0$. One molecule would exclude the center of another from the volume $b = 4\pi(2r_0)^3/3 = 8 \times 4\pi r_0^3/3$, that is, eight times the volume of a single molecule. Do not take this expression as a definitive result; it is merely suggestive and indicates the order of magnitude. The coefficient b is best determined empirically. The product Nb represents the volume from which any one molecule is excluded by all the other molecules.

2. Attraction. The attractive force at intermediate distances will decrease the system's total energy (at given temperature T) relative to the energy of an ideal gas. For any given molecule, the reduction in energy will depend on how many other molecules are within intermediate range of it. Perhaps the simplest form for the reduction is $-a \times (N/V)$, where a is a positive constant with the units of energy times volume. Then the total energy would be

$$E = \tfrac{3}{2}NkT - N \times a\frac{N}{V}. \tag{12.47}$$

According to equation (5.16), the partition function provides an energy estimate via

$$\langle E \rangle = kT^2 \frac{\partial \ln Z}{\partial T}.$$

To generate the term $-aN^2/V$ that appears in (12.47), the logarithm of Z must have a term aN^2/VkT. Thus, to incorporate both repulsion and attraction, we write the partition function Z_{vdW} as

$$\ln Z_{\mathrm{vdW}} = \ln\left\{\frac{[(V - Nb)\lambda_{\mathrm{th}}^{-3}]^N}{N!}\right\} + \frac{aN^2}{VkT}. \tag{12.48}$$

(The last term could be tucked inside the logarithm and would then look like the familiar Boltzmann factor.)

According to equation (5.19), the pressure follows as

$$P = kT\frac{\partial \ln Z_{\mathrm{vdW}}}{\partial V}$$

$$= \frac{NkT}{V - Nb} - a\left(\frac{N}{V}\right)^2. \tag{12.49}$$

This is the *van der Waals equation of state*, derived by Johannes Diderik van der Waals in his Ph. D. thesis, written in 1873 in Leiden, the Netherlands. Our route to the result is reasonably short and also provides a function, the partition function Z_{vdW}, that can be used to calculate all other equilibrium thermodynamic properties, such as entropy and chemical potential, in a mutually consistent fashion.

Appendix D offers further qualitative understanding of how the attractive and repulsive intermolecular forces affect the pressure.

Isotherms

Figure 12.9 shows some isotherms computed with the van der Waals equation. Outside the area outlined by dashes, the isotherms agree qualitatively with the experimental isotherms in figure 12.7. Within the outlined area, however, the dip on the isotherm is not physically sensible. Where the slope dP/dV is positive, the system would be unstable.

To see why, imagine that the material—liquid, gas, or mixture—is in a vertical cylinder, as was sketched in figure 1.4. The cylinder's sides are maintained at constant temperature by contact with a water bath, say, and so any evolution of the material is isothermal. Initially, the confined material is at equilibrium (however unstable) under an external pressure produced by a weight sitting on the piston. By specification, this situation corresponds to a point in figure 12.9 where the slope dP/dV is positive. Now add the weight of a postage stamp to the piston. The additional external pressure compresses the material slightly. According to figure 12.9 and the slope, the material's pressure drops. So the weighted piston compresses the material further; the material's pressure drops again, and so on. This is instability.

An analogously unstable behavior occurs if a tiny weight is removed from the piston. In the real world and its incessant perturbations, no material can exist for any length of time on an isotherm where dP/dV is positive.

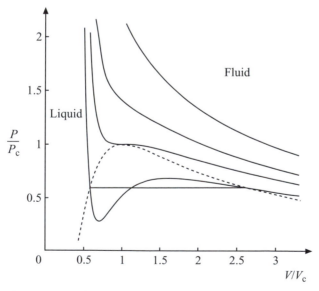

Figure 12.9 Isotherms of the van der Waals equation of state. The pressure and volume are given relative to their values at the critical point, P_c and V_c. Within the area outlined by dashes, the literal curve must be replaced by the horizontal line, as explained in the text.

On isotherms below $(27/32)T_c$, the dip is so large that the pressure goes negative. That is not wholly absurd. The sap in tall trees is believed to be pulled to the tree tops, not pushed. Moving in tiny tubes, the sap is under tension and may be said to have a negative pressure. In the lab, liquids have been put under tensions that correspond to negative pressures of order -500 atmospheres. Nonetheless, a negative pressure is not acceptable in a region where liquid and *gas* are to coexist.

The oscillatory portion of an isotherm must be replaced by a horizontal line that connects a 100 percent vapor point with a 100 percent liquid point that has the same numerical value for the chemical potential. (Recall that each horizontal *line* in figure 12.7 corresponds to a *point* on the vaporization curve of figure 12.1, and hence the chemical potential has the same numerical value all along the line segment.) Figure 12.10 and the Gibbs–Duhem relation provide a construction for the line, as follows.

Even though the van der Waals equation fails as a description of nature within the area outlined by dashes, the thermodynamic functions (like P and μ) satisfy all the usual equations because we derive them from a partition function. Thus the Gibbs–Duhem equation holds (at least in a mathematical sense). Along an isotherm, that equation reduces to

$$\Delta\mu = \frac{V}{N}\Delta P. \qquad (12.50)$$

Think of integrating an infinitesimal version of this relation, $d\mu = (V/N)dP$, along the isothermal curve $ABCDE$ in figure 12.10. From point A to point B, the differential dP is positive, and so μ increases. From B to C, the differential dP is negative, and so μ decreases. The net change in μ is given by the dashed area that is bounded by the curve ABC and the horizontal line. Perhaps you can see this most clearly by rotating the figure 90°, so that the P axis takes the usual position of the integration axis in an integral of the form $\int(V/N)dP$. (Beware, however, because then leftward is the

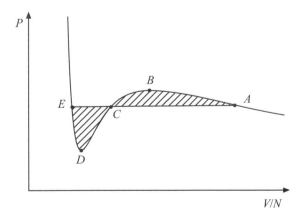

Figure 12.10 Selecting the correct horizontal line segment. To facilitate using the Gibbs–Duhem equation, the pressure P is plotted as a function of V/N, where N is the total number of molecules.

direction of increasing P.) For the integration from point C to D and thence to E, the reasoning is similar. Now the change in μ is given by (-1) times the dashed area bounded below by curve CDE. To ensure that μ at the low-volume end E equals μ at the high-volume end A, the horizontal line must be chosen so that the two dashed areas are equal.

The coefficients a and b

In the simplest version of the van der Waals equation, the coefficients a and b are constants. Certainly we treated them as constants in constructing Z_{vdW} and then in differentiating that function. Continuing in this spirit, we ask, how can one determine a and b empirically? Data at the critical point suffice, as follows.

In section 12.8, we noted that one may think of the critical point as the limit as isothermal line segments (in the coexistence region) shrink to zero length. Because $dP/dV = 0$ on the line segments, that derivative is zero at the critical point also. In the case of the van der Waals equation, we can establish another property of the critical point by studying the non-physical undulation of the isotherm. The dip is a local minimum and has zero slope at the bottom; the adjacent local maximum (at larger V/V_c) has zero slope at the top. As one shifts attention to isotherms of successively higher temperature, those points of zero slope converge toward each other. In the limit, they give a zero value to d^2P/dV^2 at the critical point. Analytically, one may think of it this way:

$$\left.\frac{d^2P}{dV^2}\right|_{\text{c.p.}} = \lim_{V_{\text{at max}} - V_{\text{at dip}} \to 0} \frac{dP/dV|_{\text{at max}} - dP/dV|_{\text{at dip}}}{V_{\text{at max}} - V_{\text{at dip}}} = 0.$$

Both terms in the numerator are always zero, and so the limit is zero.

For an easy way to determine the coefficients a and b, impose the two derivative conditions at the critical point. The equations are these:

$$\left.\frac{\partial P}{\partial V}\right|_{\text{c.p.}} = \frac{-NkT_c}{(V_c - Nb)^2} + 2a\frac{N^2}{V_c^3} = 0.$$

$$\left.\frac{\partial^2 P}{\partial V^2}\right|_{\text{c.p.}} = \frac{2NkT_c}{(V_c - Nb)^3} - 6a\frac{N^2}{V_c^4} = 0.$$

(The switch to partial derivative signs is just a notational precaution. A derivative like dP/dV along an isotherm requires that we hold T constant when differentiating P as the latter is given in terms of V and T by the van der Waals equation.) The solutions are

$$\left(\frac{V}{N}\right)_c = 3b, \tag{12.51}$$

$$kT_c = \frac{8}{27}\frac{a}{b}. \tag{12.52}$$

An additional relationship follows from the van der Waals equation itself. Substitute into (12.49) the expressions for $(V/N)_c$ and kT_c in terms of a and b; one finds

$$P_c = \frac{1}{27}\frac{a}{b^2}. \tag{12.53}$$

Table 12.5 displays some values for the coefficients a and b. Both the coefficient b and the volume over which the potential energy is negative will grow with the "volume" of a molecule; hence the trend in the depth of the attractive well is more clearly revealed by the ratio a/b, cited in electron volts, than by a alone.

Equation (12.52) tells us that the critical temperature is proportional to a/b and hence to the well depth (more or less). This echoes a remark made in section 12.1: empirically, the critical temperature is proportional to the well depth: $kT_c \cong$ (well depth). For this relationship, there was a ready interpretation: if the molecular translational kinetic energy, $\frac{3}{2}kT$, exceeds the well depth, then the molecule will not be bound by the attractive force exerted by another molecule. A distinct, high-density liquid state will not be possible.

The simple structure of the van der Waals equation provides yet another relationship. We found three conditions that relate the three intensive variables $\{T_c, P_c, (V/N)_c\}$ and the two coefficients a and b. It should be possible to solve for a and b (as we did) and to have left over one condition that relates the intensive variables among themselves. That is indeed so. Consider the dimensionless combination $(N/V)_c(kT_c/P_c)$. Equations (12.51) to (12.53) imply

$$\left(\frac{N}{V}\right)_c \frac{kT_c}{P_c}\bigg|_{vdW} = \frac{8}{3} = 2.67. \tag{12.54}$$

Regardless of how the coefficients a and b are chosen, either empirically or by theoretical computation, equation (12.54) is a prediction by the van der Waals

Table 12.5 *Empirical constants for the van der Waals equation. The coefficient b is given in units of 10^{-30} m^3 (1 cubic angstrom). The data in table 12.1 provide a and b via the expressions a = $(27/64)(kT_c)^2/P_c$ and b = $(1/8)kT_c/P_c$.*

Substance	b (10^{-30} m^3)	a (eV $\times 10^{-30}$ m^3)	a/b (eV)
Water	50.7	9.55	0.188
Sulfur dioxide	94.7	11.9	0.125
Carbon dioxide	71.3	6.30	0.0884
Oxygen	52.9	2.38	0.0451
Argon	53.8	2.36	0.0439
Nitrogen	64.3	2.36	0.0367
Hydrogen	44.3	0.428	0.009 66
Helium-4	39.4	0.0597	0.001 52

equation. Empirical values of $(N/V)_c(kT_c/P_c)$ for the gases in table 12.5 range from 3.3 for hydrogen to 4.4 for water. The van der Waals equation does a lot better than the ideal gas law would do if applied at the critical point, for it would give the value 1. Of course, such use would not be fair, for the ideal gas law claims validity only for dilute gases. Nonetheless, the van der Waals equation falls short of high accuracy.

Van der Waals himself never intended that the coefficients a and b be taken as literal constants. The title of his dissertation was "On the continuity of the gaseous and liquid states," and that title displays the focus of his effort: to show that the gaseous and liquid states have much more in common than had been thought. Perhaps the greatest triumph of his thesis was to exhibit, in one equation, the continuity along a path (in the P–V plane of figure 12.7) from the vapor phase up to the fluid phase and then down to the liquid phase. Van der Waals devoted his career to the equation of state, broadly construed; for his many insights, he received the Nobel Prize for Physics in 1910. In his acceptance speech, he complained that people repeatedly attributed to him the opinion that the coefficients a and b are constants. Not so. Rather, from the time of his thesis research onward, he sought to find the way in which they slowly varied. For us, however, there is little benefit in pursuing that investigation. Today, the van der Waals equation is best taken with constant a and b, used for semi-quantitative work, and studied as a good example of an equation of state.

12.10 Essentials

1. Coexistence of two phases requires that their chemical potentials be equal.

2. The vaporization curve—the curve along which liquid and vapor coexist—terminates at the *critical point*. At temperatures higher than T_c, the system is a single-phase "fluid" with the volume-filling character of a gas.

3. For one molecule, the *latent heat of vaporization*, denoted by L_{vap}, is

$$L_{vap} \equiv \left(\begin{array}{c} \text{energy input by heating to promote one molecule} \\ \text{from the liquid to the vapor at constant } T \text{ and } P \end{array} \right).$$

The latent heat can be expressed as

$$L_{vap} = \Delta\varepsilon + P\Delta v = \Delta(\varepsilon + Pv),$$

where lower case ε and v denote the average energy per molecule and the volume per molecule, respectively. Analogous definitions and expressions apply to melting and to sublimation.

4. The *Gibbs–Duhem relation* describes (in part) how the chemical potential depends on the intensive variables T and P:

$$\Delta\mu = -s\Delta T + v\Delta P,$$

where s is the entropy per molecule.

5. The *Clausius–Clapeyron equation* gives the slope of the vaporization curve (in the P–T plane):

$$\frac{dP}{dT} = \frac{s_{\text{vap}} - s_{\text{liq}}}{v_{\text{vap}} - v_{\text{liq}}} = \frac{L_{\text{vap}}}{(v_{\text{vap}} - v_{\text{liq}})T}.$$

Analogous expressions hold for the melting and sublimation curves.

6. Along the vaporization curve, the pressure is described—approximately but well—by the equation

$$P = \frac{P_A}{\exp(-L_{\text{vap}}/kT_A)} \exp(-L_{\text{vap}}/kT),$$

provided that L_{vap} is taken as a constant (equal to a suitable typical value of the latent heat).

7. The van der Waals equation of state is the relationship

$$P = \frac{NkT}{V - Nb} - a\left(\frac{N}{V}\right)^2.$$

The positive constants b and a represent the effects of strong repulsion at short distance and mild attraction at intermediate distance, respectively.

Further reading

A wealth of information about phases is packed into D. Tabor's *Gases, Liquids, and Solids*, second edition (Cambridge University Press, New York, 1979). The derivations are elementary; the data are copious; and the insights are wonderful. Tabor's development of the van der Waals equation is well worth reading, and so is his discussion of negative pressure in liquids.

   J. Wilks and D. S. Betts provide a splendid survey of both ^{3}He and ^{4}He in *An Introduction to Liquid Helium*, 2nd edition (Oxford University Press, New York, 1987). Another good source is E. R. Dobbs, *Solid Helium Three* (Oxford University Press, New York, 1994).

Problems

1. The temperature outside is $-5\,°\text{C}$, and the pond is frozen. Yet children skate on a thin, evanescent layer of liquid water (which provides a virtually frictionless glide).

Use figure 12.1 to explain how the liquid film arises. [The conventional wisdom in this problem is questioned by S. C. Colbeck, "Pressure melting and ice skating," *Am. J. Phys.* **63**, 888–90 (1995) and by S. C. Colbeck *et al.*, "Sliding temperatures of ice skates," *Am. J. Phys.* **65**, 488–92 (1997).]

2. *Solid–vapor equilibrium: little boxes.* For a simple model of a solid, suppose that each atom is restricted to the immediate vicinity of a lattice site. Such an atom may move only within a little box of volume v_{sol}, the volume per atom of the solid. Interatomic forces bind the atoms to the solid by an energy $-\varepsilon_0$ (per atom) relative to the gas phase, where ε_0 is a positive constant. Note: the energy difference ε_0 compares an atom in its single-particle ground state in its box with an atom at rest in the gas phase. Take ε_0 to be of order 1 electron volt.

(a) Select one of the following two contexts:

$$\lambda_{\text{th}} \gg (v_{\text{sol}})^{1/3} \text{ or } \lambda_{\text{th}} \ll (v_{\text{sol}})^{1/3}.$$

Provide physical data and numerical justification for the reasonableness of your choice. (Each context can be justified.)
(b) Calculate the chemical potential for the atoms in the solid.
(c) Derive an equation for the vapor pressure above the solid.
(d) Should the vapor pressure be an intensive or an extensive variable? Does your theoretical expression conform to your expectation?

3. *Solid–vapor equilibrium: Einstein model.* For a more sophisticated model of a solid, suppose that each atom may oscillate about its equilibrium location in the lattice; the atom behaves like a harmonic oscillator that is free to vibrate in three dimensions. All atoms vibrate with the same frequency ν_0. Thus each atom is restricted to the immediate vicinity of a lattice site—but by a linear restoring force, not by the walls of a tiny box. The interatomic forces play another role, too. They bind the atoms to the solid by an energy $-\varepsilon_0$ (per atom) relative to the gas phase, where ε_0 is a positive constant. (Note. The energy difference ε_0 compares an atom in its ground state at its equilibrium site with an atom at rest in the gas phase.)

(a) Calculate the chemical potential for an atom in the solid.
(b) Derive an equation for the vapor pressure above the solid.
(c) What form for the vapor pressure emerges when $kT \gg h\nu_0$?

4. The system consists of $N = 10^{24}$ molecules of water. The initial state is pure vapor at $T = 373$ K and atmospheric pressure. The water is heated at constant pressure to $T = 700$ K, subsequently compressed isothermally to a pressure of 230 atmospheres, next cooled to 373 K at constant pressure, and finally allowed to expand isothermally to atmospheric pressure. The final state is pure liquid.

(a) What is the net change in the system's energy?

(b) And in its chemical potential?

5. The top of Mt. Everest is 8,854 meters above sea level. Calculate the temperature at which water boils there.

6. Here are some data for ammonia (NH_3).

Vapor pressure of solid ammonia: $\ln P = 16.27 - 3{,}729/T$.

Vapor pressure of liquid ammonia: $\ln P = 12.79 - 3{,}049/T$.

The pressure is in atmospheres; the temperature, in kelvin.

(a) What is the temperature of ammonia's triple point? And its pressure?

(b) Determine the latent heat of vaporization, L_{vap}, at the triple point (in eV/molecule). Explain any approximations.

(c) At the triple point, the latent heat of sublimation is $L_{sublimation} = 0.3214$ eV/molecule. What is the value of L_{fusion} (in eV/molecule)?

7. In figure 12.1, the coexistence curve that separates the vapor phase from the other phases has a cusp at the triple point: a discontinuous change of slope. Develop an explanation for why such a discontinuity is to be expected.

At the triple point, the volume per water molecule has the following set of ratios: $v_{vapor}/v_{liq} = 2.06 \times 10^5$ and $v_{sol}/v_{liq} = 1.09$. The latent heats of vaporization, sublimation, and fusion are the following: 0.47, 0.53, and 0.062 eV/molecule.

8. Liquid water—in a sealed but flexible container—is vaporized by heating at atmospheric pressure and a temperature of 373 K. Under these conditions, the volume per molecule in the liquid phase is $v_{liq} = 31.2 \times 10^{-30}$ m^3, that is, 31.2 cubic angstroms. The volume per molecule in the vapor is much larger: $v_{vap}/v_{liq} = 1{,}600$. In the following, "water" refers to the liquid and vapor together.

When *one molecule* is promoted from the liquid to the vapor, what will be

(a) the work done by the water (in electron volts)?

(b) the change in entropy of the water?

(c) the change in the internal energy of the water (in eV)?

(d) the change in the Gibbs free energy of the water?

9. For liquid ^{3}He, the temperature and associated vapor pressure are given by the following pairs of numbers:

(T, P): $(0.2, 1.21 \times 10^{-5})$, $(0.3, 1.88 \times 10^{-3})$, $(0.4, 2.81 \times 10^{-2})$, $(0.5, 1.59 \times 10^{-1})$.

The temperature is in kelvin; the pressure is in millimeters of mercury. Atmospheric pressure corresponds to 760 millimeters of mercury (in an old fashioned mercury manometer).

(a) What is the latent heat of vaporization, L_{vap}, in the temperature range $0.2 \leqslant T \leqslant 0.5$ K?

(b) Determine the vapor pressure at $T = 0.1$ K.

10. For solid ^{3}He, calculate the entropy per atom that the lattice vibrations contribute at temperatures below 1 K. The Debye temperature is $\theta_D = 16$ K. Compare this contribution to s_{sol} with that from the nuclear spin.

11. $C_P - C_V$ *in greater generality.*

(a) Take the energy to be a function of both temperature and volume: $E = E(T, V)$. Derive the following connection between the heat capacities at constant pressure and at constant volume:

$$C_P = C_V + \left[\left(\frac{\partial E}{\partial V} \right)_T + P \right] \times \left(\frac{\partial V}{\partial T} \right)_P .$$

(b) Check that the relation reduces properly when you apply it to a classical ideal gas.

(c) Apply the relation to the van der Waals gas.

(d) In which ways does the ratio C_P/C_V for a (monatomic) van der Waals gas differ from the corresponding ratio for a monatomic classical ideal gas?

(e) Evaluate the ratio of heat capacities under the conditions $V/N = 5 \times b$ and $kT = 5 \times a(N/V)$.

12. *Comparing the van der Waals and classical ideal gas laws.*

(a) Rearrange the factors in the van der Waals equation so that each term on the right-hand side of equation (12.49) is proportional to $(N/V)kT$. What conditions on N/V and kT (relative to the coefficients a and b) will yield individual corrections of order 1 percent to the pressure as given by the ideal gas law?

(b) Use data from table 12.5 to determine the size of the corrections for air under typical room conditions. What is the net correction?

13. Compare the entropy change of a van der Waals gas with that of a classical ideal gas in the following contexts:

(a) heating from temperature T_i to T_f at constant volume;

(b) isothermal expansion from volume V_i to V_f;

(c) slow adiabatic expansion from volume V_i to V_f.

14. The critical point for the noble gas neon lies at $T_c = 44.4$ K and $P_c = 26.9$ atmospheres. Predict the volume per atom $(V/N)_c$ at the critical point. Explain your reasoning, too.

15. Investigate the *Dieterici* equation of state:

$$P = \frac{NkT}{V - Nb} \exp\left(\frac{-aN}{VkT}\right).$$

For example, does it predict a critical isotherm? Is which regions of the P–V plane does it differ significantly from the van der Waals equation of state? How does its prediction for the combination $(N/V)_c(kT_c/P_c)$ compare with experiment?

13 The Classical Limit

13.1 Classical phase space
13.2 The Maxwellian gas
13.3 The equipartition theorem
13.4 Heat capacity of diatomic molecules
13.5 Essentials

This chapter adapts the quantum results of chapter 5 to the language of classical physics. The chapter can be read any time after you have studied chapter 5. Of all the other chapters, only chapters 14 and 15 make essential use of the material developed here.

13.1 Classical phase space

Classical physics describes a collection of particles in terms of their positions and momenta. For a single particle restricted to one-dimensional motion, the variables are x and p_x. To provide a graphical representation of the particle's classical state, one uses an $x-p_x$ plane, as illustrated in figure 13.1, and plots a point in the plane. The $x-p_x$ plane is called the particle's two-dimensional *phase space*.

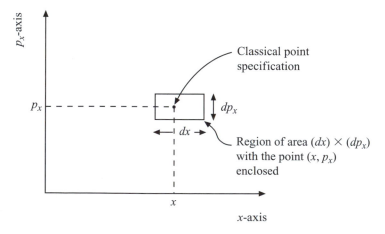

Figure 13.1 The two-dimensional phase space for a particle constrained to one-dimensional motion.

306

The Heisenberg uncertainty principle puts a bound, however, on our simultaneous knowledge of position and momentum. The principle asserts that

$$\Delta x \, \Delta p_x \geqslant \frac{h}{4\pi}, \tag{13.1}$$

provided Δx and Δp_x are the root mean square estimates of the uncertainties in position and momentum. Moreover, in quantum theory, one cannot construct a wave function that ascribes a definite position to a particle and, simultaneously, a definite momentum. The optimal wave function leads to uncertainties Δx and Δp_x that meet the lower bound implied by (13.1). This suggests that a single quantum state corresponds (at best) to a region in the two-dimensional phase space whose "area" is approximately h in size.

The preceding paragraph has two immediate implications.

1. We can afford to frame statements about probabilities in terms of *ranges* of the continuous variables x and p_x, using differentials such as dx and dp_x, rather than using a literal point in phase space: (x, p_x).
2. When we convert a sum over quantum states to an integral over the classical two-dimensional phase space, we need to reason that

$$\left(\begin{array}{c} \text{the number of states in the} \\ \text{rectangular region } dx \times dp_x \end{array} \right) = \# \frac{dx \times dp_x}{h}, \tag{13.2}$$

where $\#$ is a dimensionless number of order 1 and whose precise value remains to be determined.

Indeed, let us establish the numerical value of the constant $\#$ right now. Equation (5.32) gave us the partition function for a single particle free to move in a three-dimensional cubical box:

$$Z_1 = \sum_\alpha e^{-\varepsilon_\alpha / kT} = \frac{(2\pi m k T)^{3/2}}{h^3} V. \tag{13.3}$$

The classical analog must be

$$Z_{\text{semi-classical}} = \int \exp\left(-\frac{p^2}{2m} \bigg/ kT \right) \#^3 \frac{d^3 x \, d^3 p}{h^3}, \tag{13.4}$$

where the energy is the classical kinetic energy, $p^2 / 2m$, and where

$$d^3 x \, d^3 p = dx \, dy \, dz \times dp_x \, dp_y \, dp_z$$

in terms of Cartesian coordinates. One factor of $\#/h$ arises for each pair of position and momentum variables. [Because Planck's constant remains in a multiplicative fashion (although not in the energy expression), the integral in (13.4) is best called the "semi-classical" partition function.] The spatial integration goes over the box volume V. Hold the momentum variables fixed and integrate with respect to position; that integration will produce a factor of V. Because the energy depends on the momentum magnitude only, the momentum integration is done most easily with spherical polar coordinates. One replaces $d^3 p$ by $4\pi p^2 \, dp$, as sketched in figure 13.2. Thus

$$Z_{\text{semi-classical}} = \frac{\#^3 V 4\pi}{h^3} \int_0^\infty \exp\left(-\frac{p^2}{2m}\bigg/ kT\right) p^2 \, dp$$

$$= \#^3 \frac{(2\pi mkT)^{3/2}}{h^3} V. \tag{13.5}$$

Appendix A provides the value of the definite integral. Upon comparing equations (13.3) and (13.5), we find that the dimensionless constant $\#$ is precisely 1.

The generalization of equation (13.2) to three-dimensions is the statement,

$$\left(\begin{array}{c} \text{the number of states in the} \\ \text{six-dimensional region } d^3x \, d^3 p \end{array}\right) = \frac{d^3 x \, d^3 p}{h^3}. \tag{13.6}$$

The full phase space for a single particle has six dimensions; fortunately, there is no need to visualize that. It suffices to visualize a three-dimensional "position" space (or "configuration" space) and, separately, a three-dimensional "momentum" space.

Generalization to 2, 3, or 10^{20} particles is not difficult, but the chapter will not need that generalization in any detail, and so we skip the extra algebra.

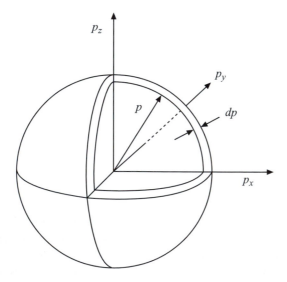

Figure 13.2 A thin shell in momentum space. Its "volume" is the surface area, $4\pi p^2$, times the radial thickness dp, that is, $4\pi p^2 \, dp$.

13.2 The Maxwellian gas

Now we turn to the classical probabilities. Consider a dilute gas, so dilute that we may ignore intermolecular forces and may treat the gas as a classical ideal gas. Classical physics usually fails when it tries to cope with molecular structure, and so we take the gas molecules to be structureless. Classical physics does, however, allow us to focus on a specific molecule and to ask for the probability that its position lies in the small volume element d^3x around position $\mathbf{x}$ and that its momentum lies in the "volume" d^3p around $\mathbf{p}$ in the momentum portion of phase space. Then we write

$$\begin{pmatrix} \text{probability that the molecule has} \\ \text{position in } d^3x \text{ around } \mathbf{x} \text{ and} \\ \text{momentum in } d^3p \text{ around } \mathbf{p} \end{pmatrix} \equiv \mathscr{P}(\mathbf{x}, \mathbf{p})d^3x\, d^3p. \tag{13.7}$$

The function $\mathscr{P}(\mathbf{x}, \mathbf{p})$ is called a *probability density*, and we work it out next.

The gas is in thermal equilibrium at temperature T, and so we may apply the canonical probability distribution. The full quantum version of the canonical distribution, from equation (5.10), is

$$P(\Psi_j) = \frac{\exp(-E_j/kT)}{Z}. \tag{13.8}$$

The symbol $P(\Psi_j)$ and the right-hand side refer to the probability for a *single* quantum state Ψ_j. In going to a classical version, we need to multiply by the number of quantum states that correspond to a given small region in the classical phase space. Thus, for one particle, the classical analog of equation (13.8) is

$$\mathscr{P}(\mathbf{x}, \mathbf{p})d^3x\, d^3p = \begin{pmatrix} \text{probability for} \\ \text{one quantum state} \end{pmatrix} \times \begin{pmatrix} \text{number of quantum} \\ \text{states in } d^3x\, d^3p \end{pmatrix}$$

$$= \frac{e^{-\varepsilon(\mathbf{x},\mathbf{p})/kT}}{Z_{\text{semi-classical}}} \times \frac{d^3x\, d^3p}{h^3}. \tag{13.9}$$

Here $\varepsilon(\mathbf{x}, \mathbf{p})$ is the molecule's energy as a function of position (in the Earth's gravitational field, say) and of momentum. By our analysis in section 13.1, the denominator $Z_{\text{semi-classical}}$ is equal to the integral of the other factors over all phase space:

$$Z_{\text{semi-classical}} = \int e^{-\varepsilon(\mathbf{x},\mathbf{p})/kT} \frac{d^3x\, d^3p}{h^3}. \tag{13.10}$$

That value for $Z_{\text{semi-classical}}$ ensures that the integral of $\mathscr{P}(\mathbf{x}, \mathbf{p})$ over all phase space yields one, for the molecule is certain to be found somewhere in its total phase space.

The probability on the left-hand side of equation (13.9) is intended to be entirely classical, as befits the definition in equation (13.7). There is to be no h and no notion of quantum states or uncertainty principle. On the right-hand side in equation (13.9), the explicit factor of $1/h^3$ cancels with the $1/h^3$ that is implicit in $Z_{\text{semi-classical}}$, and so

$\mathscr{P}(\mathbf{x}, \mathbf{p})$ is indeed independent of Planck's constant. In that sense especially, equation (13.9) is a classical limit.

If the energy $\varepsilon(\mathbf{x}, \mathbf{p})$ does not depend on $\mathbf{x}$, then appeal to equation (13.5) for the value of $Z_{\text{semi-classical}}$ gives

$$\mathscr{P}(\mathbf{x}, \mathbf{p}) \, d^3x \, d^3p = \frac{e^{-p^2/2mkT}}{V(2\pi mkT)^{3/2}} d^3x \, d^3p. \qquad (13.11)$$

For one molecule of the classical ideal gas (in the absence of forces other than those of the confining walls), this is the basic and correctly normalized probability distribution.

A note about terminology. In classical physics, most variables may take on a continuous range of values. A probability like that described on the left-hand side of equation (13.7) is proportional to one or more differentials. The function that multiplies the differentials is called the *probability density* (as noted above), and we will call the product as a whole the *probability distribution*.

Two kinds of conversions are useful. Here is a statement of the procedures; illustrations follow.

1. *Reduction.* To specialize the probability distribution, integrate over variables that you no longer care to retain. (The procedure is based on the last theorem in appendix C.)
2. *Transformation.* To transform to other independent variables, (a) compare differentials and (b) equate probabilities.

To illustrate reduction, suppose we no longer care about where the molecule may be, but we do want a probability distribution for momentum. Integration of (13.11) with respect to position gives a probability distribution for momentum alone:

$$\mathscr{P}(\mathbf{p}) d^3p = \int_{\text{volume } V} \mathscr{P}(\mathbf{x}, \mathbf{p}) d^3x \, d^3p$$

$$= \frac{e^{-p^2/2mkT}}{(2\pi mkT)^{3/2}} d^3p. \qquad (13.12)$$

To illustrate transformation, suppose we want a probability distribution for velocity, that is, $f(\mathbf{v}) \, d^3v$ is to give the probability that the molecule has velocity in the region d^3v around the point $\mathbf{v}$ in a velocity space. [The notation shifts from $\mathscr{P}$ to f for two reasons: (1) to avoid depending on the argument alone to distinguish one probability density from another; (2) for historical reasons.] Momentum and velocity are related by the equation $\mathbf{p} = m\mathbf{v}$, where m is the molecule's rest mass. To each region d^3v in velocity space, there corresponds a unique region d^3p in momentum space. Figure 13.3 illustrates this fact. The volumes are related by

$$d^3p = (m \, dv_x)(m \, dv_y)(m \, dv_z) = m^3 \, d^3v. \qquad (13.13)$$

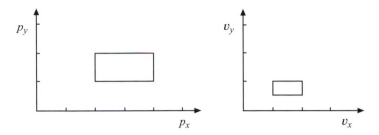

Figure 13.3 Corresponding regions in momentum and velocity space (displayed in two dimensions). The mass m was taken to have the value $m = 2$ kg, an unrealistic but convenient value.

Moreover, the probability that the molecule is to be found with velocity in the region d^3v around $\mathbf{v}$ must be the same as the probability that the molecule be found in the *uniquely corresponding* region d^3p around $\mathbf{p}$. Thus

$$f(\mathbf{v})\,d^3v = \mathscr{P}(\mathbf{p})\,d^3p \tag{13.14}$$

for the corresponding regions. Using equation (13.13) to express d^3p in terms of d^3v, we find

$$f(\mathbf{v})d^3v = \mathscr{P}(\mathbf{p})|_{\mathbf{p}=m\mathbf{v}}\,m^3\,d^3v$$

$$= \left(\frac{m}{2\pi kT}\right)^{3/2}e^{-mv^2/2kT}d^3v. \tag{13.15}$$

James Clerk Maxwell arrived at this functional form in 1860 (by a different route and with a slightly different significance). We will call the expression in (13.15) the *Maxwell velocity distribution* (which is short for a "probability distribution for a molecule's velocity"). It deserves a separate display:

$$\left(\begin{array}{c}\text{probability that molecule has}\\ \text{velocity in } d^3v \text{ around } \mathbf{v}\end{array}\right) \equiv f(\mathbf{v})d^3v = \left(\frac{m}{2\pi kT}\right)^{3/2}e^{-mv^2/2kT}d^3v.$$

$$\tag{13.16}$$

Note that the right-hand side depends on the magnitude of the velocity $\mathbf{v}$ but not on the direction. Thus all directions (at fixed magnitude) are equally likely. One says that the probability distribution is *isotropic*.

If the direction of molecular motion is irrelevant, then the reduction procedure will generate a probability distribution for the magnitude only of the velocity: the speed. Integration over direction at fixed speed amounts to asking for the probability that the velocity vector falls in a thin spherical shell of radius v and thickness dv in velocity space. The picture is akin to figure 13.2. Thus we write

$$\left(\begin{array}{c}\text{probability that molecule has}\\ \text{speed in the range } v \text{ to } v + dv\end{array}\right) \equiv \tilde{f}(v)dv = \int_{\text{thin shell}} f(\mathbf{v})d^3v$$

$$= \left(\frac{m}{2\pi kT}\right)^{3/2} e^{-mv^2/2kT} 4\pi v^2 \, dv. \qquad (13.17)$$

Figure 13.4 illustrates the probability density, $\tilde{f}(v)$, for the speed distribution. The initial rise is approximately proportional to v^2 and corresponds to the increasing volume of thin shells (of fixed radial thickness Δv) as the radius increases; the volume grows as $4\pi v^2 \Delta v$. Ultimately, the exponential takes over, and the probability density approaches zero asymptotically.

The probability distribution for speed, $\tilde{f}(v) \, dv$, depends on molecular parameters and the temperature through the combination kT/m only. That combination has the dimensions of $(\text{length})^2/(\text{time})^2$, that is, the dimensions of $(\text{speed})^2$. [You can see this quickly by recalling that the exponent in equation (13.17) must be dimensionless.] Any characteristic speed that the probability distribution provides will necessarily be proportional to $\sqrt{kT/m}$. Only the dimensionless numerical coefficient may vary. Three characteristic speeds come to mind, as follows.

1. Most probable speed. The most probable speed corresponds to the maximum in figure 13.4, where the slope is zero. Thus one solves the algebraic equation

$$\frac{d\tilde{f}(v)}{dv} = \sqrt{\frac{2}{\pi}}\left(\frac{m}{kT}\right)^{3/2}\left[2v + \left(\frac{-mv}{kT}\right)v^2\right]e^{-mv^2/2kT} = 0.$$

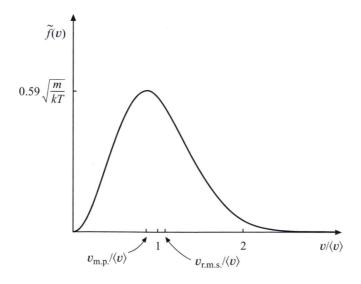

Figure 13.4 The probability density for the Maxwell speed distribution. The abscissa is marked in the dimensionless ratio $v/\langle v \rangle$.

The most probable speed, denoted $v_{\text{m.p.}}$, emerges as

$$v_{\text{m.p.}} = \sqrt{2}\sqrt{\frac{kT}{m}}.$$

(13.18)

2. Mean speed. An average or mean speed $\langle v \rangle$ is computed by weighting the speed v with its probability of occurrence $\tilde{f}(v)\,dv$ and then integrating:

$$\langle v \rangle = \int_0^\infty v\tilde{f}(v)dv = \int_0^\infty e^{-mv^2/2kT}\sqrt{\frac{2}{\pi}\left(\frac{m}{kT}\right)^{3/2}}v^3\,dv$$

$$= \sqrt{\frac{8}{\pi}}\sqrt{\frac{kT}{m}},$$

(13.19)

upon appeal to appendix A for the definite integral.

3. Root mean square speed. A calculation of $\langle v^2 \rangle$ proceeds as

$$\langle v^2 \rangle = \int_0^\infty v^2 \tilde{f}(v)\,dv = 3\frac{kT}{m},$$

(13.20)

again upon appeal to appendix A for the explicit integral. Of course, this result is equivalent to none other than the equation

$$\tfrac{1}{2}m\langle v^2 \rangle = \tfrac{3}{2}kT,$$

a relation familiar from section 1.2, but it is a nice check. Taking the square root of (13.20) yields the root mean square speed, $v_{\text{r.m.s.}}$:

$$v_{\text{r.m.s.}} = \sqrt{3}\sqrt{\frac{kT}{m}}.$$

(13.21)

The three speeds are closely similar. The mean speed $\langle v \rangle$ is 13 percent larger than $v_{\text{m.p.}}$, and $v_{\text{r.m.s.}}$ is 22 percent larger. The common proportionality to $\sqrt{kT/m}$ has two immediate implications: higher temperature implies higher speed, and larger mass implies lower speed.

For diatomic nitrogen ($m = 4.65 \times 10^{-26}$ kg) at room temperature ($T = 293$ K), the root mean square speed is

$$v_{\text{r.m.s.}} = 511 \text{ m/s}.$$

(The length of a football field is approximately 100 meters, and so you can visualize $v_{\text{r.m.s.}}$ as approximately five football fields per second.) The speed of sound in air at room temperature is $\cong 340$ m/s. Thus it is comparable to $v_{\text{r.m.s.}}$, as one would expect, because sound waves should propagate at roughly the speed of typical molecular motions.

The root mean square speed is easily measured—albeit indirectly—with a scale, a vacuum pump, and a barometer. Equation (1.6), derived from kinetic theory, can be written as

$$P = \tfrac{1}{3} \frac{Nm}{V} \langle v^2 \rangle. \tag{13.22}$$

The combination Nm/V is the mass density. For air, that density can be measured by weighing a stoppered one-liter flask both before and after it is evacuated. (The dominant constituents of air, diatomic nitrogen and oxygen, differ little in molecular mass, and so—to 15 percent accuracy—one may lump all the molecules in air together as a single species: "air molecules" of a single mass m.) A mercury barometer provides the current value of atmospheric pressure. Standard values are $Nm/V = 1.2$ kg/m^3 (at $T = 293$ K and atmospheric pressure) and $P_{\text{atmospheric}} = 1.01 \times 10^5$ N/m^2. Those values yield

$$v_{\text{r.m.s.}} \cong 500 \text{ m/s}.$$

(The literal numerical result is 502 m/s, but our lumping all molecular constituents into one average gas precludes anything beyond two-figure accuracy, at most.)

Normalization

The probability distributions in this section are all correctly normalized, that is, the integral of the probability density over the relevant domain equals 1. The probability distribution $\mathscr{P}(\mathbf{x}, \mathbf{p}) \, d^3x \, d^3p$ in (13.9) and (13.10) was constructed to have that property. Reduction and transformation preserve the normalization. Algebraic errors can creep in, however, and so *always* check that your probability distribution is correctly normalized.

If you need to construct a probability distribution from scratch, you can sometimes split the work into two pieces:

1. write down the essential variable part, perhaps as a Boltzmann factor times an appropriate differential expression;
2. determine an unknown multiplicative constant by requiring that the probability distribution be correctly normalized.

In part 1, you will need to be careful to insert the correct analog of $d^3x \, d^3p$, especially if you are using coordinates other than Cartesian.

13.3   The equipartition theorem

The kinetic energy of a particle depends quadratically on the momentum components: $(p_x^2 + p_y^2 + p_z^2)/2m$. Sometimes a potential energy is quadratic in the position variable x: $\tfrac{1}{2}k_{\text{sp}}x^2$, where k_{sp} denotes the spring constant. Or a potential energy can be expanded around the minimum of a more complicated potential well, and then $\tfrac{1}{2}k_{\text{sp}}x^2$ is the first term of interest in the expansion. In short, energy expressions that have a quadratic structure arise often. The equipartition theorem provides a short route to determining the associated energy as a function of temperature—provided that two assumptions are fulfilled.

Assumption 1. The *classical* version of the canonical probability distribution is applicable and adequate.

Assumption 2. The classical expression for the total energy splits additively into two parts: one part depends quadratically on a single variable (x, say), and the other part is entirely independent of that variable. Thus

$$E = ax^2 + E_{\text{other}}. \qquad (13.23)$$

Here a is a positive constant (but is otherwise arbitrary), and E_{other} depends on other variables (such as y, z, and $\mathbf{p}$) but not on x. The range of the variable x must be $-\infty \leqslant x \leqslant \infty$ or must stretch from zero to either positive or negative infinity.

Equipartition theorem. Given assumptions 1 and 2, the equipartition theorem states that the estimated value of the energy ax^2 is always $\frac{1}{2}kT$, regardless of the numerical value of the constant a and independent of the details of E_{other}:

$$\langle ax^2 \rangle = \tfrac{1}{2}kT. \qquad (13.24)$$

The proof is short. Assumption 1 entitles us to write

$$\langle ax^2 \rangle = \frac{\int ax^2 e^{-E/kT} \, dx \, d(\text{others})}{\int e^{-E/kT} \, dx \, d(\text{others})},$$

where $d(\text{others})$ denotes differentials of the other classical variables. Assumption 2 enables us to factor the exponentials and then the integrals:

$$\langle ax^2 \rangle = \frac{\int ax^2 e^{-ax^2/kT} \, dx \times \int e^{-E_{\text{other}}/kT} \, d(\text{others})}{\int e^{-ax^2/kT} \, dx \times \int e^{-E_{\text{other}}/kT} \, d(\text{others})}.$$

The integrals with the other variables cancel, and so those variables and E_{other} are indeed irrelevant. The substitution $q^2 = ax^2/kT$ will extract the dependence on kT from the remaining integrals. One finds

$$\langle ax^2 \rangle = kT \frac{\int q^2 e^{-q^2} \, dq}{\int e^{-q^2} \, dq}$$

$$= \tfrac{1}{2}kT.$$

Already the first line tells us that the result is independent of the constant a and is proportional to kT. The integrals in appendix A lead to the factor of $1/2$.

The outcome of the proof is appropriately called an "equipartition" theorem because energy is doled out equally to every variable whose classical energy expression is quadratic.

Note how extensive the "other" variables may be. They may include the variables for other particles or other aspects of the physical system. So long as assumptions 1 and 2 are fulfilled, the coordinate x and the energy ax^2 may be a tiny aspect of a large and complex system.

For example, one could choose the center-of-mass momentum component p_x of a molecule in a dense gas or a liquid. Intermolecular forces certainly are exerted on the molecule, but they are not coupled to p_x in the classical energy expression. Hence the equipartition theorem asserts that $\langle p_x^2/2m \rangle = \frac{1}{2}kT$. In classical theory, the intermolecular forces do not affect the estimated molecular kinetic energy.

Harmonic oscillator

The harmonic oscillator provides a good model for the vibration of the two atoms in a diatomic molecule. The general form of the vibrational energy would be

$$\varepsilon = \frac{p^2}{2m_{\text{eff}}} + \frac{1}{2}k_{\text{sp}}x^2. \tag{13.25}$$

The mass m_{eff} is the effective mass (or reduced mass) for the model, and k_{sp} represents the curvature at the bottom of the attractive potential well, displayed in figure 11.1. The variable x represents the separation of the two atoms minus their separation at minimum potential energy. Thus x is zero when the separation corresponds to the bottom of the potential well. Although x cannot meaningfully range from $-\infty$ to $+\infty$, that is not a serious difficulty. Unless the temperature is so high that the molecule is close to dissociation, the distant behavior of the potential energy curve is inconsequential. One may use the harmonic oscillator form as though the true potential extended parabolically from $-\infty$ to $+\infty$. In using the model to estimate the vibrational energy as a function of temperature, the pressing question is whether classical physics is adequate. Let us see.

The classical calculation of the total vibrational energy goes quickly. Appeal to the equipartition theorem yields

$$\langle \varepsilon \rangle_{\text{classical}} = \left\langle \frac{p^2}{2m_{\text{eff}}} \right\rangle + \langle \tfrac{1}{2}k_{\text{sp}}x^2 \rangle$$

$$= \tfrac{1}{2}kT + \tfrac{1}{2}kT = kT. \tag{13.26}$$

For the quantum estimate of the vibrational energy, we can take the result from equation (6.5) and restore to it the constant $\frac{1}{2}h\nu$ that corresponds to the energy scale implicit in equation (13.25):

$$\langle \varepsilon \rangle_{\text{quantum}} = \tfrac{1}{2}h\nu + \frac{h\nu}{e^{h\nu/kT} - 1}. \tag{13.27}$$

Here ν is the frequency of oscillation, which is given in terms of the parameters by $2\pi\nu = \sqrt{k_{\text{sp}}/m_{\text{eff}}}$. As $T \rightarrow 0$, more specifically, when $kT \ll h\nu$, the quantum estimate goes to the value $\tfrac{1}{2}h\nu$, called the *zero point energy*. Already the Heisenberg uncertainty principle ensures that the quantum estimate cannot vanish as $T \rightarrow 0$ because zero energy would require that both x^2 and p^2 be zero. Indeed, minimizing the right-hand side of equation (13.25) subject to $|x| \times |p| = Ch$, where C is a dimensionless constant of order unity, leads to a minimum energy of order $h\nu$. (In chapter 6, we chose a zero of energy so that the oscillator energy would be zero in the limit $T \rightarrow 0$. That choice is sensible for electromagnetic radiation and is practical for sound waves. Now, however, we retain the zero point energy, which does have empirical consequences in molecular physics and elsewhere.)

Consider next the high temperature limit: $kT \gg h\nu$. The exponent in (13.27) is then small, and one can expand the exponential:

$$\langle \varepsilon \rangle_{\text{quantum}} = \tfrac{1}{2}h\nu + \frac{h\nu}{1 + (h\nu/kT) + \cdots - 1}$$

$$= \tfrac{1}{2}h\nu + kT + \cdots \cong kT. \tag{13.28}$$

The last step follows already because of the strong inequality, $kT \gg h\nu$. In this limit, one recovers the classical value. [Moreover, if one takes four terms in the expansion of the exponential, the term in $\tfrac{1}{2}h\nu$ is canceled exactly. One finds

$$\langle \varepsilon \rangle_{\text{quantum}} = kT + \tfrac{1}{12}(h\nu/kT)^2 \times kT + \cdots .$$

The first correction term is quite small.]

Figure 13.5 compares the classical and quantum energies as a function of temperature.

Does the classical result apply to diatomic nitrogen at room temperature? Spectroscopic data give $\nu = 7.07 \times 10^{13}$ Hz, and so $h\nu = 0.292$ eV. For room temperature, $kT \cong 0.025$ eV. Thus

$$\frac{kT}{h\nu} \cong \frac{1}{10},$$

and the molecule is in the low temperature regime, not in the high temperature limit. No, classical physics is not adequate.

When $kT/h\nu = 1/10$ precisely, equation (13.27) yields

$$\langle \varepsilon \rangle_{\text{quantum}} = (\tfrac{1}{2} + 4.5 \times 10^{-5})h\nu = 0.146 \text{ eV},$$

relative to the bottom of the potential well. The well depth for nitrogen is -7.37 eV, relative to zero at infinite separation. Thus the estimated energy, $\langle \varepsilon \rangle_{\text{quantum}}$, leaves the vibrational system both close to its ground state and close to the bottom of the potential well. Approximating the well by a parabola that extends from $-\infty$ to $+\infty$ is entirely adequate because the vibrational motion is confined to the region near the

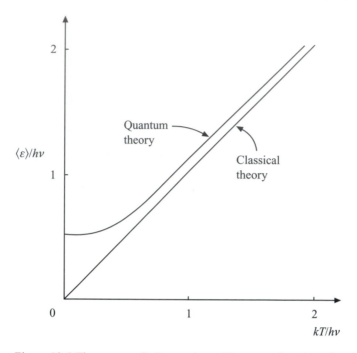

Figure 13.5 The energy of a harmonic oscillator as a function of temperature: classical and quantum estimates. In the units used for the two axes, the slope of each curve is $d\langle \varepsilon \rangle / dT$ divided by k. Thus the vibrational contribution to the molecular heat capacity, $d\langle \varepsilon \rangle / dT$, is proportional to the slope. When $kT/h\nu \leqslant 1/3$, the quantum (and true) heat capacity will be much less than the classical value.

minimum. The harmonic oscillator model is perfectly fine, but one has to use the correct physics in applying it.

13.4 Heat capacity of diatomic molecules

The heat capacity of typical diatomic molecules—at room temperature and in a dilute gas—belongs in every physicist's store of readily accessible knowledge. We can now determine the contributions to that heat capacity expeditiously. Translation, vibration, and rotation need to be considered, as follows.

1. Translation. Classical physics certainly applies to the translational motion, and the translational kinetic energy depends quadratically on the three components of momentum. Thus the equipartition theorem implies

$$\langle \varepsilon_{\text{translation}} \rangle = 3 \times \tfrac{1}{2}kT \qquad (13.29)$$

for a single molecule.

2. Vibration. Our exploration of the harmonic oscillator and diatomic nitrogen in the previous section suggests that room temperature is a very low temperature as far as vibrational motion is concerned. One can expect the vibrational energy to be merely the zero point energy (to good approximation):

$$\langle \varepsilon_{\text{vibration}} \rangle \cong \tfrac{1}{2} h\nu. \tag{13.30}$$

A characteristic vibrational temperature, θ_v, was defined in section 11.3 by the relationship $k\theta_v \equiv h\nu$. So long as the physical temperature T is much less than θ_v, equation (13.30) will hold. Moreover, when the inequality $T \ll \theta_v$ holds, the slope of the quantum curve in figure 13.5 is so nearly horizontal that vibration makes essentially no contribution to the heat capacity.

3. Rotation. A diatomic molecule has rotational symmetry about the axis joining the two nuclei. Call that axis the z-axis. Also, denote by I the moment of inertia for rotation about an axis passing through the center of mass and oriented perpendicular to the symmetry axis. Then, for our purposes, the kinetic energy of rotation has the structure

$$\varepsilon_{\text{rotation}} = \frac{J_x^2}{2I} + \frac{J_y^2}{2I}, \tag{13.31}$$

where the rotational angular momentum is denoted by **J**. Why is there no term for "rotation" about the symmetry axis? Because such "rotation" really would be excited electronic motion and is separated from the molecular ground state by such a large energy step that the possibility is negligible at room temperature.

Quantum theory assigns to the numerators in equation (13.31) characteristic values of order $\hbar^2$. Section 11.3 defined a characteristic rotational temperature, θ_r, by the relationship $k\theta_r \equiv \hbar^2/2I$. When the physical temperature is much greater than θ_r, classical physics applies. Table 11.2 indicates that room temperature exceeds θ_r by a wide margin for typical diatomic molecules, and so we may apply the equipartition theorem to the rotational energy:

$$\langle \varepsilon_{\text{rotation}} \rangle = 2 \times \tfrac{1}{2} kT. \tag{13.32}$$

(The rotational partition function, Z_{rot}, evaluated in section 11.3 in the context $T \gg \theta_r$, yields the same estimated energy. Thus it corroborates the claim that, when the thermal inequality holds, then classical physics applies.)

Table 13.1 summarizes the contributions to the total molecular energy and to the heat capacity at constant volume, C_V. (Recall the context: a dilute gas, in which intermolecular forces are negligible and do not contribute to C_V.) Thus, at room temperature, C_V typically has the value

$$C_V = \tfrac{5}{2} Nk. \tag{13.33}$$

By equation (1.18) and the preceding result, the theoretical ratio of heat capacities is

$$\gamma \equiv \frac{C_P}{C_V} = \frac{C_V + Nk}{C_V} = \frac{7}{5} = 1.4. \tag{13.34}$$

Table 13.1 *Energies and contributions to the heat capacity for a diatomic molecule. The thermal domain $\theta_r \ll T \ll \theta_v$ is specified, and room temperature usually satisfies both strong inequalities.*

	$\langle \varepsilon \rangle$	$d\langle \varepsilon \rangle / dT$
Translation	$\frac{3}{2}kT$	$\frac{3}{2}k$
Vibration	$\frac{1}{2}hv$	0
Rotation	kT	k
Sum	$\frac{5}{2}kT + \frac{1}{2}hv$	$\frac{5}{2}k$

How well do theory and experiment agree? Taking N_2, O_2, CO, and HCl as typical diatomic molecules, one finds agreement to within 0.3 percent.

13.5 Essentials

1. In classical physics, a particle's state is represented by a point in its *phase space*, a space whose axes are position variables and momentum variables.

2. The correspondence between a region in phase space and quantum states is the following:

$$\left(\begin{array}{c} \text{the number of states in the} \\ \text{six-dimensional region } d^3x\,d^3p \end{array} \right) = \frac{d^3x\,d^3p}{h^3}.$$

3. Specify a structureless classical ideal gas in thermal equilibrium. Then the basic probability relations are these:

$$\left(\begin{array}{c} \text{probability that the molecule has} \\ \text{position in } d^3x \text{ around } \mathbf{x} \text{ and} \\ \text{momentum in } d^3p \text{ around } \mathbf{p} \end{array} \right) \equiv \mathscr{P}(\mathbf{x},\,\mathbf{p})d^3x\,d^3p = \frac{e^{-\varepsilon(\mathbf{x},\mathbf{p})/kT}}{Z_{\text{semi-classical}}} \times \frac{d^3x\,d^3p}{h^3},$$

where

$$Z_{\text{semi-classical}} = \int e^{-\varepsilon(\mathbf{x},\mathbf{p})/kT} \frac{d^3x\,d^3p}{h^3}.$$

4. Other probabilities follow by two kinds of conversions:

1. *Reduction.* To specialize the probability distribution, integrate over variables that you want to eliminate.

2. **Transformation.** To transform to other independent variables, (a) compare differentials and (b) equate probabilities.

5. The *Maxwell velocity distribution* (which is short for a "probability distribution for a molecule's velocity") follows from item 3 as

$$\left(\begin{array}{c} \text{probability that molecule has} \\ \text{velocity in } d^3v \text{ around } \mathbf{v} \end{array} \right) \equiv f(\mathbf{v})d^3v = \left(\frac{m}{2\pi kT} \right)^{3/2} e^{-mv^2/2kT} d^3v.$$

6. The speed distribution follows from item 5 as

$$\left(\begin{array}{c} \text{probability that molecule has} \\ \text{speed in the range } v \text{ to } v + dv \end{array} \right) \equiv \tilde{f}(v)dv = \left(\frac{m}{2\pi kT} \right)^{3/2} e^{-mv^2/2kT} 4\pi v^2 dv.$$

7. Alone for dimensional reasons, all characteristic speeds derived from item 6 are proportional to $\sqrt{kT/m}$. The three common characteristic speeds are the following.

Most probable speed: $v_{\text{m.p.}} = \sqrt{2}\sqrt{\dfrac{kT}{m}}$.

Mean speed: $\langle v \rangle = \sqrt{\dfrac{8}{\pi}}\sqrt{\dfrac{kT}{m}}$.

Root mean square speed: $v_{\text{r.m.s.}} = \sqrt{3}\sqrt{\dfrac{kT}{m}}$.

8. *Equipartition theorem.* Given

1. that the *classical* version of the canonical probability distribution is applicable and adequate and
2. that the classical expression for the total energy splits additively into two parts, one part dependent *quadratically* on a single variable (x, say) that has an *infinite range* and the other part *entirely independent* of that variable,

then the equipartition theorem states that the estimated value of the energy ax^2 is always $\frac{1}{2}kT$, regardless of the numerical value of the constant a and independent of the details of the other part of the energy:

$$\langle ax^2 \rangle = \tfrac{1}{2}kT.$$

The weakness in this theorem is the first assumption.

9. For diatomic molecules in the thermal domain $\theta_r \ll T \ll \theta_v$, translation and rotation contribute classical values to the energy, but vibrational motion is frozen into merely the zero point energy. Thus, when present as a dilute gas, such molecules have a heat capacity at constant volume given by

$$C_V = \tfrac{5}{2}Nk.$$

The ensuing ratio of heat capacities is $\gamma = \frac{7}{5} = 1.4$. For typical diatomic molecules, room temperature satisfies both strong inequalities.

Further reading

Maxwell's derivation appeared in J. C. Maxwell, "Illustrations of the dynamical theory of gases.—Part I. On the motions and collisions of perfectly elastic spheres," *Phil. Mag.* **19**, 19–23 (1860), a portion of a series of papers.

Problems

1. *The single-particle density of states.* A semi-classical expression like equation (13.6) provides a convenient route to the single-particle density of states.

(a) Integrate over space (of volume V) and over the direction of the momentum **p** to determine $D(p)dp$, where $D(p)$ denotes the number of states per unit interval of momentum magnitude.

(b) Adopt the non-relativistic relationship between kinetic energy and momentum, $\varepsilon = p^2/2m$, and determine the number of states per unit energy interval, $D(\varepsilon)$. Do you find agreement with our previous result?

(c) Consider the relativistic relationship between total energy and momentum, $\varepsilon_{rel} = (p^2c^2 + m^2c^4)^{1/2}$. Determine the number of states per unit interval of total energy, $D(\varepsilon_{rel})$.

2. *Probability distribution for a velocity component.* Reduce the Maxwell velocity distribution to a probability distribution for the velocity component v_x only. Sketch your probability density and confirm that your probability distribution is correctly normalized (to 1). Calculate $\langle v_x \rangle$ and $\langle v_x^2 \rangle$.

3. Use the Maxwell velocity distribution to calculate $\langle |v_z| \rangle$, the mean of the absolute value of the z-component of velocity. Compare this with $\langle v \rangle$.

4. Calculate the probability that a gas molecule has velocity in the following range:

$$\sqrt{kT/m} \leqslant v_z \leqslant 1.01\sqrt{kT/m} \qquad \text{and} \qquad \sqrt{v_x^2 + v_y^2} \geqslant \sqrt{kT/m}.$$

The temperature and number density are such that the gas acts like a classical ideal gas. Give a numerical answer.

5. *Probability distribution for energy.* Transform the Maxwell speed distribution to a probability distribution for the kinetic energy ε. Provide a qualitatively faithful sketch of the probability density. What value do you find for $\langle \varepsilon \rangle$? What is the most probable value of the kinetic energy? Compare the latter with $\frac{1}{2}m(v_{m.p.})^2$. Comments?

6. *Doppler shift.* If an atom is moving when it emits light, the frequency ν that is observed in the lab will be Doppler shifted relative to the frequency ν_0 in the atom's rest frame. For light that travels along the x-axis, the lab frequency is given by

$$\nu = \nu_0 \times \left(1 + \frac{\upsilon_x}{c}\right),$$

correct to first order in inverse powers of the speed of light c. Here υ_x is the atom's x-component of velocity.

Take the light source to be sodium vapor at $T = 10^3$ K, and take $\nu_0 = 5.09 \times 10^{14}$ Hz. The mass of a sodium atom is 3.82×10^{-26} kg. Imagine that a spectroscope is oriented along the x-axis.

(a) Determine the probability distribution $\mathscr{P}(\nu)d\nu$ for receiving a photon in the frequency range ν to $\nu + d\nu$. The probability distribution is to be normalized to unity. (For convenience, allow $|\upsilon_x|$ to exceed the speed of light, but provide justification for that mathematical approximation.)

(b) What is the root mean square spread of lab frequencies? That is, what is $\langle(\nu - \nu_0)^2\rangle^{1/2}$? How large is that spread in comparison with the natural frequency ν_0?

7. *Relative motion.* The probability that two atoms will react when they collide (perhaps to form a diatomic molecule) depends on their energy in their center-of-mass reference frame. In the following, take the two atoms to have the same mass.

(a) Express the energy in the lab frame in terms of the velocity of the center of mass, $\mathbf{V_{CM}}$, and the relative velocity, $\Delta\mathbf{v} = \mathbf{v}_1 - \mathbf{v}_2$.

(b) Take the probability distribution for the two atoms to be the product of two Maxwell velocity distributions. Transform that probability distribution to a probability distribution for $\mathbf{V_{CM}}$ and $\Delta\mathbf{v}$.

(c) Reduce your answer in part (b) to a probability distribution for $\Delta\mathbf{v}$ alone. What is the most probable value for the magnitude of the relative velocity? How does that value compare with the most probable value of a single atom's speed?

(d) Calculate the mean value of the relative speed, $\langle|\Delta\boldsymbol{v}|\rangle$. By what numerical factor does this differ from the mean speed of one atom, $\langle\upsilon\rangle$?

(e) By transformation and reduction, convert your probability distribution for $\Delta\mathbf{v}$ to a probability distribution for the energy $\varepsilon_{\text{rel.motion}}$ in the CM frame. (Here $\varepsilon_{\text{rel.motion}}$ denotes the energy of the relative motion, which is the energy in the CM frame.) What value do you find for $\langle\varepsilon_{\text{rel.motion}}\rangle$?

8. *Sound waves in air.* A sound wave consists of correlated variations in the pressure P, the mass density ρ, and the bulk motion of gas molecules. The characteristic speed υ_{ch} is determined by the relation

$$v_{ch} = \left[\left(\frac{\partial P}{\partial \rho} \right)_{adiabatic} \right]^{1/2}.$$

The subscript indicates that the partial derivative corresponds to an adiabatic change, not an isothermal change.

(a) Compute v_{ch} for a classical ideal gas (which may be diatomic) in terms of kT, the molecular rest mass m, and the heat capacities.

(b) Evaluate v_{ch} for diatomic nitrogen at room temperature and then compare with the measured speed of sound in air.

9. *Heat capacities of diatomic molecules.* Determine γ, the ratio of heat capacities, for a dilute gas in the following cases.

(a) Carbon monoxide when $T = 300$ K.
(b) Iodine when $T = 500$ K.
(c) Hydrogen when $T = 10$ K.

10. *Classical solid.* Consider a solid whose atoms form a cubical lattice; common sodium chloride will do nicely. An individual atom may vibrate independently in the x, y, and z directions of Cartesian axes.

(a) If the N atoms of the solid are considered as N *independent* harmonic oscillators, what does classical physics predict for C/N, the heat capacity per atom?

The classical value of C/N is known as the *Dulong and Petit value.* In 1819, P. L. Dulong and A. T. Petit found experimentally that C/N is approximately the same for many metals at the relatively high temperatures they used. (Dulong and Petit expressed their experimental results in the units of that era, different from our units, but the essential content was the same. They determined heat capacities by measuring cooling rates from initial temperatures of order 300 to 500 K.)

(b) If you have read section 6.5, would you expect Dulong and Petit to have found the classical value for C/N when they studied the metals silver, gold, copper, and iron?

11. For a particle constrained to one-dimensional motion (on $-\infty \leqslant x \leqslant \infty$), suppose the energy expression is

$$\varepsilon(x,\, p) = \frac{p^2}{2m} + bx^{2n},$$

where b is a positive constant and n is a positive integer. Compute the estimated values of the kinetic and potential energies as functions of the temperature T and constants. You will need to consider some integrals, but aim to calculate the estimates without evaluating any integral explicitly.

12. *Intramolecular speed*. For diatomic nitrogen at room temperature, estimate the speed with which each atom moves with respect to the molecular center of mass. Estimate also the percentage uncertainty in your estimate of the speed.

13. *Rotating dust grain*. Consider an interstellar dust grain whose shape is like a cigar. The diameter is 10^{-7} meter, and the length is 10 times as long. The grain is electrically neutral and is in thermal equilibrium with dilute gaseous hydrogen. Typically, the grain will have some angular momentum relative to its center of mass. Would you expect the angular momentum to lie primarily parallel to the long axis or perpendicular to it? Be at least semi-quantitative in your response, please.

14. *Paramagnetism classically*. Treat the spatially fixed paramagnetic particle of section 5.3 classically, allowing the magnetic moment $\mathbf{m_B}$ to take on any orientation relative to the fixed external magnetic field $\mathbf{B}$. The orientational energy is

$$-\mathbf{m_B} \cdot \mathbf{B} = -m_B B \cos \theta$$

where θ is the angle between the magnetic moment and the field.

(a) Determine the classical probability distribution $\mathscr{P}(\theta)d\theta$ that the orientation angle θ lies in the range $d\theta$ around θ. Recall that the expression for an infinitesimal solid angle in spherical polar coordinates is $\sin \theta d\theta d\varphi$, where φ is the azimuthal angle. Confirm the normalization.
(b) Calculate the estimate ⟨component of $\mathbf{m_B}$ along $\mathbf{B}$⟩ as a function of temperature T and field magnitude B. The result is called the *Langevin* expression (after the French physicist Paul Langevin).
(c) Plot your result from (b) as a function of $m_B B/kT$ and compare it with the quantum result (for spin $\frac{1}{2}\hbar$), displayed in section 5.3. For a good comparison, superimpose the two graphs, classical and quantum.

15. *Isothermal atmosphere revisited*. A monatomic classical gas is held in a vertical cylinder of height H in the Earth's gravitational field g. There are N atoms, each of mass m, in thermal equilibrium at temperature T.

(a) Determine the probability density $\mathscr{P}(\mathbf{x}, \mathbf{p})$ for a single atom of the gas. Be sure that the integral of $\mathscr{P}(\mathbf{x}, \mathbf{p})d^3x\, d^3p$ over the relevant domain equals one.
(b) Calculate the average energy ⟨E⟩$/N$.
(c) Graph the quotient ⟨E⟩$/NmgH$ as a function of kT/mgH. (Both of these variables are dimensionless and hence advantageous.)
(d) By inspection, for which value of kT/mgH is the heat capacity largest?
(e) What is ⟨E⟩$/N$ in the limits of small and large values of kT/mgH (relative to 1)?

16. *Effusion*. Air leaks into an evacuated chamber through a pin-hole of area A.

(a) Adapt the analysis of section 1.2 to calculate the rate at which molecules enter the chamber (in terms of A, T, N/V, and the mass m of an "air molecule"). Take the wall to be infinitesimally thin so that transit through the hole is not a complication. Moreover, ignore the hole's effect on the local conditions in the air.

(b) If the pin-hole has the extent $A = 10^{-12}$ m^2 and if the air is under typical room conditions, what is the entry rate (in molecules per second)?

(c) Consider the molecules that leak through the hole in one second. Qualitatively, how does their average kinetic energy compare with $\frac{3}{2}kT$? Explain your reasoning.

(d) Calculate in detail the average kinetic energy discussed in part (c). Express it as a multiple of kT.

17. *A sobering thought in this day of energy awareness.* Suppose you turn on the furnace to raise the temperature of the air in a room from 19 °C to 22 °C. The furnace consumes energy in achieving this increment. Does the energy in the room increase? Why? (Here I am thinking strictly of the energy of the air in the room.) If your answer is no, where has the energy gone? (Recall that, in a realistic house, the pressure in the room remains atmospheric pressure, that is, remains equal to the pressure outdoors.)

14 Approaching Zero

14.1 Entropy and probability

14.2 Entropy in paramagnetism

14.3 Cooling by adiabatic demagnetization

14.4 The Third Law of Thermodynamics

14.5 Some other consequences of the Third Law

14.6 Negative absolute temperatures

14.7 Temperature recapitulated

14.8 Why heating increases the entropy. Or does it?

14.9 Essentials

This chapter explores in greater depth the connections among entropy, temperature, and energy transfer by heating (or cooling). The sections are diverse but related, albeit loosely. The first section develops a general expression for entropy in terms of probabilities. The next section provides an example and also prepares the ground for section 14.3, which discusses a classic method of reaching low temperatures. The section entitled "The Third Law of Thermodynamics" explores the constraints on methods for approaching absolute zero. In section 14.6, we discover that temperatures numerically below absolute zero have been produced in the lab, and we explore that context. Finally, the last three sections reflect on the notions of temperature, heating, and entropy.

In short, the chapter's title, "Approaching zero," highlights the central four sections, sections 14.3 to 14.6. The sections that precede and follow those sections are loosely related to the central sections and are also valuable in their own right.

14.1 Entropy and probability

Chapter 2 used the idea of multiplicity and its evolution to state the Second Law of Thermodynamics. A quantitative study of how multiplicity changes during a slow isothermal expansion led to entropy, defined as $S \equiv k \ln (\text{multiplicity})$. In chapter 5, we used the entropy concept to derive the canonical probability distribution; thereby we introduced an explicit expression for probabilities at thermal equilibrium. Now we complete the loop by learning how the probabilities can be used to calculate the entropy.

First we marshal the essential equations. Equation (5.10) gave the canonical probability distribution as

$$P(\Psi_j) = \frac{\exp(-E_j/kT)}{Z}, \tag{14.1}$$

and (5.15) gave the energy estimate $\langle E \rangle$ as

$$\langle E \rangle = \sum_j E_j \, P(\Psi_j). \tag{14.2}$$

Finally, equation (5.25) gave the entropy as

$$S = \frac{\langle E \rangle}{T} + k \ln Z \tag{14.3}$$

in the context of the canonical probability distribution.

To write the entropy in terms of the probabilities, eliminate E_j and $\langle E \rangle$ among the three equations. First take the logarithm of equation (14.1) and then rearrange terms as

$$E_j = -kT \, [\ln P(\Psi_j) + \ln Z].$$

Insert this expression for E_j into the right-hand side of (14.2):

$$\langle E \rangle = -kT \sum_j [\ln P(\Psi_j) + \ln Z] P(\Psi_j)$$

$$= -kT \left[\ln Z + \sum_j P(\Psi_j) \ln P(\Psi_j) \right].$$

Because the probabilities sum to 1, the term in $\ln Z$ reduces to merely $-kT \ln Z$. Now use this expression for $\langle E \rangle$ in (14.3), finding

$$S = -k \sum_j P(\Psi_j) \ln P(\Psi_j). \tag{14.4}$$

The entropy emerges as a sum of $-P(\Psi_j) \ln P(\Psi_j)$; each logarithm is weighted by the probability that the state Ψ_j is the correct state to use. Because the probability $P(\Psi_j)$ is less than 1 (or equal to 1), $\ln P(\Psi_j)$ is negative (or zero), and the entropy is a positive quantity (or zero).

What happens when Ψ_j corresponds to a state of high energy E_j and consequently small probability? Mathematically, as $P_j \to 0$, the limit of the function $P_j \ln P_j$ is zero (as you can confirm with l'Hôpital's rule). Thus states of vanishing probability do not contribute to the entropy. That makes sense, for such states do not contribute to the multiplicity.

The structure of equation (14.4),

$$\left[\begin{array}{c} \text{sum of} \\ -\text{probability} \times \ln(\text{probability}) \end{array} \right], \tag{14.5}$$

is the uniquely correct way to assess the logarithm of a "multiplicity" in a general sense of that notion. When Claude Shannon invented the field of Information Theory (in 1948), he found himself led to the structure in (14.5). Working for Bell Telephone Laboratories, he asked himself, how can one encode messages most efficiently? In turn, that led him to the question, how uncertain is one about a message when one knows merely its length? For example, in a message of eight characters written in correct English, the number of possibilities is *not* $(26 + 2)^8$, where 26 is the number of letters in the alphabet and where the "+2" accounts for "space" and "period." (The distinction between capital and lower case letters is ignored here, as it is in telegraph messages, which are sent in all caps.) No English word has four q's in a row: qqqq. There are many other exclusions. Moreover, certain combinations are more probable than others; for example, the pair "ab" is more probable than "bb." If the first seven characters spell out "St Loui," the probability is high that the last letter is "s." Shannon set up three general criteria that any reasonable measure of uncertainty about a message ought to satisfy. Then he found that only a structure like (14.5) would meet the criteria.

To pursue the example further, let $n \equiv (26 + 2)^8$. If all n sequences of 8 characters were equally probable as a message, then each sequence would have a probability of $1/n$. According to (14.5), the equation

$$\left[\begin{array}{c} \text{sum of} \\ -\text{probability} \times \ln\,(\text{probability}) \end{array} \right] = n \times \left(-\frac{1}{n} \times \ln \frac{1}{n} \right) = \ln n \tag{14.6}$$

would hold. The uncertainty would be $\ln n$. The form is like the logarithm of a multiplicity as we developed that idea in chapter 2. The structure in (14.5) and in (14.4) generalizes from a situation in which the microstates are equally probable to a situation in which the microstates differ in probability among themselves. In short, we are *not* replacing the old idea that "entropy is the logarithm of the multiplicity (times Boltzmann's constant)." Rather, we are retaining that idea but are generalizing the notion of multiplicity. Microstates that differ in probability contribute differently to the multiplicity.

Section 5.5 was devoted to choosing the energy range δE in the quantum form of multiplicity, which we had expressed in section 4.2 in terms of the density of states $D(E)$ for the entire system: $S = k \ln [D(E)\delta E]$. I characterized the choice as "plausible and even natural" and promised evidence that the selection is also "uniquely correct." Although more details could be filled in, the correspondence between entropy as expressed in equation (14.4) and Shannon's uncertainty measure is evidence that section 5.5 selected δE correctly.

A physical example of how to use equation (14.4) may be welcome; the next section provides it.

14.2 Entropy in paramagnetism

We return to the single paramagnetic atom of section 5.3. Its spin is $\frac{1}{2}\hbar$, and its location is absolutely fixed in some crystal lattice. The system has two energy eigenstates:

magnetic moment parallel (∥) or anti-parallel (anti-∥) to an external magnetic field **B**. Section 5.3 gave the probabilities as

$$P_\parallel = \frac{e^{+m_\mathrm{B} B/kT}}{2\cosh\left(m_\mathrm{B} B/kT\right)}, \tag{14.7}$$

$$P_{\text{anti-}\parallel} = \frac{e^{-m_\mathrm{B} B/kT}}{2\cosh\left(m_\mathrm{B} B/kT\right)}. \tag{14.8}$$

Equation (14.4) gives the entropy (divided by k) as

$$S/k = -\left(P_\parallel \ln P_\parallel + P_{\text{anti-}\parallel} \ln P_{\text{anti-}\parallel}\right), \tag{14.9}$$

and so substitution yields

$$S/k = -\frac{m_\mathrm{B} B}{kT}\tanh\left(m_\mathrm{B} B/kT\right) + \ln\left[2\cosh\left(m_\mathrm{B} B/kT\right)\right]. \tag{14.10}$$

[This result is, of course, derivable also from equation (14.3) and the partition function $Z = 2\cosh(m_\mathrm{B} B/kT)$.] Figure 14.1 displays the entropy as a function of temperature at fixed external field.

In this ideal situation—no interaction with neighboring magnetic moments—the entropy depends on B and T through the ratio B/T only. In the limit $T \to \infty$ at fixed field B, the probabilities (14.7) and (14.8) become equal. Their limiting value is $\frac{1}{2}$, and so equation (14.9) implies

$$\lim_{T\to\infty} S/k = \ln 2 \tag{14.11}$$

at fixed field B. At temperatures such that $kT \gg m_\mathrm{B} B$, the energy difference between

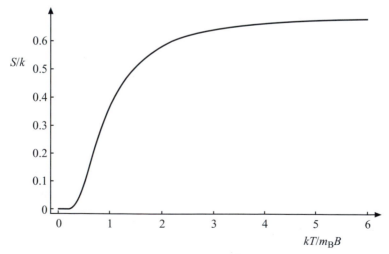

Figure 14.1 The entropy of an ideal spin $\frac{1}{2}\hbar$ paramagnet. The high temperature asymptote is $S/k \to \ln 2$, which equals 0.693.

the two orientations is insignificant; the two orientations are equally probable; and the multiplicity has the simple value 2.

For the opposite limit, $T \to 0$ at fixed field, one can first factor the large exponential in $P_\parallel$ out of the numerator and denominator and then expand:

$$P_\parallel = \frac{1}{1 + e^{-2m_B B/kT}} \cong 1 - e^{-2m_B B/kT}. \tag{14.12}$$

The sum of the probabilities remains equal to 1, and so

$$P_{\text{anti-}\parallel} \cong e^{-2m_B B/kT}. \tag{14.13}$$

At low temperature, the magnetic moment is overwhelmingly more likely to point parallel to the external field, the orientation of low energy, than anti-parallel.

Substitution of the probabilities (14.12) and (14.13) into equation (14.9), expansion of the first logarithm, comparison of the several terms, and retention of only the largest term yield

$$S/k \cong 2\,\frac{m_B B}{kT}\,e^{-2m_B B/kT} \tag{14.14}$$

when $kT \ll m_B B$. In the limit as $T \to 0$ at fixed field, the exponential dominates over any mere inverse power of T, and so one finds

$$\lim_{T \to 0} S/k = 0, \tag{14.15}$$

as figure 14.1 illustrated. The system settles into its ground state; the multiplicity becomes 1; and the entropy goes to zero.

While we are on the subject of entropy, an adjective that will be useful later should be defined: *isentropic* means "at constant entropy."

14.3 Cooling by adiabatic demagnetization

In 1926, William Francis Giauque and (independently) Peter Debye suggested that adiabatic demagnetization would provide a practical way to reach extremely low temperatures. Seven years later, having overcome immense technical problems, Giauque cooled gadolinium sulfate to 0.25 K, a new record for low temperature. For understanding the process, figure 14.2 displays the essential information: the graphs of entropy versus temperature for low and high values of magnetic field, graphs based on figure 14.1 and equation (14.10). The entropy expression in (14.10) is a function of the dimensionless ratio $m_B B/kT$. Consequently, for a given value of the entropy, lower field B requires lower temperature T, and so the low field curve lies to the left of the high field curve.

The system consists of N paramagnetic particles together with other particles in a crystalline solid. To see how the cooling arises, start with the system at point a: in low magnetic field and at initial temperature T_i. (Note. Think of T_i as 1 K or less. Then the initial temperature is so low that the entropy associated with the lattice vibrations is

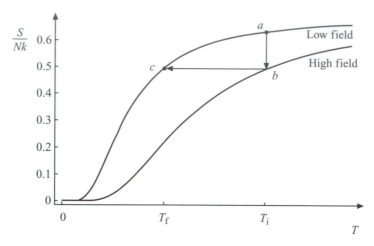

Figure 14.2 Entropy versus temperature for two values of the magnetic field. The path a to b describes an isothermal increase in the magnetic field; the path b to c, an adiabatic (and reversible) decrease in the field.

negligible in comparison with the orientational entropy of the paramagnetic particles, both initially and throughout the steps.) Increase the field isothermally to the higher value, taking the system to point b. Now isolate the system thermally, so that no energy transfer by conduction or radiation can occur; this provides the environment for an adiabatic process. Slowly reduce the field to its original low value. The adiabatic and reversible process holds the entropy constant. The isentropic process from point b to point c leaves the system with a final temperature T_f that is substantially lower than the initial temperature. The paramagnetic sample has been cooled.

What is the quantitative relationship for the amount of cooling? Recall that the entropy expression in (14.10) depends on B and T through the ratio $m_B B / kT$ only. The ratio must be preserved during the isentropic demagnetization: $T_f / B_{\text{low field}} = T_i / B_{\text{high field}}$. Consequently, the final temperature is

$$T_f = \frac{B_{\text{low field}}}{B_{\text{high field}}} T_i. \tag{14.16}$$

[Bear in mind that—so far—we have neglected the mutual magnetic interactions of the N paramagnets. When $B_{\text{low field}}$ becomes as small as the magnetic field that the paramagnets themselves produce, the relationship in (14.16) needs to be augmented. More about that comes shortly.]

Cooling from another point of view

You may wonder, where is the microscopic kinetic energy whose change I can use as a touchstone for a change in temperature? There is no need to grasp for microscopic kinetic energy. Figure 14.3 shows graphs of entropy versus energy for low and high

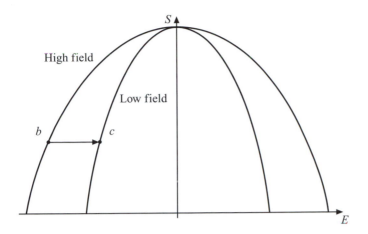

Figure 14.3 Entropy versus energy for low and high magnetic field.

magnetic fields. The construction is as follows. The energy (as a function of T and B) can be computed from the partition function as

$$\langle E \rangle = kT^2 \frac{\partial \ln Z}{\partial T}$$

$$= -m_{\mathrm{B}} BN \tanh \left(m_{\mathrm{B}} B / kT \right), \tag{14.17}$$

where $\ln Z = N \ln \left[2 \cosh \left(m_{\mathrm{B}} B / kT \right) \right]$, that is, N times the logarithm of the partition function for a single paramagnet of spin $\frac{1}{2}\hbar$. Given $T > 0$, large field implies a large but negative value for $\langle E \rangle$. The maximum negative value is $-m_{\mathrm{B}} BN$ and corresponds to all magnetic moments being aligned parallel to the field. This physical situation corresponds to the system's ground state and hence to a multiplicity of 1 and an entropy of zero. Equation (14.10) gave the entropy for a single paramagnet; multiplication by N gives the entropy for the present system. One can solve equation (14.17) for T in terms of $\langle E \rangle$ and then substitute for T in the entropy expression (14.10). These steps provide S as a function of $\langle E \rangle$, which is written here as merely E. The value B of the magnetic field remains in the relationship as a parameter. The graph for high field will extend farther along the energy axis than the curve for low field.

The isentropic demagnetization corresponds to the line b to c in figure 14.3. Recall now the general quantitative definition of temperature,

$$\frac{1}{T} = \left(\frac{\partial S}{\partial E} \right)_{\text{fixed external parameters}}. \tag{14.18}$$

In section 4.3, we derived this relationship in a context where the volume V was the sole external parameter. In the present situation, the external magnetic field B is the external parameter. To the eye, the slope $(\partial S / \partial E)_B$ is larger at point c than at point b.

The general definition says that larger slope means lower temperature; the system has been cooled.

Indeed, the graphs for low and high field have the same analytic form; the low field curve is merely narrower by the ratio $B_{\text{low field}}/B_{\text{high field}}$. Thus we infer that the slope at point c is larger by the ratio $B_{\text{high field}}/B_{\text{low field}}$. By keeping track of all the ratios and reciprocals, one finds that the final temperature is given by the expression already derived in equation (14.16). The graphical analysis, however, has the merit of taking us back to the general definition of temperature. Recall, from section 1.1, that temperature is "hotness measured on some definite scale." In turn, "hotness" means readiness to transfer energy by conduction or radiation. For that, the rate of change of entropy with energy provides the crucial measure. In situation c, the system is colder because its entropy grows more rapidly with an increase in its energy. The system absorbs energy more readily, a sure sign of coldness.

It is hard to divorce oneself from the notion that temperature is invariably and inextricably bound up with microscopic kinetic energy. In reality, temperature is a deeper notion than any such connection with irregular microscopic motion. The reasoning from "hotness measured on some definite scale" to the general calculational prescription in (14.18) is a much better view to take.

Further illustration

Figure 14.4 illustrates the situations a, b, and c of figure 14.2 by displaying typical distributions of magnetic moments. The estimated magnetic moment along **B** is

$$\langle \text{magnetic moment along } \mathbf{B}\rangle = N m_{\text{B}} \tanh \frac{m_{\text{B}} B}{kT},$$

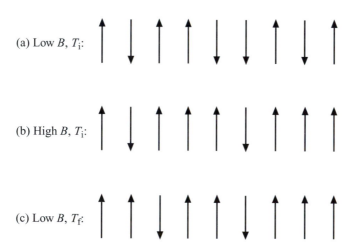

(a) Low B, T_i:

(b) High B, T_i:

(c) Low B, T_f:

Figure 14.4 Illustrations of the stages in cooling by adiabatic demagnetization. The arrows represent individual magnetic moments; the external magnetic field **B** points up the page. What matters is the fraction of arrows pointed upward, not which arrows point upward.

a result derived in section 5.3. Like the entropy, the estimated total moment depends on the ratio $m_B B/kT$. Therefore processes in the $S-T$ plane in figure 14.2 can be readily illustrated with the average alignment of the moments.

Experiments again

We return now to the experiments. Figure 14.5 sketches the apparatus of the 1930s. Liquid helium provides the initial temperature T_i. By pumping away helium vapor, one can cool the liquid—by evaporation—to 1 K. The paramagnetic sample is suspended in a chamber immersed in the liquid helium. Initially, the chamber contains dilute gaseous helium also; the gas provides thermal contact between the paramagnetic sample and the liquid helium while the external magnetic field is raised to the value $B_{\text{high field}}$. Thus the gas ensures an isothermal magnetization, the process a to b of figure 14.2. Next, the gaseous helium is pumped out, leaving the paramagnetic sample thermally isolated. Slow, adiabatic demagnetization takes the sample isentropically from point b to point c in figure 14.2. Apparatus of this nature enabled Giauque to cool to 0.25 K. Lower values for the initial temperature and other choices of paramagnetic salt dropped the final temperature by two orders of magnitude. The salt cerium magnesium nitrate yields $T_f \cong 3$ millikelvin (3 mK).

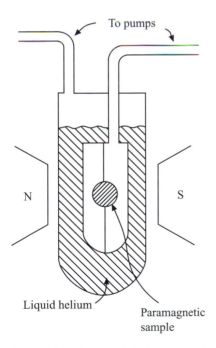

To pumps

N   S

Liquid helium

Paramagnetic sample

Figure 14.5 The essentials for cooling by adiabatic demagnetization. The pole faces of an external magnet are labeled N and S.

The lower limit

What sets the limit to the temperature that isentropic demagnetization can achieve? The mutual interactions of the magnetic moments are central to the answer. The separation of neighboring magnetic moments is typically 5×10^{-10} m or so. At that distance, a single electronic magnetic moment makes a field of 0.01 tesla (within a factor of 2). In comparison with the high external field $B_{\text{high field}}$, which may be 8 tesla, the field from a neighboring paramagnet is truly insignificant. When the external field is reduced to zero, however, the mutual interactions come into their own. The interactions correlate the orientations of adjacent magnetic moments and thus reduce the multiplicity and entropy below the values that they would have in the absence of all interactions. If the moments are associated with angular momenta of $J\hbar$, then each moment has $2J + 1$ possible orientations relative to a fixed axis. The maximum orientational entropy would be $Nk \ln(2J + 1)$. The mutual interactions ensure that S/Nk never gets as large as $\ln(2J + 1)$. (An exception is the limit $T \rightarrow \infty$, but we are far more interested in low temperatures.)

Equation (14.10) gave the entropy of an *ideal* spin $\frac{1}{2}\hbar$ paramagnet, and we noted that S/Nk is a function of B/T only. To include the mutual interactions in an approximate way, we may use the same functional form but replace B/T by the quotient

$$\frac{(B_{\text{loc}}^2 + B^2)^{1/2}}{T}, \tag{14.19}$$

where B_{loc} is a "local" field that arises from the mutual interactions and has a magnitude of order 0.01 tesla. No direction is associated with the "field" B_{loc}. Rather, the square B_{loc}^2 represents the correlating effects of the local mutual interactions in a scalar fashion. [A fully quantum calculation when kT is large relative to all magnetic energies does generate the combination displayed in (14.19); so the expression has a theoretical basis as well as being a natural interpolating form.] Isentropic demagnetization preserves the numerical value of the ratio in (14.19). Demagnetization from $B_{\text{high field}}$ to a zero value for the external field yields a final temperature

$$T_{\text{f}} = \frac{B_{\text{loc}}}{(B_{\text{loc}}^2 + B_{\text{high field}}^2)^{1/2}} T_{\text{i}} \cong \frac{B_{\text{loc}}}{B_{\text{high field}}} T_{\text{i}}. \tag{14.20}$$

In the initial, high field situation, one may omit B_{loc}, and so the denominator simplifies to merely $B_{\text{high field}}$.

In short, isentropic demagnetization reduces the temperature by the factor $B_{\text{loc}}/B_{\text{high field}}$. The mutual interactions, represented by B_{loc}, ensure that the factor is greater than zero.

By the way, you may have wondered why cooling by adiabatic demagnetization employs such esoteric compounds: gadolinium sulfate, cesium titanium alum, cerium magnesium nitrate, and iron ammonium alum, to name just four. In each such compound, one metallic ion produces the magnetic moment; in the short list, the relevant ions are gadolinium, titanium, cerium, and iron. The remainder of each compound serves to separate adjacent magnetic ions so that their mutual interactions are relatively small. The large spatial separation translates into a small value for B_{loc} and, as equation (14.20) displays, leads to a low final temperature.

In order of magnitude, nuclear magnetic moments are smaller by a factor of 10^{-3} than electronic or orbital moments (because the proton and neutron masses are roughly 2,000 times the electron's mass). Thus B_{loc} is smaller by the same factor, and so isentropic demagnetization of nuclei should reach temperatures of order 10^{-3} times the electronic limit.

One can also repeat the process of adiabatic demagnetization (as described more fully in the next section). Repeated adiabatic demagnetization has cooled the conduction electrons in copper to 10^{-5} K. (Here the conduction electrons and the nuclei are at the same temperature.) In the metal rhodium, the nuclear spins—considered as a system themselves—have been cooled to 3×10^{-10} K.

14.4 The Third Law of Thermodynamics

According to the preceding section, a single stage of isothermal magnetization and subsequent isentropic demagnetization will not cool a sample to absolute zero. The mutual interactions of the paramagnets impose a nonzero lower limit to the temperature attained.

Could a sequence of such stages reach absolute zero? The first stage reaches a low temperature and is used to cool the (thermally smaller) sample of the second stage, which reaches a lower temperature and is used to cool the (still smaller) sample of the third stage, and so on. Equation (14.20) implies that each stage reduces the temperature by some nonzero fraction; no finite number of such reductions will drive the temperature to zero. Figure 14.6 attempts to show the futility of such a sequence.

The paramagnetic example illustrates one of three alternative formulations of the *Third Law of Thermodynamics*. The three versions are the following.

1. *Unattainability form.* No process can lead to $T = 0$ in a finite number of steps.
2. *Absolute entropy form.* The entropy goes to zero as $T \to 0$.
3. *Entropy change form.* The entropy change in any isothermal process goes to zero as $T \to 0$.

In stating the three versions, I have presumed that the physical system is in complete thermal equilibrium, so that thermodynamics may properly be applied to the entire system.

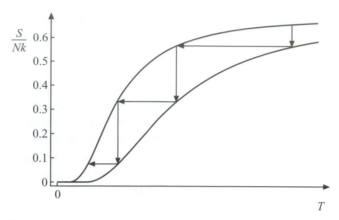

Figure 14.6 Successive stages of isothermal magnetization and isentropic demagnetization. Each decrement in temperature is less than the previous decrement. Even more importantly, each decrement is the same fraction of the current stage's starting temperature. Thus repeated stages fail to take the system to absolute zero.

Figure 14.6 illustrates each of these formulations. First, the unattainability form is a statement that we have already noted. Second, each entropy curve goes to zero as $T \to 0$. For paramagnetism, an analytic statement, $S/k \to 0$ as $T \to 0$, was derived earlier as equations (14.14) and (14.15). Third, the vertical distance between the two curves represents the entropy change during the isothermal magnetization. That distance goes to zero as $T \to 0$ and so the entropy difference in the isothermal process goes to zero. For an analytic expression, one can differentiate $S(T, B)$, as given by equation (14.14), with respect to B. Then one finds that $(\Delta S)_T \equiv (\partial S/\partial B)_T \Delta B \to 0$ as $T \to 0$ at fixed B and ΔB.

Historically, the entropy change form came first. The German physical chemist Walther Nernst introduced it in 1906, having boldly inferred it from the behavior of certain chemical reactions at low temperatures. Controversy about the formulation led him (in 1912) to offer the unattainability form as an alternative. Once the appropriate route has been pointed out, a little effort suffices to show that these two forms imply each other and thus are equivalent. In the interval between Nernst's two announcements, Max Planck proposed the absolute antropy form. Planck's form is stronger. It implies the other two forms, but they imply only that the entropy must approach a constant value, independent of external parameters and phase (such as liquid versus solid phase). For example, if the graphs in figure 14.6 were shifted upward by a constant amount, they would still illustrate the unattainability and entropy change forms but not Planck's absolute entropy form.

To be sure, a few physicists question the equivalence of the unattainability and the entropy change forms. References to their views appear in the further reading section at the end of the chapter.

"Deriving" the Third Law

In a sense, Planck's form follows readily from the canonical probability distribution and the expression (14.4) for entropy in terms of those probabilities. The canonical probability distribution,

$$P(\Psi_j) = \frac{\exp(-E_j/kT)}{Z},$$

implies that the system settles into its ground state as $T \to 0$. The probability for any and all states of higher energy vanishes. If the ground state is unique, then its probability approaches 1, and equation (14.4) implies that the entropy goes to zero. If the ground state is n-fold degenerate, then the probability for each of the degenerate states approaches $1/n$, and $S \to k \ln n$. Comparison must be made with the typical entropy at much higher temperatures, which is of order Nk in its dependence on the number of particles N. Even if the degeneracy integer n were of order N, the ratio of entropies would be of order $(\ln N)/N$, which is virtually zero when N is of order 10^{20}. For all practical purposes, the entropy would approach zero.

This kind of reasoning, although technically correct, fails to provide an adequate foundation for the Third Law. To see why, let us take as our system the conduction electrons in 10 cubic centimeters of copper. According to section 9.1, the ground state has all single-particle states filled out to the Fermi energy ε_F. The total energy is $E_{g.s.} = \frac{3}{5}N\varepsilon_F$. To construct the first excited state, we promote one electron from energy ε_F to a single-particle state just beyond the Fermi energy. To determine the increment $\Delta\varepsilon$ in energy, we reason that

$$D(\varepsilon_F)\,\Delta\varepsilon = 1, \tag{14.21}$$

that is, the density of single-particle states times $\Delta\varepsilon$ gives one new single-particle state. Taking $D(\varepsilon_F)$ from (9.13), we find

$$\Delta\varepsilon = \frac{1}{D(\varepsilon_F)} = \frac{2}{3}\frac{\varepsilon_F}{N}. \tag{14.22}$$

Thus the first excited state has energy

$$E_{\text{first excited}} = E_{g.s.} + \Delta\varepsilon = E_{g.s.} + \frac{2}{3}\frac{\varepsilon_F}{N}. \tag{14.23}$$

The Boltzmann factors in $P(\Psi_j)$ imply that the ratio of probabilities is

$$\frac{P_{\text{first excited}}}{P_{g.s.}} = \frac{\exp(-E_{\text{first excited}}/kT)}{\exp(-E_{g.s.}/kT)} = e^{-\Delta\varepsilon/kT}. \tag{14.24}$$

If this ratio of probabilities is to be small, then one needs a temperature low enough that $\Delta\varepsilon/kT$ is much greater than 1:

$$T \ll \frac{\Delta\varepsilon}{k} = \frac{2}{3}\frac{\varepsilon_F}{Nk} = \frac{2}{3}\frac{T_F}{N}. \tag{14.25}$$

For copper, the experimental Fermi temperature is $T_F = 6.0 \times 10^4$ K. Ten cubic centimeters of copper contain $N = 8.5 \times 10^{23}$ atoms, and so the requirement is

$$T \ll 4.7 \times 10^{-20} \text{ K}. \tag{14.26}$$

Yet when Nernst proposed his first version of the Third Law, attempts to liquefy helium still failed (because the critical temperature of ^{4}He, 5.2 K, could not yet be reached), and 20 K was considered a low temperature. The Third Law, as a part of empirical physics, applies already at temperatures regularly attained in the twentieth century. The law does not require the outrageously low temperatures, like that displayed in (14.26), that are needed if one is to make the first excited state of a macroscopic system improbable. In the next paragraphs, we find a more satisfactory perspective.

Table 14.1 displays the entropy at "low" temperature of several systems that we have studied. The entropies were calculated as

$$S(T) = \int_0^T \frac{C_V(T')}{T'} \, dT'. \tag{14.27}$$

In turn, the heat capacities were calculated from the relation $C_V = (\partial \langle E \rangle / \partial T)_V$ and from expressions for $\langle E \rangle$ derived in sections 6.5, 9.1, and 9.4. Each system has a characteristic temperature: the Debye temperature θ_D, the Fermi temperature T_F, or the Bose temperature T_B. For all three systems, the entropy at high temperature is of order Nk. Perhaps equation (5.41) provides the most convincing evidence for that claim: in the semi-classical limit, the entropy of an ideal gas is Nk times a logarithm, and "all logarithms are of order 1" is a rule of thumb in physics. The fourth column in table 14.1 gives a ratio of entropies: the numerator is the entropy at a temperature of one-hundredth of the characteristic temperature; the denominator is Nk, which is the value of the entropy at high temperature, in order of magnitude. The entropy has already become small. At $T = 10^{-4} T_{\text{char}}$, the entropy is much smaller still.

For copper, the characteristic temperatures are $\theta_D = 343$ K and $T_F = 6.0 \times 10^4$ K. The Bose temperature for ^{4}He, if it were an ideal gas at the temperature and number density where it becomes a superfluid, would be $T_B = 3.1$ K. These are reasonable, laboratory-scale temperatures. In part, the empirical content of the Third Law is this:

Table 14.1 *Examples of how entropy and heat capacity approach zero.*

System	C_V	Entropy	S/Nk when $T = 0.01 T_{\text{char}}$	$T = 10^{-4} T_{\text{char}}$
Lattice vibrations	$\dfrac{12\pi^4}{5} Nk \left(\dfrac{T}{\theta_D}\right)^3$	$\dfrac{4\pi^4}{5} Nk \left(\dfrac{T}{\theta_D}\right)^3$	7.8×10^{-5}	7.8×10^{-11}
Conduction electrons	$\dfrac{\pi^2}{2} Nk \dfrac{T}{T_F}$	$\dfrac{\pi^2}{2} Nk \dfrac{T}{T_F}$	4.9×10^{-2}	4.9×10^{-4}
Bose gas: $T < T_B$	$0.770 \times \dfrac{5}{2} Nk \left(\dfrac{T}{T_B}\right)^{3/2}$	$0.770 \times \dfrac{5}{3} Nk \left(\dfrac{T}{T_B}\right)^{3/2}$	1.3×10^{-3}	1.3×10^{-6}

physical systems have characteristic temperatures (that depend on intensive variables like N/V but not on the system's size); those temperatures have reasonable, laboratory-scale size; and when $T \leqslant 0.01\,T_{\text{char}}$ or so, the entropy is already small and approaches zero rapidly (relative to its value at high temperature).

14.5   Some other consequences of the Third Law

A glance at the heat capacities C_V in table 14.1 shows that they, too, go to zero as $T \to 0$. That is true of heat capacities in general. [If heat capacities did not go to zero, integrals like that in equation (14.27) would fail to converge, but the entropy is finite near absolute zero.] Moreover, many other quantities in thermal physics go to zero, or their slopes go to zero, as $T \to 0$. Two additional examples will suffice.

  1. $(dP/dT)_{coexistence}$. The helium isotopes are unique in that the liquid phase—as well as the solid phase—persists in the limit of absolute zero. Indeed, the solid forms only under relatively high pressure: approximately 25 atmospheres for ^{4}He. Figure 12.6 shows the solid–liquid coexistence curve for ^{3}He.

  Along a coexistence curve that joins two phases, such as a melting curve, the Clausius–Clapeyron equation asserts that

$$\frac{dP}{dT} = \frac{\Delta s}{\Delta v}.$$

The symbol Δs denotes the change in entropy per particle between the two phases, and Δv is the change in volume per particle. Now specialize to helium on its melting curve. The quantity Δs is an "entropy change in an isothermal process," here the transfer of an atom from the solid to the liquid. According to the entropy change form of the Third Law, $\Delta s \to 0$ as $T \to 0$. The volume change Δv remains nonzero, and so

$$\left(\frac{dP}{dT}\right)_{\text{coexistence}} \to 0 \quad \text{as} \quad T \to 0. \tag{14.28}$$

  Along the melting curve for ^{4}He, measurements by Sir Francis Simon and C. A. Swenson show that $(dP/dT)_{\text{coexistence}}$ goes to zero as T^7 in the interval $1.0 \leqslant T \leqslant 1.4$ K, a remarkably swift decline. The melting curve flattens out as it approaches absolute zero.

  2. *Coefficient of thermal expansion.* The combination of derivative and division by volume,

$$\frac{1}{V}\left(\frac{\partial V}{\partial T}\right)_P,$$

describes the rate of expansion with temperature (at constant pressure) in a fashion that is independent of the system's size. The combination is called the *coefficient of thermal expansion*:

$$\left(\begin{array}{c} \text{coefficient of} \\ \text{thermal expansion} \end{array} \right) \equiv \frac{1}{V} \left(\frac{\partial V}{\partial T} \right)_P. \tag{14.29}$$

Although there is no obvious connection with entropy, we can derive one. Equations (10.29) and (10.30) give the volume V and entropy S as partial derivatives of the Gibbs free energy:

$$V = \left(\frac{\partial G}{\partial P} \right)_{T,N},$$

$$S = -\left(\frac{\partial G}{\partial T} \right)_{P,N}.$$

Differentiate the first of these equations with respect to T; the second, with respect to P. The mixed second derivatives of G are equal, and so

$$\left(\frac{\partial V}{\partial T} \right)_P = -\left(\frac{\partial S}{\partial P} \right)_T. \tag{14.30}$$

The derivative on the right is again an entropy change in an isothermal process and must go to zero as $T \rightarrow 0$. Consequently, the coefficient of thermal expansion must go to zero as T goes to zero. Figure 14.7 provides an example from research on solid helium.

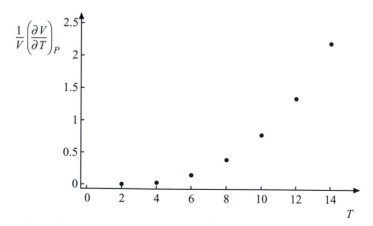

Figure 14.7 The coefficient of thermal expansion versus temperature for solid ^{4}He. The coefficient is given in units of 10^{-3} per kelvin; the temperature, in kelvin. The points descend toward zero approximately as T^3. Data were taken at a number density corresponding to 6×10^{23} atoms (one mole) per 12 cm^3 (under high pressure, of order 10^3 atmospheres). [*Source*: J. S. Dugdale and J. P. Franck, *Phil. Trans. R. Soc.* **257**, 1–29 (1964).]

14.6 Negative absolute temperatures

We return now to paramagnetism. Figure 14.8 shows a graph of entropy versus energy for a system of N ideal spin $\frac{1}{2}\hbar$ paramagnets; thus it reproduces a curve from figure 14.3. Below the S versus E curve is the temperature T that follows from the general quantitative definition of temperature, which takes here the form

$$\frac{1}{T} = \left(\frac{\partial S}{\partial E}\right)_B. \tag{14.31}$$

When the energy is greater than zero, the slope of the entropy graph is negative, and so the absolute temperature is negative. That may be all well and good as far as

(a)

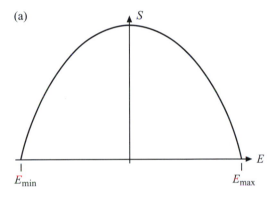

(b)

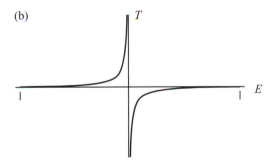

Figure 14.8 (a) Entropy versus energy for a system of N ideal spin $\frac{1}{2}\hbar$ paramagnets. The energy $E_{\min}$ arises when all magnetic moments point parallel to the magnetic field **B**. In contrast, $E_{\max}$ arises when all moments point anti-parallel to **B**. (b) The corresponding absolute temperature, calculated from the slope of the curve in part (a) according to the relationship $1/T = (\partial S/\partial E)_B$. As E approaches its extreme values in graph (a), the slopes become infinite, and so T approaches zero. The difference in the sign of the slopes implies an approach to absolute zero from opposite sides. [Note. The slope steepens toward infinity in an extraordinarily narrow interval at the extremes of E: an interval of less than one part in 10^{14} of the entire energy range. That interval is less than the width of the line used to draw the curve, and so the approach to an infinite slope is not visible (in this mathematically faithful drawing).]

mathematics goes, but is a negative absolute temperature physically realizable? The answer is a resounding "yes." In 1950, Edward M. Purcell and Robert V. Pound produced such a temperature in the lab. In the intervening decades, their technique has become routine.

For Purcell and Pound, the paramagnetic system was the assembly of nuclei in a crystal of lithium fluoride. (The fluorine isotope ^{19}F has a nuclear spin of $\frac{1}{2}\hbar$; the lithium isotope ^{7}Li has a nuclear spin of $\frac{3}{2}\hbar$. The entropy versus energy curve for the composite system is qualitatively like that in figure 14.8.) The experiment commenced at room temperature, where the lattice vibrations remain significant, both as an energy source and as a contributor to the overall entropy. The exchange of energy between the magnetic moments and the lattice vibrations, however, was so slow that, for short times (of order 10 seconds), the crystal acts as though the orientational and magnetic aspects of the nuclei were entirely decoupled from the vibrational aspects of the same nuclei. One can meaningfully speak of a system of nuclear paramagnets "isolated" from the lattice vibrations. The trick, then, is to promote the nuclear paramagnets to a macro-state where they have positive energy: the right-hand side of figure 14.8.

The essence of the technique used by Purcell and Pound is a sudden reversal of an external field. Initially, the paramagnets are at thermal equilibrium at a positive temperature in a field of 0.01 tesla. The magnetic moments point predominantly parallel to the external field and have negative potential energy. Then the field is reversed extremely quickly, in a time of order 2×10^{-7} second. The reversal is so fast that the spins cannot change their orientation during the reversal. They find themselves pointed predominantly anti-parallel to the new field, an orientation of positive potential energy. In a very short time, they come to a new thermal equilibrium, both with respect to the new external field and also with respect to their own mutual interactions. The paramagnets now have a negative absolute temperature.

After the sudden reversal, all subsequent changes in external magnetic field occur relatively slowly. To good approximation, the paramagnets can follow the changes, continuously and isentropically. Their temperature may change continuously in magnitude but remains negative in sign. This conclusion follows from our analysis in section 14.3; in an isentropic process, the combination

$$\frac{\left(B_{\text{loc}}^2 + B^2\right)^{1/2}}{T}$$

remains constant even as the external field B changes slowly.

To test for a negative temperature, Purcell and Pound removed their crystal from the 0.01 tesla field and placed it in a much larger field: 0.64 tesla. No matter how the crystal was carried from one magnet to the other and no matter how it was oriented in the larger field, the paramagnetic system exhibited a large amount of energy, characteristic of moments lining up anti-parallel to the large external field, the orientation of positive potential energy. Purcell and Pound applied electromagnetic radiation at a frequency chosen so that ^{7}Li nuclei could flip from an orientation of low potential energy to an orientation of high energy by absorbing a photon. In earlier experiments, when both lattice and spins were at room temperature, such radiation was invariably absorbed. Now, instead of absorption (and as they had hoped), Purcell and Pound found stimulated *emission* of radiation at that frequency. The spin system had so much energy that spins predominantly flipped to low-energy orientations and simultaneously emitted a photon. From the direct observation of stimulated emission of radiation one infers that the moments are predominantly anti-parallel to the field and that the nuclear spin system is at a negative temperature.

Hotter and colder

When the crystal was placed in the large external field and stimulated with radio waves, the crystal radiated as would be expected for a paramagnetic system at $T = -350$ K. At any positive temperature, the paramagnets would have absorbed the radio waves. Precisely because the paramagnets emitted energy (rather than absorbed it), they were hotter than paramagnets at any positive temperature. We conclude that a temperature of $T = -350$ K is hotter than any positive temperature. Indeed, all negative temperatures are hotter than all positive temperatures.

For an analytic way to see this conclusion, return to equation (14.31) and figure 14.8. Where the slope of the S versus E curve is negative (and hence where T is negative), a *decrease* in energy produces an increase in entropy. In contrast, at positive temperatures, an increase in energy is required to produce an increase in entropy. If two systems, one at a negative temperature and the other at a positive temperature, interact thermally, then the growth of entropy (at fixed total energy) will be largest if the system at negative temperature gives up energy and the system at positive temperature accepts energy. According to the Second Law, this is the route to a new thermal equilibrium. The energy donor is, by definition, the hotter system, and so the system at negative temperature is the hotter system.

Purcell and Pound probed the crystal periodically with radio waves. As the nuclear spin system cooled down slowly (with a time constant of about 5 minutes), its temperature went from -350 K to $-1,000$ K and thence to $-\infty$ K, which is physically equivalent to $+\infty$ K, and then continued through $+1,000$ K to $+300$ K, which is ordinary room temperature (in round figures). The spin system did *not* pass through absolute zero. Not only was this sequence found experimentally; we can see it theoretically in figure 14.8. The spin system loses energy as it heats up the lattice vibrations and as it emits photons in response to radio-frequency stimulation. As the energy of the spin system decreases, the point on the temperature graph moves leftward, and the temperature heads toward $-\infty$ K. Recall that the numerator of the canonical probability distribution is the Boltzmann factor, $\exp(-E_j/kT)$. As $T \to -\infty$, the exponent goes to zero for all E_j, and so the probability distribution becomes perfectly flat. Such flatness arises whatever the sign of T (when $|T| = \infty$), and so there is continuity in the probability distribution as the temperature makes the discontinuous switch from $-\infty$ K to $+\infty$ K. The spin system passes smoothly from negative to positive temperature through the back door of infinite temperature. The unattainability form of the Third Law becomes the colloquial expression, "You can *not* get to zero from *either* side."

Although there is no need to distinguish between plus and minus infinity for the temperature, there is a vast difference between approaching absolute zero from positive and from negative temperatures. As the temperature goes to zero from the positive side (denoted $T \rightarrow +0$), the system settles into its ground state. In contrast, as the temperature goes to zero from the negative side (denoted $T \rightarrow -0$), the system is pushed into its highest energy state. This distinction carries at least two consequences. For one, the coldest temperatures are just above 0 K on the positive side, and the hottest temperatures are just below 0 K on the negative side. Figure 14.9 displays this. We explore another consequence of the distinction in the next subsection.

Conditions that permit a negative absolute temperature

If we imagine a negative absolute temperature, then the Boltzmann factor, $\exp(-E_j/kT)$, in the canonical probability distribution grows as the energy eigenvalue E_j increases. The states with high energy are more probable than those with low energy, exactly the reverse of the situation at positive temperature. In a sense, the basic experimental problem of achieving a negative temperature is to make the high-energy states more probable than the low-energy states. If the system has energy eigenstates of arbitrarily high energy (as does a gas), then no finite supply of energy can promote the system to a negative temperature; an extremely hot positive temperature is the best one can do.

Indeed, three conditions must be met if a system is to be capable of being at a negative temperature.

1. The system must be macroscopic and in thermal equilibrium, so that the very notion of temperature is applicable.

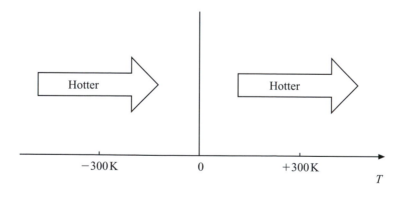

Figure 14.9 A depiction of hotness as a function of the absolute temperature. The vertical line at $T = 0$ represents both a barrier for the arrows and an unattainable value. One cannot cool a system to absolute zero from above nor heat it to zero from below.

2. The possible values of the system's energy must have a finite upper limit.
3. The system must be isolated (effectively, at least) from systems not satisfying the second condition.

The finite upper bound on the sequence of energies E_j ensures that a finite amount of energy can produce a Boltzmann factor that grows with increasing E_j.

Effective isolation is necessary because systems in thermal contact reach the same temperature in thermal equilibrium. For two systems in thermal contact, a negative temperature is possible for either only if it is possible for both.

Note that perfect isolation is not required. Effective isolation means "little interaction during the time needed for the system of interest to come to internal thermal equilibrium, for all practical purposes." This permits one to regard the system of nuclear spins in lithium fluoride as effectively isolated from the lattice vibrations (which cannot attain a negative temperature). To be sure, some early skeptics of negative absolute temperatures insisted that the spins first come into equilibrium with the vibrational motion of the nuclei before one would talk about a temperature for the spins. The demand was unrealistic and unnecessary. Indeed, such insistence was equivalent to demanding that the nurse not take my temperature until I have come into equilibrium with the ubiquitous 2.7 K cosmic blackbody radiation—by which time the doctor is not going to help me.

Since 1950, when the nuclei of the insulator lithium fluoride were promoted to a negative temperature, physicists have generated negative nuclear spin temperatures in silver and rhodium, both electrical conductors. In particular, the nuclear spin temperature of rhodium has been made as hot as -750 picokelvin, that is, -7.5×10^{-10} K.

14.7 Temperature recapitulated

Now is a good time to recapitulate some key aspects of temperature.

General qualitative definition. We can order objects in a sequence that tells us which will gain energy by heating (and which will lose energy by cooling) when we place them in thermal contact. Of two objects, the object that loses energy is the "hotter" one; the object that gains energy is the "colder" one. *Temperature* is hotness measured on some definite scale. That is, the goal of the "temperature" notion is to order objects in a sequence according to their "hotness" and to assign to each object a number—its temperature—that will facilitate comparisons of "hotness."

General quantitative definition. Section 4.3 led us to the relationship

$$\frac{1}{T} = \left(\frac{\partial S}{\partial E}\right)_{\text{fixed external parameters}}. \tag{14.32}$$

The rate of change of entropy with energy (at fixed external parameters) gives a quantitative definition of temperature and one that can be applied to an individual system. The rate $(\partial S/\partial E)_{\text{f.e.p.}}$ (where f.e.p. denotes "fixed external parameters") serves also a comparative function. Recall that the Second Law of Thermodynamics implies evolution to increased multiplicity and hence to increased entropy (in the context of isolation). Thus, when two systems are placed in thermal contact (but are otherwise isolated), the system with the greater $(\partial S/\partial E)_{\text{f.e.p.}}$ rate will gain energy (by heating) from the other system (so that the total entropy will increase). The system with the greater value of $(\partial S/\partial E)_{\text{f.e.p.}}$ will be the colder system, and so $(\partial S/\partial E)_{\text{f.e.p.}}$ provides a way to order systems according to their hotness.

Energy per particle or per mode. If classical physics is sufficient, then the equipartition theorem states that every quadratic energy expression has an estimated value of $\frac{1}{2}kT$. (For the precise conditions under which the theorem holds, see section 13.3.) Thus translational kinetic energy has an estimated value of $3 \times \frac{1}{2}kT$. The kinetic and potential energies of a harmonic oscillator are each $\frac{1}{2}kT$. In a loose but useful sense, these results lead to a physicist's rule of thumb.

> If classical physics suffices, then the energy per particle or per mode is approximately kT.   (14.33)

Valuable though this rule is, it is not the whole story. As soon as quantum theory becomes essential, the energy estimate changes dramatically. One way to understand the change is this: introducing Planck's constant enables one to construct a characteristic energy (that is independent of temperature); then a whole new range of estimated energies can be constructed from kT and the characteristic energy.

A good example of a characteristic energy is the Fermi energy ε_{F}:

$$\varepsilon_{\text{F}} = \left(\frac{3}{8\pi}\right)^{2/3} \frac{h^2}{2m} \left(\frac{N}{V}\right)^{2/3}, \tag{14.34}$$

correct for fermions with spin $\frac{1}{2}\hbar$. The Fermi energy depends on h, m, and the intensive variable N/V.

When the temperature is low (relative to the Fermi temperature T_{F}), the estimated energy per fermion is

$$\frac{\langle E \rangle}{N} = \frac{3}{5}\varepsilon_{\text{F}} + \frac{\pi^2}{4}\left(\frac{kT}{\varepsilon_{\text{F}}}\right) kT, \tag{14.35}$$

a result that follows from equations (9.12), (9.13), and (9.18). This expression differs greatly from "approximately kT." The characteristic energy (here ε_{F}) and the associated characteristic temperature (here $T_{\text{F}} = \varepsilon_{\text{F}}/k$) enable one to construct energy estimates that are much more complicated than merely kT. Nature has seized the opportunity and pursued it vigorously. The implication is this: when quantum theory is essential, there is no general rule for the energy per particle or per mode. Rather, look for the characteristic energy or temperature, and expect the energy estimate to be a function of them as well as of kT. The most that one could write in any generality is the expression

$$\frac{\langle E \rangle}{N} = kT \times \text{function of} \left(\frac{kT}{\varepsilon_{\text{characteristic}}} \right), \tag{14.36}$$

but, unless one knows the function, the formal relationship is barren.

Temperature is deeper than average kinetic energy. It is easy to come away from introductory physics with the impression that "absolute temperature is a measure of average kinetic energy," where the word "is" has the significance of a definition or an explanation. Such an impression would be wrong—for several reasons, among which are the following.

1. Consider a gaseous system that is in the quantum domain and yet is nearly classical. For this context, section 8.5 showed that fermions and bosons—when manifestly at the same temperature—have different average translational kinetic energies. By itself, temperature does *not* determine average kinetic energy.
2. Section 14.6 showed that the absolute temperature may be negative, both theoretically and experimentally. Kinetic energy is never negative.
3. No one would go to the trouble of an elaborate definition such as $1/T = (\partial S/\partial E)_{\text{f.e.p.}}$ if temperature were merely a constant times average translational kinetic energy.

The misconception that introduces this subsection is propagated with the best of intentions: to make absolute temperature easier to understand. The root of the conceptual error lies in this: a belief that the purpose of "temperature" is to tell us about a physical system's *amount* of energy. That is not the purpose of the temperature notion. Rather, temperature is intended to tell us about a system's *hotness, its tendency to transfer* energy (by heating). All physical systems are capable of heating or cooling others. The purpose of temperature is to rank the systems with respect to their ability to heat one another.

14.8 Why heating increases the entropy. Or does it?

You may have wondered, why does energy transfer by heating increase the entropy but energy transfer by work done does not? Here both transfers are prescribed to be done slowly; experience indicates that the physical system then remains close to equilibrium. The connection between entropy and uncertainty, developed in section 14.1, provides a key to understanding why a difference arises.

Take first the case of work done. An external parameter, such as volume or magnetic field, is changed slowly in a known way. The external parameter appears in the Schrödinger equation for the system. If a quantum state $\Psi(t_0)$ was the appropriate state to use to describe the system initially (at time $t = t_0$), then the state $\Psi(t)$ that evolves (deterministically) according to the Schrödinger equation is the appropriate state later. The same is true for other quantum states that, initially, have a nonzero probability of being the right state to use. The quantum states may change, but no new uncertainty is introduced; hence the multiplicity and entropy remain constant.

In section 2.5, we saw how the entropy can remain constant during a slow adiabatic expansion. The spatial part of the multiplicity increases (because of the increasing volume). The momentum part of the multiplicity decreases (because the gas molecules do work, lose energy, and develop smaller momenta, on the average). Countervailing effects can balance, and the entropy can remain constant as the external parameter (here the volume) is changed.

In the case of heating, the situation is significantly different. There must be an interaction between the system of interest and the source of energy. The former might be the fried eggs of section 1.1, and the latter the hot metal of the frying pan. At the interface, there are electrical interactions between organic molecules in the eggs and metal atoms in the pan. These interactions accomplish the transfer of energy by heating, but we do not know them in detail, and so the Schrödinger equation itself contains terms that are uncertain. We cannot follow the evolution of quantum states for the eggs deterministically, and so our uncertainty grows. Because entropy can be conceived as a measure of uncertainty about the appropriate quantum state to use, it makes sense that the entropy increases.

That paragraph, however, cannot be the whole story. For example, how does one understand the decrease in entropy when a system is cooled? After all, similar uncertainty arises about the microscopic details of the interaction. Why doesn't the entropy increase for cooling, too?

Cooling a dilute gas at constant volume entails extracting energy from the gas. The momentum part of the multiplicity decreases, but the spatial part remains constant. Effects do *not* balance and cancel; hence the multiplicity and entropy decrease.

In heating and cooling, the uncertainty about the microscopic details implies that entropy *need not* remain constant. What does determine the entropy change is, formally, the partial derivative $(\partial S/\partial E)_{\text{f.e.p.}}$, itself multiplied by the amount of energy transferred. In turn, the derivative depends on the system's constituents, their inter-actions, the environment, and the current macrostate. Usually the derivative is positive, and so heating increases the entropy, and cooling decreases it. We have found, however, that $(\partial S/\partial E)_{\text{f.e.p.}}$ can be negative for a nuclear spin system. (Recall the entropy versus energy curve in figure 14.8.) When the derivative is negative, heating decreases the entropy, and cooling increases it.

In short, the uncertainty in the microscopic interactions of heating and cooling provides permission for an entropy change but does not prescribe its sign. What gives definitive information is the derivative $(\partial S/\partial E)_{\text{f.e.p.}}$.

The evolution of entropy under various circumstances has occupied some of the keenest minds in physics for over a century. Questions abound. Not all the answers are known—or, at least, not all are agreed upon. Some sources for exploring the issues further are given in the references for this chapter.

14.9   Essentials

1. Entropy can be expressed directly in terms of probabilities:

$$S = -k \sum_j P(\Psi_j) \ln P(\Psi_j).$$

2. In cooling by adiabatic demagnetization, the final temperature equals a ratio of magnetic fields times the initial temperature:

$$T_f \cong \frac{B_{\text{loc}}}{B_{\text{high field}}} T_i,$$

where B_{loc} is a "local" field that arises from the local interactions among the paramagnets and with their crystalline environment. No direction is associated with the "field" B_{loc}. Rather, the square B_{loc}^2 represents the correlating effects of the local interactions in a scalar fashion.

3. The Third Law of Thermodynamics has three versions.

   1. *Unattainability form.* No process can lead to $T = 0$ in a finite number of steps.
   2. *Absolute entropy form.* The entropy goes to zero as $T \rightarrow 0$.
   3. *Entropy change form.* The entropy change in any isothermal process goes to zero as $T \rightarrow 0$.

4. In part, the empirical content of the Third Law is this: physical systems have characteristic temperatures (that depend on intensive variables like N/V but not on the system's size); those temperatures have reasonable, laboratory-scale size; and when $T \leqslant 0.01 T_{\text{char}}$ or so, the entropy is already small and approaches zero rapidly (relative to its value at high temperature).

5. The Third Law, especially the entropy change form, can be used to deduce the limiting behavior (as $T \rightarrow 0$) of surprisingly diverse quantities, such as the slope of the solid–liquid coexistence curve and the coefficient of thermal expansion.

6. Temperatures below absolute zero exist both theoretically and experimentally. According to the general relationship

$$\frac{1}{T} = \left(\frac{\partial S}{\partial E} \right)_{\text{fixed external parameters}},$$

such negative absolute temperatures arise whenever the entropy S decreases with increasing energy E.

7. All negative temperatures are hotter than all positive temperatures. Moreover, the coldest temperatures are just above 0 K on the positive side, and the hottest temperatures are just below 0 K on the negative side.

8. Rule of thumb: if classical physics suffices, then the energy per particle or per mode is approximately kT.

Valuable though this rule is, it is only a small part of the whole story. As soon as quantum theory becomes essential, the energy estimate changes dramatically. Introducing Planck's constant enables one to construct a characteristic energy (that is independent of temperature); then a whole new range of estimated energies can be constructed from kT and the characteristic energy.

Further reading

Kurt Mendelssohn tells a marvelous story in his *Quest for Absolute Zero: the Meaning of Low Temperature Physics*, 2nd edition (Halsted Press, New York, 1977).

More technical but highly recommended is David S. Betts, *An Introduction to Millikelvin Technology* (Cambridge University Press, New York, 1989).

J. Wilks, *The Third Law of Thermodynamics* (Oxford University Press, New York, 1961), provides a comprehensive and accessible exposition.

The issue of equivalence among various statements of the Third Law is nicely discussed by Mark W. Zemansky and Richard H. Dittman, *Heat and Thermodynamics*, 6th edition (McGraw-Hill, New York, 1981), section 19-6. (Of course, Wilks also examines this issue.)

To be sure, the equivalencies are questioned by John C. Wheeler, *Phys. Rev. A*, **43**, 5289–95 (1991) and **45**, 2637–40 (1992). Wheeler's view is that the unattainability and entropy change versions are not equivalent thermodynamic statements and that they make independent statements about the kinds of interactions that are found in nature. See also P. T. Landsberg, *Am. J. Phys.* **65**, 269–70 (1997).

A fully quantum mechanical derivation of the effective field B_{loc} is provided in chapter 12 of Ralph Baierlein, *Atoms and Information Theory* (W. H. Freeman, New York, 1971).

The behavior of $(dP/dT)_{coexistence}$ is reported in F. E. Simon and C. A. Swenson, "The liquid-solid transition in helium near absolute zero," *Nature* **165**, 829–31 (1950).

Edward M. Purcell and Robert V. Pound report the first negative absolute temperature in "A nuclear spin system at negative temperature," *Phys. Rev.* **81**, 279–80 (1951). A fine theoretical discussion is provided by Norman F. Ramsey, "Thermodynamics and statistical mechanics at negative absolute temperatures," *Phys. Rev.* **103**, 20–8 (1956).

An example of negative temperatures as—by now—a standard aspect of physics is provided by R. T. Vuorinen *et al.*, "Susceptibility and relaxation measurements on rhodium metal at positive and negative spin temperatures in the nanokelvin range,"

J. Low Temp. Phys. **98**, 449–87 (1995). A review article from the same laboratory is A. S. Oja and O. V. Lounasmaa, "Nuclear magnetic ordering in simple metals at positive and negative nanokelvin temperatures," *Rev. Mod. Phys.* **69**, 1–136 (1997).

An engaging introduction to Shannon's uncertainty measure is provided by Claude E. Shannon and Warren Weaver, *The Mathematical Theory of Communication* (University of Illinois Press, Urbana, IL, 1964).

Seminal papers on entropy by Ludwig Boltzmann, Ernst Zermelo, and others are reprinted by Stephen G. Brush in *Kinetic Theory, Vol. 2: Irreversible Processes* (Pergamon Press, New York, 1966). Joel L. Lebowitz's article, "Boltzmann's entropy and Time's arrow" [in *Physics Today* **46**, 32–8 (September 1993)] provides an accessible introduction to many issues. Letters in response to Lebowitz's presentation appeared in *Physics Today* **47**, 11–15 and 115–17 (November 1994).

Edwin Jaynes has spoken out clearly on the issues of entropy and its evolution. Particularly noteworthy is his paper "Gibbs vs Boltzmann Entropies" in *Am. J. Phys.* **33**, 391–8 (1965). That paper and some other relevant papers appear in the collection *E. T. Jaynes: Papers on Probability, Statistics, and Statistical Physics* (Reidel, Dordrecht, 1983), edited by R. D. Rosenkrantz.

Problems

1. *Entropy for fermions.*

(a) Consider a statement that is necessarily either true or false, such as the statement, "The moon contains more than 10^{14} kilograms of gold." On the basis of available terrestrial and lunar evidence, one assigns (somehow) numerical values to the probabilities of the statement's being true, P_{yes}, and being false, P_{no}. Information theory and equation (14.5) imply that one's uncertainty about whether the moon contains more than 10^{14} kilograms of gold is proportional to the expression

$$- (P_{yes} \ln P_{yes} + P_{no} \ln P_{no}).$$

Graph the expression as a function of P_{yes}, which has the range $0 \leqslant P_{yes} \leqslant 1$. Is the behavior qualitatively reasonable? Why?

(b) Use the structure in part (a) to construct the entropy for an ideal quantum gas of N fermions in terms of the estimated occupation numbers. (If you have not done problem 8.1, read that problem over first.) The entropy is to be taken as a measure of one's uncertainty of the true quantum state (which is equivalent to one's uncertainty about which single-particle states are occupied). Then compare your result with the expression in problem 8.6.

2. *Relative heat capacities.* At low temperature, the influx of energy from stray sources of heating is a perennial problem. Cooling by adiabatic demagnetization will provide a useful "cold reservoir" only if the heat capacity of the paramagnetic material is relatively large (so that the reservoir does not heat up rapidly). Consider a cold

reservoir formed by N spatially fixed paramagnets of spin $\frac{1}{2}\hbar$ and magnetic moment m_{B}.

(a) Use figure 5.3 or 14.1 to estimate where the heat capacity is largest (as a function of $kT/m_{\mathrm{B}}B$). If $B = B_{\mathrm{loc}} = 0.01$ tesla, how low is the temperature?
(b) Compute the heat capacity per paramagnet at constant magnetic field, C_B/N, as a function of T and B. Plot C_B/Nk versus $kT/m_{\mathrm{B}}B$.
(c) Evaluate C_B/Nk for the value of the quotient $kT/m_{\mathrm{B}}B$ that you estimated in part (a).
(d) Compare your value of C_B/Nk in part (c) with C_V/Nk for metallic copper at $T = 0.01$ K (in the absence of any magnetic field). How many copper atoms must be collected to give as large a heat capacity as one paramagnet?

3. *Another limit.*

(a) Construct a line of reasoning to show that $(\partial P/\partial T)_V$ should go to zero as $T \to 0$.
(b) Test this proposition against some system for which you can determine the derivative and the limit explicitly.

4. *Hotter, graphically.* Sketch the graph of entropy versus energy for a system of N spatially fixed paramagnets in an external magnetic field. Then draw arrows, labeled "hotter," to indicate the direction along the energy axis in which one should move to find a hotter system.

5. *Heat engines at negative temperatures.* Consider using two heat reservoirs to run an engine (analogous to the Carnot cycle of chapter 3), but specify that both temperatures, T_{hot} and T_{cold}, are negative temperatures. The engine is to run reversibly.

(a) If the engine is to do a positive net amount of work, from which reservoir must the energy be extracted?
(b) Under the same conditions, what is the engine's efficiency?
(c) If you drop the requirement of reversibility, what is the maximum efficiency, and how would you achieve it?

6. *Magnetization as $T \to 0$.* The magnetic moment per unit volume (along the direction of an external magnetic field $\mathbf{B}$) is denoted by M and can be computed from equation (5.20):

$$M = \frac{kT}{V}\left(\frac{\partial \ln Z}{\partial B}\right)_T.$$

(a) Develop a Maxwell relationship to relate $(\partial M/\partial T)_B$ to an isothermal change in the entropy per unit volume.
(b) What can you infer about the limiting behavior of $(\partial M/\partial T)_B$ as $T \to 0$?
(c) Marshal support for your inference by citing evidence from the explicitly known behavior of spatially fixed paramagnets.

7. *Extremum at negative temperatures.* In the context of a fixed positive temperature (and fixed external parameters), the Helmholtz free energy F attains a minimum at thermal equilibrium. What is the corresponding property for F at a negative temperature? Justify your response.

8. In a hypothetical system, the single-particle states φ_α have the following energies:

φ_1: ε_0;

φ_2, φ_3, φ_4, φ_5, φ_6: $3\varepsilon_0$.

Altogether, there are only six states. The constant ε_0 is positive.

For one particle in thermal equilibrium at temperature T, compute the following:

(a) partition function Z,
(b) energy estimate $\langle E \rangle$, and
(c) entropy S.

Express your answers in terms of k, T, and ε_0.

(d) Then, *from your answer to part* (c), determine the limiting values of S when $kT \ll \varepsilon_0$ and $kT \gg \varepsilon_0$. Show your work. If the limiting values do *not* seem reasonable, explain why.

15 Transport Processes

15.1 Mean free path
15.2 Random walk
15.3 Momentum transport: viscosity
15.4 Pipe flow
15.5 Energy transport: thermal conduction
15.6 Time-dependent thermal conduction
15.7 Thermal evolution: an example
15.8 Refinements
15.9 Essentials

Two paragraphs set the scene for the entire chapter. Recall that the molecules of a gas are in continual, irregular motion. Individual molecules possess both energy and momentum, and they transport those quantities with them. What is the net transport of such quantities? In thermal equilibrium, it is zero. If, however, the system has macroscopic spatial variations in temperature, being hotter in some places than in others, then net transport of energy may arise (even from irregular molecular motion). Or if the locally averaged velocity is nonzero and varies spatially (as it does in fluid flow through a pipe), then net transport of momentum may arise.

Our strategy is first to examine irregular molecular motion in its own right, then to study transport of momentum, and finally to investigate the transport of energy.

15.1 Mean free path

Our context is a classical gas. The conditions of temperature and number density are similar to those of air under room conditions. We acknowledge forces between molecules, but we simplify to thinking of the molecules as tiny hard spheres, so that they exert mutually repulsive forces during collisions. Generalization will come later.

When any given molecule wanders among its fellow molecules, its path is a random sequence of long and short "free" paths between collisions. Figure 15.1 illustrates the irregular, broken path. There is a meaningful *average* distance that the molecule travels between collisions. We call this distance the *mean free path* and denote it by lower case script ell: ℓ. The adjective "mean" emphasizes that ℓ is the average distance between collisions. The process of random free paths and random collisions is how a given molecule diffuses through the others.

The mean free path ought to decrease if the number density increases. What is the

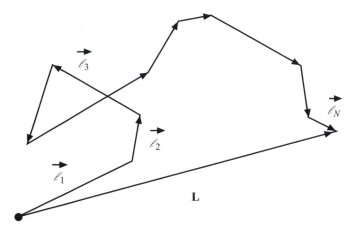

Figure 15.1 A view of the "free" paths, $\vec{\ell}_i$, as a molecule pursues a straight-line trajectory between collisions. The vector **L** describes the net displacement and is equal to the vectorial sum of the N free paths.

quantitative connection? Figure 15.2 illustrates a collision. For a collision to occur, the *edge* of another molecule must lie within *one* molecular *radius* of our molecule's centerline-of-motion.

Equivalently, the *center* of another molecule must lie within *two* molecular *radii* of the centerline. That amounts to *one* molecular *diameter*, which is denoted by d. Figure 15.3 illustrates this and goes further. For the purpose of calculating collisions, it is as though our molecule carried a shield of radius d and every other molecule whose center touched the shield caused a collision. The area of the shield is πd^2. (If you are familiar with the idea of a "cross-section," then you will recognize πd^2 as the cross-section for collision and hence for scattering.)

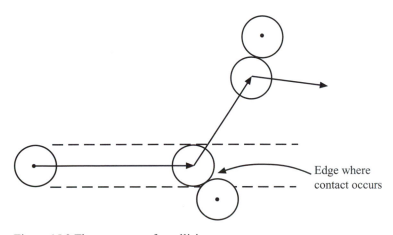

Figure 15.2 The geometry of a collision.

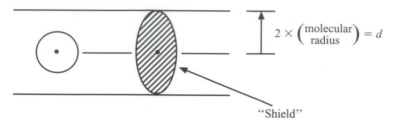

Figure 15.3 Collision as determined by the molecular diameter d.

As illustrated in figure 15.4, our molecule sweeps out "stove-pipe sections" of volume, each of which terminates in a collision. Thus each stove-pipe section contains precisely one molecule other than our own, namely, the molecule that caused the collision. On the average, the stove-pipe sections are one mean free path long. The volume of the average stove-pipe section is (length) $\times$ (cross-sectional area) $=$ $\ell \times \pi d^2$ because the radius of the shield is d. Because each stove-pipe section contains one molecule (other than our traveler), we may write

$$\left(\begin{array}{c} \text{average volume of} \\ \text{stove-pipe section} \end{array} \right) \times \left(\begin{array}{c} \text{number of molecules} \\ \text{per unit volume} \end{array} \right) = 1, \qquad (15.1)$$

that is,

$$\ell \times \pi d^2 \times n = 1. \qquad (15.2)$$

The number density arises frequently in this chapter, and so the short symbol n denotes the number of molecules per unit volume. Solving (15.2) for the mean free path, we find

$$\ell = \frac{1}{n \times \pi d^2}. \qquad (15.3)$$

The mean free path decreases as the reciprocal of the number density. Also, the larger the molecular diameter, the shorter the mean free path, which also makes sense.

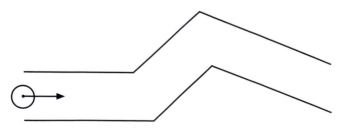

Figure 15.4 The stove-pipe sections of volume swept out by the molecule's shield.

Some numerical estimates for air

We proceed to some numerical estimates for air under typical room conditions. The dominant constituents of air are diatomic nitrogen and oxygen: N_2 and O_2. Both molecules are much more nearly spherical than dumbbell-shaped. (To be sure, the two nuclei form an exemplary dumbbell because they are separated by a distance much larger than nuclear size. The electron cloud that surrounds the nuclei, however, is nearly spherical, and that cloud determines the molecular shape and size, as they are seen in molecule–molecule collisions.) We regard molecules of N_2 and O_2 as spheres of approximately the same mass and size: mass m and diameter d.

The tactics are first to note some experimental values for mass density, next to rewrite those densities in terms of molecular and gas parameters, and finally to combine equations so as to extract ratios of characteristic lengths.

The mass density of air is readily measured; chapter 13 cited the value

$$\rho_{air} = 1.2 \text{ kg/m}^3, \tag{15.4}$$

where the letter rho, ρ, denotes mass density. The density of liquid nitrogen (at 77 K and atmospheric pressure) is measured even more easily (by weighing a beaker of it):

$$\rho_{liq\ N_2} = 800 \text{ kg/m}^3. \tag{15.5}$$

Let us regard liquid nitrogen as providing a situation where the molecules are in constant contact. Then there is one molecule in each cube of edge length d, and so the mass density is

$$\rho_{liq\ N_2} = \frac{m}{d^3} \tag{15.6}$$

in terms of the molecular parameters.

In air, the average separation of molecules determines the volume that a molecule may call its own. That volume extends to each side of a molecule by a distance equal to half the molecule-to-molecule separation. In short, there is one molecule in a volume equal to (separation)^3. Consequently, the mass density of air may be written as

$$\rho_{air} = \frac{m}{(\text{separation})^3}. \tag{15.7}$$

Now to extract some length ratios. Form the ratio of equations (15.6) and (15.7); then take a cube root:

$$\frac{(\text{separation})}{d} = \left(\frac{\rho_{liq\ N_2}}{\rho_{air}}\right)^{1/3} = 8.7 \cong 10. \tag{15.8}$$

The average center-to-center separation of nearby molecules is approximately 10 times the molecular diameter.

The number density n can be expressed in terms of d and the ratio of mass densities. Form the cube of equation (15.8) and rearrange to find

$$n = \frac{1}{(\text{separation})^3} = \frac{1}{d^3}\left(\frac{\rho_{\text{air}}}{\rho_{\text{liq N}_2}}\right). \tag{15.9}$$

Substitute this expression for n into equation (15.3) and then rearrange to produce the ratio

$$\frac{\ell}{d} = \frac{1}{\pi}\left(\frac{\rho_{\text{liq N}_2}}{\rho_{\text{air}}}\right) = 212 \cong 200. \tag{15.10}$$

Taking the round figures, one can say that the mean free path is approximately 20 times the average separation of the molecules in air.

15.2 Random walk

Glance again at figure 15.1 as we pose the question, given some starting point for our molecule, how far from that point will the molecule get in N free paths, on the average? The vector location $\mathbf{L}$ after N free paths has the form

$$\mathbf{L} = \vec{\ell}_1 + \vec{\ell}_2 + \vec{\ell}_3 + \cdots + \vec{\ell}_N, \tag{15.11}$$

where $\vec{\ell}_i$ denotes the ith free path. To extract the magnitude of the distance, take the scalar product of $\mathbf{L}$ with itself:

$$\mathbf{L}\cdot\mathbf{L} = (\vec{\ell}_1 + \vec{\ell}_2 + \vec{\ell}_3 + \cdots + \vec{\ell}_N)\cdot(\vec{\ell}_1 + \vec{\ell}_2 + \vec{\ell}_3 + \cdots + \vec{\ell}_N)$$

$$= \vec{\ell}_1\cdot\vec{\ell}_1 + \vec{\ell}_2\cdot\vec{\ell}_2 + \cdots + 2\vec{\ell}_1\cdot\vec{\ell}_2 + 2\vec{\ell}_1\cdot\vec{\ell}_3 + \cdots.$$

Now take the average over many sets of N free paths. The cross terms (such as $\vec{\ell}_1\cdot\vec{\ell}_3$) may be either positive or negative and hence will average to zero. The average of $\vec{\ell}_i\cdot\vec{\ell}_i$ will be the same for all values of i: $i = 1, 2, 3, \ldots, N$. Thus

$$\langle\mathbf{L}\cdot\mathbf{L}\rangle = N\langle\vec{\ell}_1\cdot\vec{\ell}_1\rangle. \tag{15.12}$$

The average of the magnitude of the first free path, $\langle|\vec{\ell}_1|\rangle$, is precisely the mean free path that we denoted by ℓ in section 15.1. Let us ignore the distinction between the root mean square value, $\langle\vec{\ell}_1\cdot\vec{\ell}_1\rangle^{1/2}$, and the value $\langle|\vec{\ell}_1|\rangle$, particularly since the factor that relates them is close to unity and is difficult to determine. Proceeding in that spirit, we take the square root of equation (15.12) and find

$$\langle\mathbf{L}\cdot\mathbf{L}\rangle^{1/2} \equiv L_{\text{r.m.s.}} = \sqrt{N}\ell, \tag{15.13}$$

where $L_{\text{r.m.s.}}$ denotes the root mean square of the net distance traveled. Equation (15.13) tells us that $L_{\text{r.m.s.}}$ grows not as N but only as $\sqrt{N}$. The randomly oriented free paths tend to cancel one another; the residual distance grows only as a consequence of incomplete cancellation.

To determine how far the molecule will get from a starting point in a given elapsed time t, we need the number of steps N that can be taken in that time. If we were to stretch a string along the tortuous path of figure 15.1, we would get a length $\langle v \rangle t$, where $\langle v \rangle$ denotes the mean speed. So the number N of free paths is

$$N = \frac{\text{length of path}}{\ell} = \frac{\langle v \rangle t}{\ell}.$$ (15.14)

Substituting this expression into (15.13) yields

$$L_{\text{r.m.s.}} = \sqrt{\frac{\langle v \rangle t}{\ell}}\,\ell = \sqrt{\langle v \rangle t \ell}.$$ (15.15)

The net distance traveled grows only as $\sqrt{t}$. Figure 15.5 illustrates the reason why.

Numerical estimates

How far does a chosen air molecule diffuse in various lengths of time? The diameter of both N_2 and O_2 is approximated well by $d \cong 3 \times 10^{-10}$ meter. Then equation (15.10) implies a mean free path of $\ell \cong 6 \times 10^{-8}$ meter. Chapter 13 gave the root mean square speed as $v_{\text{r.m.s.}} \cong 500$ m/s, and so $\langle v \rangle \cong 460$ m/s. This information

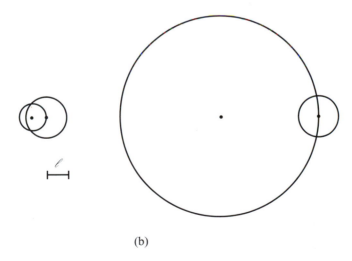

(a)   (b)

Figure 15.5 Geometry to illustrate the growth as $\sqrt{t}$. The two sketches show where the molecule could be after one more free path (of average length). Less of the circle of radius ℓ is outside the already-achieved distance in sketch (b) than in sketch (a). Thus, the larger the net distance already traveled, the smaller the probability that the next free path will increase the distance. As soon as L is much larger than ℓ, the probability of an increase in L barely exceeds $1/2$. The molecule is almost as likely to regress as to progress.

provides the product $\langle v \rangle \ell$ that appears in equation (15.15), and table 15.1 lists some times and distances. If diffusion alone were responsible for apprising you that onions are frying on the stove, you could wait a long time in the next room before you became aware that dinner was cooking. Convection is much more effective than diffusion at moving the aroma of cooking food or the scent of flowers through the air. Usually, convection is responsible for bringing to us whatever we smell.

15.3 Momentum transport: viscosity

When fluid flows slowly through a pipe, the velocity profile has the shape shown in figure 15.6. The flow is fastest at the pipe's central axis. As one approaches the walls, the velocity drops to zero. In fluid dynamics, this property is called the *no slip*

Table 15.1 *Net distance,* $L_{r.m.s.}$, *diffused by an air molecule. The calculation is based on* $\ell = 6 \times 10^{-8}$ *meter and* $\langle v \rangle = 460$ *m/s.*

Time t	$L_{r.m.s.}$
1 s	0.53 cm
10 s	1.7 cm
1 min	4.1 cm
10 min	13 cm
1 hour	32 cm
1 day	1.5 m

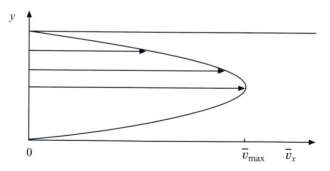

Figure 15.6 Velocity profile for slow fluid flow in a pipe. The velocity component $\bar{v}_x$ is the velocity component along the pipe; the overbar denotes an average of molecular velocity components taken over a small volume (that is, small relative to the pipe radius R). The y-axis represents a slice through the pipe: from one interior wall through the central axis and to the diametrically opposite wall.

condition. Experiment abundantly confirms its existence, and we can think of it this way: the molecular attraction between wall atoms and fluid molecules holds the latter at rest in an infinitesimal layer adjacent to the walls.

The major purpose of the diagram, however, is merely to give us a concrete instance in which a component of fluid velocity (here the x-component) varies in size in an orthogonal direction (here the y-direction):

$$\bar{v}_x = \bar{v}_x(y). \tag{15.16}$$

In such a context, molecular diffusion transports x-momentum in the y-direction. To understand how this process arises, we specialize to a dilute classical gas, as follows.

Part (b) of figure 15.7 shows a small portion of the velocity profile, and part (a) shows two molecules that will cross the plane $y = y_0$ from opposite sides. On average, the lower molecule has a larger value of v_x and hence transports more x-component of momentum upward across the plane than the upper molecule transports in the opposite direction. Thus diffusion produces a net upward transport of x-momentum.

Now we quantify the transport. The goal is to determine the dependence on temperature T, number density n, mean free path ℓ, and so on. In setting up the calculation, we will not concern ourselves with factors of 2 or π.

Of the molecules immediately below the plane $y = y_0$, half have a positive value for v_y, the y-component of velocity. The flux of such molecules upward across the plane is $\frac{1}{2}nv_y$, where v_y denotes now some typical (positive) value. Those molecules carry, on average, an x-momentum of size $m\bar{v}_x(y_0 - \ell)$, characteristic of the fluid velocity one mean free path back along their trajectory. (Their last scattering could not, on average, be farther back than one mean free path.) So we write

$$\left(\begin{array}{c} \text{flux of } x\text{-momentum} \\ \text{upward across } y = y_0 \end{array} \right) \cong \frac{1}{2}nv_y m\bar{v}_x(y_0 - \ell). \tag{15.17}$$

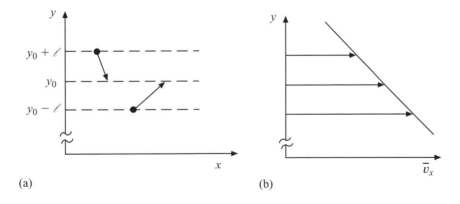

(a) (b)

Figure 15.7 Local analysis for momentum transport by diffusion. (a) Two molecules about to cross the plane $y = y_0$. (b) The local velocity profile of the fluid flow. Over a small range in y, which need be only a few mean free paths long, the mean molecular velocity $\bar{v}_x$ varies in a nearly linear fashion.

The downward flux is similar but is proportional to $\bar{v}_x(y_0 + \ell)$, the fluid velocity one mean free path above the plane $y = y_0$.

For the net flux of x-momentum, we have

$$\begin{pmatrix} \text{net flux of } x\text{-momentum} \\ \text{upward across unit area} \end{pmatrix} \cong \tfrac{1}{2} n v_y m \bar{v}_x(y_0 - \ell) - \tfrac{1}{2} n v_y m \bar{v}_x(y_0 + \ell)$$

$$\cong -n v_y m \ell \frac{\partial \bar{v}_x}{\partial y}. \tag{15.18}$$

A Taylor series expansion of $\bar{v}_x(y_0 \mp \ell)$ about $y = y_0$ generates the last line. For v_y, we can take a substantial fraction of $\langle v \rangle$, the average value of the magnitude of the irregular molecular velocity. (The irregular part of the velocity is $\mathbf{v} - \bar{\mathbf{v}}$, that is, the true molecular velocity minus the local average velocity, which is associated with the macroscopic fluid motion.) The typical trajectory crosses the plane $y = y_0$ obliquely, and so ℓ is an over-estimate of the y-distance where the last collision occurred. All in all, to replace v_y in the second line of (15.18) with $\tfrac{1}{3}\langle v \rangle$ is reasonable. (More detail in setting up the calculation leads to precisely the factor $1/3$. It would be only fair, however, to add that even more sophisticated calculations modify the $1/3$ to a different but similar fraction.)

The structure of equation (15.18) is this:

$$\begin{pmatrix} \text{net flux of } x\text{-momentum} \\ \text{in } y\text{-direction} \end{pmatrix} = -\eta \frac{\partial \bar{v}_x}{\partial y}, \tag{15.19}$$

where the coefficient eta, η, is called the *coefficient of (dynamic) viscosity* and where we have an approximate expression for the coefficient,

$$\eta \cong \frac{1}{3} n \langle v \rangle m \ell, \tag{15.20}$$

valid for a dilute classical gas. The structure in (15.19) holds for most fluids, not just a dilute gas. The velocity profile and its spatial variation appear solely in the derivative $\partial \bar{v}_x / \partial y$. The coefficient of viscosity is characteristic of the fluid itself, independent of any average velocity.

To see how the coefficient η for a dilute gas depends on intrinsic parameters, we first eliminate the mean free path ℓ in terms of such parameters. Substituting from equation (15.3), we find

$$\eta \cong \frac{1}{3} n \langle v \rangle m \frac{1}{n \times \pi d^2} = \frac{1}{3} \frac{m}{\pi d^2} \langle v \rangle$$

$$= \frac{1}{3} \frac{m}{\pi d^2} \sqrt{\frac{8}{\pi} \frac{kT}{m}}. \tag{15.21}$$

Thus the coefficient of viscosity depends on the temperature T, the molecular mass m, and the molecular diameter d. Once this dependence is recognized, one can say that the coefficient η does *not* depend on the number density or the pressure. When James Clerk Maxwell derived this implication, it surprised him. He put it to experimental test (in 1866) and found his derivation vindicated.

To be sure, there is a natural limit to such independence. If the number density drops so low that the mean free path becomes comparable to the size of the container or to the distance over which the average velocity changes greatly, then the local analysis in equations (15.17) and (15.18) fails, and so the entire calculation needs to be reformulated.

Equation (15.21) implies that the viscosity of a dilute gas increases with temperature. The molecules move faster and hence can transport momentum more readily. Experiment confirms the increase.

The behavior is in dramatic contrast, however, to the way that viscosity changes for a liquid. For the latter, an increase in temperature brings a decrease in viscosity. Molecules in a liquid are virtually in constant contact with their neighbors, and strong intermolecular forces operate. To a surprisingly large extent, the molecules are locked into their locations relative to their neighbors. Increasing the temperature makes it easier for molecules to surmount the potential energy barriers that inhibit shifts in location. Thus an increase in temperature enables the liquid molecules to slide past one another more easily, and so the fluid becomes less viscous.

Table 15.2 lists the coefficient of viscosity for some common substances.

Table 15.2 *The coefficient of (dynamic) viscosity. The ambient pressure is 1 atmosphere.*

Fluid	Temperature	Viscosity ($N \cdot s/m^2$)
Argon	300 K	2.3×10^{-5}
Helium	300 K	2.0×10^{-5}
Hydrogen	300 K	0.90×10^{-5}
Nitrogen	300 K	1.8×10^{-5}
Water	0 °C	1.8×10^{-3}
	20 °C	1.0×10^{-3}
	60 °C	0.47×10^{-3}
Glycerin	−20 °C	130
	0 °C	12
	20 °C	1.5

Source: *CRC Handbook of Chemistry and Physics*, 71st edn, edited by David R. Lide (Chemical Rubber Publishing Company, Boston, 1992).

15.4 Pipe flow

We are all familiar with viscosity as a physical phenomenon: maple syrup from the refrigerator pours slowly; motor oil "thickens" in cold weather; and suntan lotion may feel viscous. Yet students in physics or astronomy rarely see a calculation where viscosity is prominent, and so this section provides an example.

Turn to the context of figure 15.8: steady fluid flow through a length of pipe. Diffusion transports x-momentum to the walls, where the walls absorb it. The fluid between points a and b would slow to a halt if momentum were not continually supplied from outside. Pressure at the upstream end (point a), acting to the right, adds momentum to the fluid. A lesser pressure at the downstream end (point b), acting to the left, subtracts momentum from the fluid. The question we address is this: for fixed flow rate (in cubic meters per second), how does the pressure difference $P_a - P_b$ scale when the radius R of the pipe is varied?

Specify that the fluid is incompressible. That is surely a good approximation for water or blood, and it is a good approximation even for air at subsonic flow speeds. The detailed shape of the velocity profile remains unknown (to us), but—for purposes of scaling—the qualitative shape suffices.

Our plan is this: determine the rates of momentum loss and gain; equate them; extract the pressure difference $P_a - P_b$; and then incorporate the condition of fixed flow rate.

We use equation (15.19) to estimate the momentum loss rate:

$$\begin{pmatrix} \text{momentum} \\ \text{loss rate} \end{pmatrix} = \begin{pmatrix} \text{momentum flux across} \\ \text{unit area of wall} \end{pmatrix} \times \begin{pmatrix} \text{total wall} \\ \text{area} \end{pmatrix}$$

$$\cong \eta \frac{\bar{v}_{\max}}{R} \times 2\pi R L_{ab}. \tag{15.22}$$

The derivative $\partial \bar{v}_x / \partial y$ at the wall must scale as the change in $\bar{v}_x$ between center and wall, which is $(\bar{v}_{\max} - 0)$, divided by the pipe's radius.

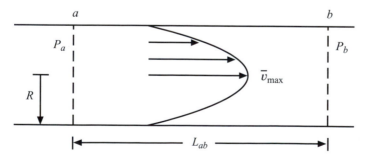

Figure 15.8 Momentum balance for steady flow through a pipe: some context.

The pressure at point a acts on an area πR^2 and so exerts a force $P_a\pi R^2$ to the right. At point b, a force $P_b\pi R^2$ acts to the left. Then Newton's second law implies

$$\left(\begin{array}{c}\text{momentum}\\\text{gain rate}\end{array}\right) = (P_a - P_b)\pi R^2. \tag{15.23}$$

Equating the gain and loss rates and acknowledging the approximate nature of equation (15.22), we find

$$P_a - P_b \cong \eta \bar{v}_{\max}\frac{2L_{ab}}{R^2}. \tag{15.24}$$

The flow rate will scale as $\bar{v}_{\max}$ and as the cross-sectional area πR^2:

$$(\text{flow rate}) \cong \bar{v}_{\max}\pi R^2. \tag{15.25}$$

We incorporate the fixed flow rate by using equation (15.25) to eliminate $\bar{v}_{\max}$ in the momentum balance equation. Dropping all pretense about numerical factors of order unity, we find

$$P_a - P_b \propto \eta \frac{L_{ab}}{R^4} \times (\text{flow rate}). \tag{15.26}$$

The dependence on the inverse fourth power of the radius is well-established experimentally (for smooth, non-turbulent flow). It is one aspect of *Poiseuille's law*, as equation (15.26) is called. You can see vividly why clogging of the human arteries puts such a strain on the heart. When blood returns to the heart, it is essentially at atmospheric pressure. Decreasing the arterial radius by a mere 10 percent (throughout the vascular system) increases the required blood pressure (as blood leaves the heart) by more than 50 percent. Medicine that reduces the viscosity η can help, but it is difficult to fight a fourth power.

If you want to determine the detailed shape of the velocity profile, problem 7 provides a good start.

15.5 Energy transport: thermal conduction

Now we turn to the transport of energy by molecular diffusion. Again we begin with a dilute classical gas and employ a local analysis, as sketched in figure 15.9. The average energy per molecule, $\bar{\varepsilon}$, is a function of temperature, and the temperature decreases along the y-direction. We can expect diffusion to transport energy from the region of high temperature to that of low temperature.

Following the pattern of reasoning that we used in section 15.3, we write

$$\left(\begin{array}{c}\text{net flux of energy}\\\text{upward across unit area}\end{array}\right) \cong \tfrac{1}{2}nv_y\bar{\varepsilon}(y_0 - \ell) - \tfrac{1}{2}nv_y\bar{\varepsilon}(y_0 + \ell)$$

$$\cong -nv_y\ell\frac{d\bar{\varepsilon}}{dT}\frac{\partial T}{\partial y}. \tag{15.27}$$

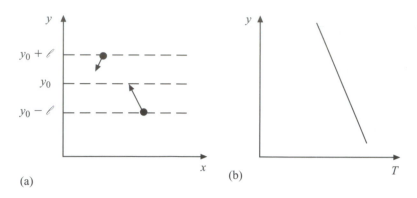

Figure 15.9 Local analysis for energy transport by diffusion. (a) Two molecules about to cross the plane $y = y_0$. (b) The profile of the temperature distribution: the temperature decreases along y and approximately linearly so in the short interval that is displayed. (The fractional change in T over a distance 2ℓ is specified to be small, and so the y and T axes cross at some finite T, not at absolute zero.)

The chain rule gives the derivative of $\bar{\varepsilon}$ with respect to position. You might wonder whether the product nv_y should be considered a function of position also and evaluated at $y_0 \mp \ell$. The answer is "no." The goal is to describe a situation wherein there is no net flux of molecules *per se*, and so nv_y must be the same (in magnitude) below and above the plane $y = y_0$.

Again we can extract a general structure:

$$\begin{pmatrix} \text{net energy} \\ \text{flux} \end{pmatrix} = -K_T \,\mathrm{grad}\, T, \tag{15.28}$$

where the positive coefficient K_T is called the *coefficient of thermal conductivity* and where we have an approximate expression,

$$K_T \cong \frac{1}{3} n \langle v \rangle \ell \frac{d\bar{\varepsilon}}{dT}, \tag{15.29}$$

valid for a dilute classical gas and based on the same approximation as for the viscosity. (There is no standard notation for the coefficient of thermal conductivity. Unadorned K is sometimes used, but that can be confused with the K for "kelvin," and so a subscript "T" for "thermal" has been appended: K_T.) The gradient, grad T, generalizes the geometry. For any smooth function, the gradient points in the direction of maximal spatial rate of change. Thus, in equation (15.28), the combination $-\mathrm{grad}\, T$ says that the energy flux points in the direction in which the temperature decreases fastest. The magnitude of a spatial gradient is always the maximal spatial rate of change, and so $|\mathrm{grad}\, T|$ generalizes the derivative of T in (15.27).

Viscosity and thermal conductivity compared

We can profitably compare the coefficients of viscosity and thermal conductivity. For a dilute classical gas, equations (15.20) and (15.29) imply that the ratio of the coefficients is the following:

$$\frac{K_T}{\eta} \propto \frac{1}{m}\frac{d\bar{\varepsilon}}{dT}. \tag{15.30}$$

Both the mean free path and the molecular speed cancel out.

If the gas is monatomic, such as argon or helium, then $\bar{\varepsilon}$ is strictly the average translational kinetic energy, $\frac{3}{2}kT$, and so $d\bar{\varepsilon}/dT = 3k/2$. The quantity mK_T/η should be the same for all monatomic gases. Experiments with helium, argon, and neon (near room temperature) give mutual agreement within 5 percent.

If the gas is diatomic, then there will be a contribution of k to $d\bar{\varepsilon}/dT$ from classical rotational motion. The vibrational motion is typically "frozen" into the ground state; it does not contribute to $d\bar{\varepsilon}/dT$. Again, the quantity mK_T/η should be the same for all diatomic gases (but should differ from the value for monatomic gases). A more cautious comparison avoids any assumptions about $d\bar{\varepsilon}/dT$ and focuses on the combination

$$\frac{mK_T}{\eta(d\bar{\varepsilon}/dT)}, \tag{15.31}$$

which should have a single value for all gases. Experimental comparison among H_2, CO, N_2, NO, and O_2 finds that the values for the combination lie close together—within 5 percent. That is the good news. The bad news is that, although the combination in (15.31) should have the same value for both diatomic and monatomic gases, it does not. The two sets differ by approximately 25 percent, and the situation is worse for polyatomic molecules (such as CO_2 and C_2H_4). The transport of energy associated with internal motions turns out to be more complicated than a simple classical "billiard ball" picture can describe. Despite this blemish, even the simple kinetic theory of transport is a notable success.

Before we go on, let me emphasize one point. In a broad sense, equation (15.28) provides a general structure for energy transport by thermal conduction. Although we derived the equation in the context of a dilute classical gas, the gradient form remains valid for most solids and liquids (as well as for dense gases).

Also, let me note that coefficients like η and K_T are called *transport coefficients* because, together with certain spatial derivatives, they describe the transport of momentum and energy.

15.6   Time-dependent thermal conduction

The local analysis in section 15.5 provides an expression for the energy flux. To solve a large-scale problem of how the temperature distribution evolves in time, one needs an equation for the time derivative $\partial T/\partial t$ in terms of the energy flux. Conservation of energy enables us to construct such an equation, as follows.

Consider the molecules that are in a small volume V_{small} in the physical system. Conservation of energy implies

$$\begin{pmatrix} \text{rate of change} \\ \text{of their energy} \end{pmatrix} + \begin{pmatrix} \text{rate at which work is done} \\ \text{by them in an expansion} \end{pmatrix} = \begin{pmatrix} \text{rate of energy input} \\ \text{by thermal conduction} \end{pmatrix}.$$

(15.32)

Fundamentally, this equation is equation (1.13) after its right-hand and left-hand sides have been interchanged and after it has been expressed as rates of change with respect to time. (Note that the number of molecules is kept fixed, but the size of V_{small} may change.)

When thermal conduction delivers energy to the molecules, the local temperature changes, the pressure changes, and the volume that the molecules occupy changes. A lot is going on, and we need to find a simplifying context. For the molecules in question, regard their energy as a function of the temperature and pressure. The volume V_{small} that they occupy is likewise a function of T and P. Then split the discussion into solids and liquids on the one hand and gases on the other.

Most solids and liquids are remarkably insensitive to changes in the ambient pressure. For example, if the pressure applied to water (at 20 °C) is increased from 1 atmosphere to 10 atmospheres, the fractional change in volume is only 5×10^{-4}. A pressure change affects the energy of the molecules much less than a temperature change does (when the changes in pressure and temperature are typical of those induced by thermal conduction). Thus, for most solids and liquids, we may approximate the entire left-hand side of (15.32) by the heat capacity at constant pressure (as though the pressure really were constant) times the rate at which the temperature is changing.

The situation with gases is quite different. A 1 percent change in pressure affects the volume as much as a 1 percent change in temperature. In expressing the left-hand side of equation (15.32) in terms of various coefficients, one must retain the change in pressure that accompanies the thermal conduction. Usually the gas is the simpler physical system, but that is not so here.

For the remainder of this section, we restrict our attention to typical solids and liquids. Then the left-hand side of (15.32) may be approximated in terms of the heat capacity at constant pressure:

$$C_P \frac{\partial T}{\partial t} = -\int_{\text{surface of small volume}} \begin{pmatrix} \text{energy flux vector} \\ \text{for thermal conduction} \end{pmatrix} \cdot d\mathbf{A}.$$

(15.33)

The vector $d\mathbf{A}$ denotes a patch of surface area and points along the outward normal. Thus the integral represents the outward flow of energy by thermal conduction. The minus sign converts to the inward flow, as required by equation (15.32).

Next, divide both sides by V_{small} and pass to the limit as $V_{\text{small}} \to 0$. The right-hand side becomes the divergence of the energy flux (times a minus sign). Thus

$$(C_P)_{\text{p.u.v.}} \frac{\partial T}{\partial t} = -\text{div}\left(\begin{array}{c} \text{energy flux vector} \\ \text{for thermal conduction} \end{array} \right). \tag{15.34}$$

The subscript "p.u.v." denotes "per unit volume." Taking the energy flux vector from equation (15.28), we arrive at the relationship

$$(C_P)_{\text{p.u.v.}} \frac{\partial T}{\partial t} = \text{div}(K_T \, \text{grad} \, T). \tag{15.35}$$

This equation is quite general; it applies to most solids and liquids (that are macroscopically at rest). As noted earlier, however, it would need to be augmented before it could be applied to gases. The equation is often called the *heat equation*, another relic of the nineteenth century. A modern view sees it as the evolution equation for the temperature distribution. (Other names are the *equation of thermal conduction* and *Fourier's equation*.)

If the thermal conductivity K_T is spatially constant (or if that is a good approximation), one may pull K_T outside the divergence operation and write

$$\frac{\partial T}{\partial t} = \frac{K_T}{(C_P)_{\text{p.u.v.}}} \, \text{div} \, \text{grad} \, T. \tag{15.36}$$

The pair of operations, first forming the gradient and then taking the divergence, produces the Laplacian operator, often denoted by ∇^2, and so the right-hand side could be rewritten with $\nabla^2 T$. The ratio of parameters on the right-hand side is called the *thermal diffusivity*:

$$\left(\begin{array}{c} \text{thermal} \\ \text{diffusivity} \end{array} \right) \equiv D_T \equiv \frac{K_T}{(C_P)_{\text{p.u.v.}}}. \tag{15.37}$$

Table 15.3 lists representative values of K_T, $(C_P)_{\text{p.u.v.}}$, and D_T. In its most succinct form, the heat equation is

$$\frac{\partial T}{\partial t} = D_T \, \text{div} \, \text{grad} \, T. \tag{15.38}$$

In this form, the equation is often called the *diffusion equation*.

Dimensional balance in the succinct form implies that the units of D_T are m^2/s. If the physical system has a characteristic length λ, then one can expect a characteristic time to emerge. On dimensional grounds, the time scale must be proportional to λ^2/D_T. Thus large thermal diffusivity implies a short time scale and hence rapid changes in temperature. The next section provides an example of how the combination λ^2/D_T arises.

Table 15.3 *Thermal conductivity K_T, heat capacity per unit volume (at constant pressure) $(C_P)_{p.u.v.}$, and thermal diffusivity D_T. The ambient pressure is 1 atmosphere, where relevant.*

Substance	Temperature	K_T (J/s·m·K)	$(C_P)_{p.u.v.}$ (J/K·m³)	D_T (m²/s)
Argon	300 K	0.018	830	0.22×10^{-4}
Helium	300 K	0.16	830	1.9×10^{-4}
Hydrogen	300 K	0.19	1,200	1.6×10^{-4}
Nitrogen	300 K	0.026	1,200	0.22×10^{-4}
Water	20 °C	0.60	4.2×10^6	1.4×10^{-7}
Aluminum	25 °C	240	2.4×10^6	0.98×10^{-4}
Copper	25 °C	400	3.5×10^6	1.2×10^{-4}
Silver	25 °C	430	2.5×10^6	1.7×10^{-4}

Source: *CRC Handbook of Chemistry and Physics*, 71st edn, edited by David R. Lide (Chemical Rubber Publishing Company, Boston, 1992).

15.7   Thermal evolution: an example

To gain some sense of how to use the diffusion equation, we work out an example in one spatial dimension. At time $t = 0$, the temperature profile is

$$T(x, 0) = T_0 + \Delta T \sin\left(\frac{2\pi x}{\lambda}\right),$$

where T_0 and ΔT are positive constants and where the strong inequality $\Delta T \ll T_0$ holds. The constant λ denotes the wavelength of the sinusoidal variation in the initial temperature profile. Figure 15.10 illustrates the initial profile and provides a prelude of an evolved profile.

You may wonder, how realistic are these initial conditions? Although not common, they can be produced in the lab. Pass a pulse of laser light through a pair of slits, and let the interference pattern fall on a strip of copper foil, painted black for good absorption. The double-slit interference pattern will generate a sinusoidal spatial variation in the foil's temperature (together with an overall increase in temperature equal to the amplitude of the sinusoidal variation). (Note. The wavelength λ of the spatial variations in temperature differs from the wavelength of the laser light. The λ here depends on the slit-spacing and the distance from slits to foil, although it is proportional to the literal wavelength of the laser light.) Another example will be discussed after the primary calculation, to which we turn now.

How does the temperature distribution evolve in time? The differential equation for $T(x, t)$ is

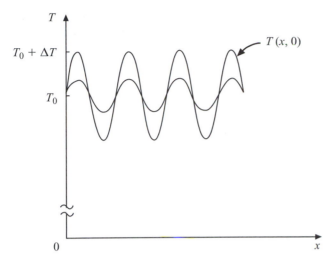

Figure 15.10 The temperature profile $T(x, t)$ at two times: $t = 0$ and $t = (\lambda/2\pi)^2/D_T$.

$$\frac{\partial T}{\partial t} = D_T \frac{\partial^2 T}{\partial x^2};$$

no derivatives with respect to y or z need appear. Because the second derivative of a sine function is proportional to the sine function, the spatial temperature variation will remain sinusoidal and have the same wavelength. Thus we try the form

$$T(x, t) = T_0 + \Delta T \sin\left(\frac{2\pi x}{\lambda}\right) \times f(t)$$

for some function $f(t)$. Substitute into the differential equation and then cancel common factors on the two sides to find an equation for $f(t)$:

$$\frac{df}{dt} = -\left(\frac{2\pi}{\lambda}\right)^2 D_T f.$$

The solution is a decaying exponential. Therefore the solution for $T(x, t)$ is

$$T(x, t) = T_0 + \Delta T \sin\left(\frac{2\pi x}{\lambda}\right) \times \exp\left[-\left(\frac{2\pi}{\lambda}\right)^2 D_T t\right].$$

The spatial variation in temperature decays away as energy diffuses from the peaks (high T) to the troughs (low T) and "fills them." As the time scale of this process, we take the e-folding time, which yields

$$\left(\begin{array}{c} \text{time scale for} \\ \text{thermal evolution} \end{array}\right) = \left(\frac{\lambda}{2\pi}\right)^2 \frac{1}{D_T}. \tag{15.39}$$

The longer the wavelength, the longer the time—and quadratically so. Long wavelength implies a shallow thermal gradient and hence a small energy flux; that is the key insight. (The spatial variation in the flux is then small also, and that variation is what dumps energy into a region or extracts it.)

Application to sound waves in air

You may have learned that audible sound waves in air constitute an adiabatic process (rather than, say, an isothermal process), at least to good approximation. The wavelengths are so long that the time required for significant energy diffusion is much longer than the oscillation period of the sound wave. Hence there is almost no energy diffusion, and the wave proceeds adiabatically. Long wavelength requires low frequency (because the speed of sound is roughly equal to the root mean square molecular speed and hence is largely independent of wavelengths or frequency). Hence we find a pleasant surprise: the sound frequency is *low* enough—*rather than high* enough—for sound propagation to be an adiabatic process.

For some detail, let us first estimate the thermal diffusivity for a dilute classical gas. Equations (15.29) and (15.37) yield

$$D_T \cong \left(\frac{1}{3} n \langle v \rangle \ell \frac{d\bar{\varepsilon}}{dT} \right) \bigg/ (C_P)_{\text{p.u.v.}}. \tag{15.40}$$

The product $n d\bar{\varepsilon}/dT$ is the rate at which internal energy per unit volume, $n\bar{\varepsilon}$, changes with temperature if the volume and hence n are kept fixed. Thus $n d\bar{\varepsilon}/dT$ is the heat capacity per unit volume at constant volume. Section 1.5 introduced the symbol γ for the dimensionless ratio of heat capacities:

$$\gamma \equiv \frac{C_P}{C_V}.$$

Using this notation, we write the thermal diffusivity as

$$D_T \cong \left[\frac{1}{3} \langle v \rangle \ell (C_V)_{\text{p.u.v.}} \right] \bigg/ (C_P)_{\text{p.u.v.}} = \frac{1}{3} \frac{\langle v \rangle \ell}{\gamma}. \tag{15.41}$$

The diffusivity depends primarily on kinematic quantities: the mean speed and the mean free path.

Next, we turn to equation (15.39) and find

$$\begin{pmatrix} \text{time scale for} \\ \text{thermal evolution} \end{pmatrix} = \left(\frac{\lambda}{2\pi} \right)^2 \frac{3\gamma}{\langle v \rangle \ell}$$

$$\cong \frac{3\gamma}{(2\pi)^2} \frac{\lambda}{\ell} \times \begin{pmatrix} \text{period of} \\ \text{sound wave} \end{pmatrix}. \tag{15.42}$$

The relations, (period of sound wave) $= \lambda/v_{\text{sound}} \cong \lambda/\langle v \rangle$, lead to the second line. At the upper end of the audible range, where the frequency is 20,000 Hz, the wavelength of sound has shrunk to approximately 2 cm. The mean free path in air is approximately 10^{-5} cm. So, in the least favorable situation, $\lambda/\ell \cong 2 \times 10^5$. The time scale for thermal evolution greatly exceeds the period of the sound wave.

Section 15.6 noted that equation (15.35) needs to be augmented before it can be applied to gases. That is true, but the part of the thermal evolution that depends on thermal conduction continues to be described by the term $\mathrm{div}(K_T \, \mathrm{grad}\, T)$. Hence the time scale for that evolution is set by D_T and the characteristic length, as reasoned at the end of section 15.6 and as displayed in equation (15.39). The analysis is consistent.

For our purposes, a gas differs from a solid or liquid fundamentally in that changes in pressure produce relatively large changes in temperature. That reaction plays a major role in determining how the temperature profile in a sound wave oscillates in time. Problem 9 explores the couplings that lead to such oscillations.

15.8   Refinements

This section discusses some refinements of the previous calculation for the mean free path.

All molecules in motion

The derivation of the relationship

$$\ell = \frac{1}{n \times \pi d^2} \tag{15.43}$$

in section 15.1 assumed, implicitly, that all molecules—other than the one we follow—are at rest. In that context, collisions occur only on the forward-facing hemisphere of our molecule. Now we acknowledge that the other molecules are in motion also. A molecule moving rapidly or obliquely may "pursue" our molecule and collide with its backward-facing hemisphere. This additional opportunity for collision increases the collision rate and reduces the mean free path relative to our earlier assessment, displayed in equation (15.43). In the following paragraphs, we determine the reduction factor.

Consider first the original situation, where the other molecules are at rest. Assign our chosen molecule a velocity $\mathbf{v}_1$ (as observed in the laboratory reference frame), but imagine riding with the molecule. In the molecule's rest frame, the other molecules stream toward it with velocity $-\mathbf{v}_1$ and hence with speed v_1. The collision rate will be proportional to v_1 because the flux of other molecules is nv_1. (After each collision, we re-start our molecule with velocity $\mathbf{v}_1$ as observed in the lab.) Next, average the collision rate over the various possible values for $\mathbf{v}_1$, using the Maxwell velocity distribution. The average collision rate is then proportional to $\langle v_1 \rangle$:

$$\left(\begin{array}{c} \text{collision rate when other} \\ \text{molecules are at rest} \end{array} \right) \propto \langle v_1 \rangle. \tag{15.44}$$

Turn now to the situation of interest: all molecules are in motion. Again assign our chosen molecule the velocity $\mathbf{v}_1$ and ride with it. The other molecules stream toward it with the *relative* speed $|\mathbf{v}_2 - \mathbf{v}_1|$, where $\mathbf{v}_2$ denotes the velocity of a subset of other molecules and where we need to average the relative speed over all possible values of $\mathbf{v}_2$. (In the preceding calculation, $\mathbf{v}_2$ was zero, and so the relative speed was merely $|0 - \mathbf{v}_1| = v_1$.) The sole difference from the preceding calculation is that we need to average the relative speed $|\mathbf{v}_2 - \mathbf{v}_1|$ with respect to both $\mathbf{v}_1$ and $\mathbf{v}_2$. In short,

$$\begin{pmatrix} \text{collision rate when other} \\ \text{molecules are in motion also} \end{pmatrix} \propto \begin{pmatrix} \text{average of } |\mathbf{v}_2 - \mathbf{v}_1| \\ \text{with respect to } \mathbf{v}_1 \text{ and } \mathbf{v}_2 \end{pmatrix} = \sqrt{2}\langle v_1 \rangle.$$

$$(15.45)$$

Problem 7 of chapter 13 outlines a route to the conclusion that the required double average is $\sqrt{2}\langle v_1 \rangle$.

Comparing equations (15.44) and (15.45) tells us that the collision rate increases by a factor of $\sqrt{2}$, and therefore the mean free path decreases by the same factor. Thus

$$\begin{pmatrix} \text{refined assessment of} \\ \text{mean free path } \ell \end{pmatrix} = \frac{1}{\sqrt{2}n \times \pi d^2}.$$

$$(15.46)$$

Intermolecular attraction

So far, the picture has been of molecules as hard spheres. That view approximates well the intermolecular repulsion at very short distances—at what one may consider to be "contact." Figure 12.8 indicated, however, that an attractive force acts when the intermolecular separation exceeds a minimum (which is close to the "contact" separation). The range of the attractive force is relatively short, extending only out to two molecular diameters or so. The attractive force will alter the trajectory of a passing molecule (even though there is no contact). It is as though the chosen molecule of section 15.1 carried a shield larger than πd^2. Although quantum theory is required to calculate the effective size of the shield, one can—as a first approximation—just replace πd^2 in a phenomenological way by $\sigma_{\text{scattering}}$, the effective shield area. (More technically, $\sigma_{\text{scattering}}$ is called the *total scattering cross-section*.) The algebraic relations are just as before; for example,

$$\begin{pmatrix} \text{refined assessment of} \\ \text{mean free path } \ell \end{pmatrix} = \frac{1}{\sqrt{2}n \times \sigma_{\text{scattering}}}.$$

$$(15.47)$$

The numerical value of $\sigma_{\text{scattering}}$ can be inferred from data on mean free paths or from the transport coefficients, η and K_T.

Molecules that move slowly will be affected more by the attractive forces than fast molecules. Thus the effective shield area is a function of the average relative speed of the molecules. (In distinction, the hard sphere area, πd^2, is independent of molecular speeds.) If the temperature decreases, the relative speed decreases, and so $\sigma_{\text{scattering}}$ increases. Because $\sigma_{\text{scattering}}$ appears in the denominator of the expression for ℓ, the mean free path decreases when the temperature decreases.

There are corresponding implications for the coefficients of viscosity and thermal conductivity. Each coefficient has a temperature dependence through the mean speed: $\langle v \rangle \propto \sqrt{T}$. Now each coefficient acquires additional temperature dependence through

$\ell(T)$. There is, however, no simple theoretical expression for the entire temperature dependence—a disappointment, perhaps.

15.9 Essentials

1. The *mean free path*, denoted by ℓ, is the average distance that a molecule travels between collisions.

2. The simplest theory yields the relationship

$$\ell = \frac{1}{n \times \pi d^2},$$

where d denotes a molecular diameter and n is the number density.

3. In a random walk, the root mean square of the *net vectorial* distance traveled is well-defined and is denoted by $L_{\text{r.m.s.}}$. After N steps of average length ℓ, the net displacement is

$$L_{\text{r.m.s.}} = \sqrt{N}\ell;$$

the distance grows only as the square root of N.

In terms of elapsed time and the mean speed, the relationship takes the form

$$L_{\text{r.m.s.}} = \sqrt{\langle v \rangle t \ell}.$$

4. The transport of momentum (in most fluids) is described by the equation

$$\left(\begin{array}{c} \text{net flux of } x\text{-momentum} \\ \text{in } y\text{-direction} \end{array} \right) = -\eta \frac{\partial \bar{v}_x}{\partial y},$$

where the *coefficient of viscosity* η is given approximately by

$$\eta \cong \frac{1}{3} n \langle v \rangle m \ell$$

for a dilute classical gas.

5. *Thermal conduction* is described by the equation

$$(\text{net energy flux}) = -K_T \, \text{grad} \, T,$$

where the *coefficient of thermal conductivity* K_T is given approximately by

$$K_T \cong \frac{1}{3} n \langle v \rangle \ell \frac{d\bar{\varepsilon}}{dT}$$

for a dilute classical gas.

6. Time-dependent thermal conduction is governed by the *diffusion equation*,

$$\frac{\partial T}{\partial t} = D_T \operatorname{div} \operatorname{grad} T,$$

where the *thermal diffusivity* D_T is defined by

$$(\text{thermal diffusivity}) \equiv D_T \equiv \frac{K_T}{(C_P)_{\text{p.u.v.}}}.$$

The differential equation applies to typical solids and liquids (but must be augmented for gases).

Further reading

A classic in the field is Sir James Jeans, *An Introduction to the Kinetic Theory of Gases* (Cambridge University Press, New York, 1940). As of its publication date, the book was authoritative, and it remains an excellent resource, but one has to pick and choose in order to find easy reading.

The "Adiabatic assumption for wave propagation" is the title of a brief and instructive exposition by N. H. Fletcher, *Am. J. Phys.* **42**, 487–9 (1974) and **44**, 486–7 (1976).

Problems

1. *Collision rate.* For air under room conditions, estimate the number of collisions that a chosen molecule makes in one second.

2. A glass disk (of radius 2 cm) is to be "aluminized" for use as a mirror. Aluminum atoms emerge from an oven through a small hole and fly 40 cm to the glass. All this takes place in the "vacuum" under a bell jar. If most of the aluminum atoms are to make the trip (from aperture to glass) without a collision, how low must be the pressure of the air that remains in the bell jar?

3. Estimate the mean free path of air molecules at a height in the atmosphere equal to 1 percent of the Earth's radius. (Recall the original definition of the meter: a length such that the distance between equator and north pole is 10^7 meters.)

4. Gases A and B are in separate containers with pressures and temperatures in the ratios $P_B/P_A = 1/6$ and $T_B/T_A = 2$. Gas B has a mean free path that is three times as long as that for gas A: $\ell_B = 3\ell_A$. Are the molecules of gas B actually smaller than those of gas A? Defend your response by determining their size, relative to the molecules of gas A.

5. *Diffusion of photons.* Within the sun, electromagnetic radiation is continually emitted and absorbed by electrons and ions. For some purposes, one may consider the

radiation to be a gas of photons whose mean free path is approximately $\ell = 1$ millimeter. (The use of a single value for ℓ ignores the variation in temperature and mass density within the sun and ignores also the Planck spectrum even at a single location, but it's a good start.)

(a) Graph the average net distance, $L_{\text{r.m.s.}}$, diffused by a photon in the following time intervals: 1, 5, 10, 50, 200, 500, and 1000 seconds.
(b) According to this model, how many years does it take a photon to diffuse from the solar center to the surface? (It is not the same photon, of course, after an absorption and emission event, but the picturesque language is harmless here.)

6. *Pipe flow and dimensional analysis.* The flow rate in a pipe (in cubic meters per second) can plausibly depend on the viscosity coefficient η, the pipe radius R, and the pressure gradient $(P_a - P_b)/L_{ab}$. How can you combine these three parameters to construct a quantity with the dimensions of a flow rate? Is there any freedom of choice? What can you conclude about the final expression for the flow rate that a correct, exact calculation will yield?

7. *Velocity profile in steady pipe flow.* Section 15.4 provides the context. That section and section 15.3 imply

$$\left(\begin{array}{c}\text{x-momentum transferred through a} \\ \text{cylindrical surface of radius r per second}\end{array}\right) = -\eta \frac{d\bar{v}}{dr} \times 2\pi r L_{ab}. \tag{1}$$

Here $\bar{v}$ denotes the local macroscopic fluid velocity, which is along the x-direction. Consider now an annular cylindrical volume: r to $r + \Delta r$ in radius and L_{ab} in length. A steady state for the momentum in that volume requires

$$(P_a - P_b)2\pi r \Delta r = \frac{d}{dr}\left(-\eta \frac{d\bar{v}}{dr} \times 2\pi r L_{ab}\right)\Delta r. \tag{2}$$

(a) Give a verbal interpretation of the expression on each side of equation (2).
(b) Integrate equation (2) and thus determine the velocity profile $\bar{v}(r)$. What is $\bar{v}_{\max}$ in terms of the viscosity coefficient η, the pipe radius R, and the pressure gradient $(P_a - P_b)/L_{ab}$?
(c) Determine the fluid flow rate in terms of R and $\bar{v}_{\max}$. Compare your answer with the estimate in section 15.4.

8. *Diffusion equation.*

(a) Return to the geometry of figure 15.9, but specify that the number density varies, $n = n(y)$, and that the temperature is uniform. Derive an approximate expression for the net particle flux; then generalize it to gradient form.
(b) Invoke conservation of particles to derive the *diffusion equation*:

$$\frac{\partial n}{\partial t} = D \operatorname{div} \operatorname{grad} n,$$

where the symbol D (without any subscript) is called the *diffusion constant*. What expression do you find for D (in terms of gas parameters)?

(c) Consider the function

$$n(\mathbf{r},\ t) = n_0 + \Delta N \frac{1}{(4\pi Dt)^{3/2}} \exp[-r^2/(4Dt)],$$

where n_0 and ΔN are positive constants.

(i) Show that the function satisfies the diffusion equation for $t > 0$.

(ii) Interpret the expression in the limit $t \to 0$ (through positive values).

(iii) Compare the behavior of $n(\mathbf{r},\ t)$ as a function of both space and time with the random walk analysis in section 15.2. Include sketches of n versus r at various times.

(iv) Consider the integral $\int[n(\mathbf{r},\ t) - n_0]d^3x$ taken over all space. What is its physical significance? Is its value constant in time? Cite evidence for your claim.

9. *Damped sound waves in air.* A good set of approximate equations for sound waves in a one-dimensional column of air are the following:

$$\frac{\partial T}{\partial t} - \left(\frac{\gamma - 1}{\gamma}\right)\frac{T_0}{P_0}\frac{\partial P}{\partial t} = D_T \frac{\partial^2 T}{\partial x^2},$$

$$\frac{\partial n}{\partial t} = -n_0 \frac{\partial \bar{v}_x}{\partial x},$$

$$mn_0 \frac{\partial \bar{v}_x}{\partial t} = -\frac{\partial P}{\partial x}.$$

The first equation is the augmented heat equation; the second expresses conservation of molecules; and the last is Newton's second law. Variables with a subscript zero denote values in the absence of the sound wave: the equilibrium values. The velocity component $\bar{v}_x$ is the local value of the macroscopic fluid velocity. (Thus $\bar{v}_x$ is an average over molecular velocities within a small volume.)

(a) Use the ideal gas law to eliminate the number density n in terms of the temperature and pressure. Expand about the equilibrium values.

(b) To get your bearings, set the thermal diffusivity to zero; assume that all deviations from equilibrium values depend on x and t through the form $\exp[i(bx - \omega t)]$, where b and ω are constants; and solve for the wave speed. The wave speed v_{wave} is the "phase velocity" of a wave crest that moves through the gas. The crest may represent a maximum in the temperature T, the pressure P, or the local macroscopic fluid velocity $\bar{v}_x$. (Thus v_{wave} is distinct from $\bar{v}_x$.)

(c) Now go back; include the thermal diffusivity, but work to first order only in $b^2 D_T/\omega$; and solve for ω when the wavelength (and hence b) are fixed, as they would be for a standing wave in a closed column of fixed length. At what rate is the wave damped?

10. *Drag at extremely low density.* A thin macroscopic disk moves with constant velocity $\mathbf{v}_0$ through an extremely dilute classical gas. The velocity $\mathbf{v}_0$ is perpendicular to the plane of the disk. The mean free path of the gas molecules is much longer than the disk's radius r_0. The molecules have mass m and mean speed $\langle v \rangle$; the inequality $\langle v \rangle \gg |\mathbf{v}_0|$ holds.

(a) Explain qualitatively why the time-averaged force $\mathbf{F}$ exerted on the disk by the gas is nonzero. Determine the dependence of the force on $\mathbf{v}_0$, $\langle v \rangle$, m, r_0, and the number density n. Viewing things from the disk's rest frame may help you.
(b) Determine the dimensionless numerical coefficient that completes the calculation of the force.

11. *Atomic and molecular size.* Use relationships and data from this chapter to estimate the size of argon atoms and hydrogen molecules. Argon has a mass 40 times as large as that of atomic hydrogen. State your line of reasoning clearly.

12. Consider two panes of glass separated by an air-filled gap of 0.5 cm. The panes are held at two slightly different temperatures, and so thermal conduction produces an energy flux proportional to the coefficient of thermal conductivity. (Ignore radiation here; focus on the transport of energy by molecular diffusion.) Initially, the air is at atmospheric pressure. Then air is slowly pumped out, so that the pressure drops continuously.

Describe the behavior of the energy flux as a function of pressure in the gap. Be quantitative where you can. (Qualitatively, there are two regimes.)

16 Critical Phenomena

16.1 Experiments
16.2 Critical exponents
16.3 Ising model
16.4 Mean field theory
16.5 Renormalization group
16.6 First-order versus continuous
16.7 Universality
16.8 Essentials

Here, in brief outline, is the run of the chapter. Section 16.1 presents two experiments to introduce the topic of "critical phenomena," as that phrase is used in physics. The next section illustrates the mathematical behavior of certain physical quantities near a critical point. Section 16.3 constructs a theoretical model: the Ising model. The following two sections develop methods for extracting predictions from the model. The last three sections draw distinctions, outline some general results, and collect the chapter's essentials.

Achieving a sound theoretical understanding of critical phenomena was the major triumph of thermal physics in the second half of the twentieth century. Thus the topic serves admirably as the culmination of the entire book.

16.1 Experiments

Liquid–vapor

We begin with an experiment that you may see as a lecture demonstration, performed either with carbon dioxide or with Freon, a liquid once used in refrigerators. A sealed vertical cylinder contains a carefully measured amount of CO_2 (say) under high pressure. To begin with, the system is in thermal equilibrium at room temperature, as sketched in part (a) of figure 16.1. A distinct meniscus separates the clear liquid phase from the vapor phase above it.

Now heat the system with a hair dryer. The meniscus rises (because the liquid expands), and gentle boiling commences. Next the meniscus becomes diffuse, and then it disappears. The CO_2 has become spatially homogeneous in density, as illustrated in figure 16.1 (b).

Figure 16.2 (a) shows the system's trajectory in the $P-V$ plane. The amount of CO_2

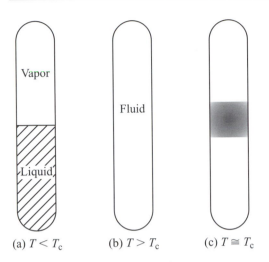

(a) $T < T_c$ (b) $T > T_c$ (c) $T \cong T_c$

Figure 16.1 Critical behavior in a liquid–vapor system. (a) Well below the critical temperature, liquid and vapor form readily distinguishable phases, and a meniscus separates them. (b) Well above the critical temperature, only a single, homogeneous "fluid" phase exists. (c) As the system is cooled down to the critical temperature, a band of turbulence appears suddenly.

was chosen so that the rise in pressure at constant volume takes the system through the critical point as it makes its way to the "fluid" region above the critical isotherm. In part (b), we see the trajectory in the P–T plane. The system moves along the vaporization curve to the critical point and then enters the fluid region.

Now turn off the heater and turn on a fan; blowing room temperature air over the cylinder will cool it—but relatively slowly (in comparison with the earlier heating). For a long time, nothing seems to happen. Then—suddenly!—a wide vertical band of turbulence appears, as sketched in figure 16.1 (c). In reflected light, the turbulent region appears bluish white; in transmitted light, it is rusty brown. Soon a meniscus becomes faintly visible in the center of the band. Droplets of mist swirl above it; bubbles gyrate in the liquid below it. The band is almost opaque because its droplets and bubbles scatter light so efficiently. (The rusty transmitted light is like a sunset; the blue end of the spectrum has been preferentially scattered out of the beam.) When a liquid is far from the critical point, fluctuations in density occur primarily on an atomic scale. Near the critical point, as bubbles and droplets form and disappear, density fluctuations occur on much larger scales, scales comparable to the wavelength of visible light. Fluctuations on that scale scatter light strongly. The phenomenon is called *critical opalescence*. Still closer to the critical point, the fluctuations grow to the size of the container; they eliminate the two-phase system as it is heated to the critical point or produce the two-phase system as the fluid CO_2 is cooled to the critical point.

You may wonder about the seeming asymmetry in behavior as the CO_2 is heated and then cooled. If one looks closely during the heating stage, one can discern similar

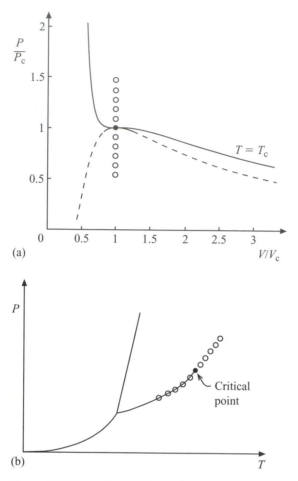

Figure 16.2 The trajectory, shown by open circles, as the liquid–vapor system is heated. The critical point is marked by a smaller, filled circle. (a) P–V plane. The increase in temperature takes the system from one isotherm in the coexistence region to another. Ultimately, the system passes through the critical point and into the fluid region. (b) P–T plane. The system moves along the vaporization curve, passes through the critical point, and enters the fluid region.

turbulence and density fluctuations. They occur on a smaller vertical scale because (presumably) the relatively rapid heating rushes the system through the critical point.

Basically, *critical phenomena* are whatever happens when a system is near a critical point. In turn, a *critical point* is a point where the existence or coexistence of phases changes qualitatively. For CO_2, the qualitative change is from the coexistence of distinct liquid and vapor phases to the existence of merely one phase, a "fluid" phase.

What might one want to know about near carbon dioxide's critical point? Several items come to mind:

1. the difference in mass density between liquid and vapor: $\rho_{liq} - \rho_{vap}$;
2. the heat capacity at constant volume, C_V;

3. some measure of the maximum length scale for the density fluctuations;
4. the isothermal compressibility, $-(1/V)(\partial V/\partial P)_T$.

Apropos of the compressibility, recall from section 12.8 and from figure 12.7 that $(\partial P/\partial V)_T$ goes to zero as T approaches T_c from above. Thus the compressibility becomes infinite, and that behavior is surely worth studying.

Section 16.2 will begin to develop general mathematical forms for behavior near a critical point, but now we turn to a second experimental example.

Ferromagnetism

Bar magnets made of iron are probably familiar to you. The iron rod creates a magnetic field outside itself, and you can use that field to pick up nails and paper clips. If one heats the bar magnet until it is red hot, specifically, to a temperature of 1,043 K, the iron loses its outside field. A phase transition has occurred. If the iron is allowed to cool, it regains the capacity to be a bar magnet.

Those are some obvious external effects and changes. One interprets them, in part, in terms of the magnetic moment per unit volume, **M**, inside the iron bar. The vector **M** is called the *magnetization*. In principle, to determine **M** one adds vectorially the individual magnetic moments in a tiny volume and then divides by the size of the volume. Thus **M** is proportional to the local average of the magnetic moments. Not all the moments in the tiny volume need point in the direction of **M**, but "most" do. The volume itself must be small when compared with a macroscopic length scale but large enough to contain millions of individual magnetic moments.

If we could look inside the once heated and then cooled iron bar, we would find a situation like that sketched in figure 16.3. The magnetization points in one direction in one region but in different directions in adjacent regions. Each region is called a *domain*, is so small that it requires a microscope to be seen, and yet contains an immense number of electronic magnetic moments, all of them aligned in a single direction, more or less. (Typical domains contain 10^{12} to 10^{18} atoms.) The alignment *within* a domain occurred *spontaneously* as the iron cooled below 1,043 K. To some extent, just which direction was chosen for the alignment was a matter of chance and

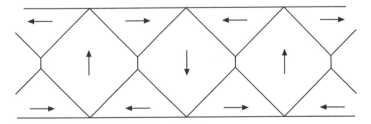

Figure 16.3 Magnetic domains in iron. The sample is an iron "whisker," a long single crystal (only 0.05 mm across a short side). The arrows indicate the direction of "macroscopic" magnetization. [*Source*: R. W. DeBlois and C. D. Graham, Jr., "Domain observations on iron whiskers," *J. Appl. Phys.* **29**, 931–40 (1958).]

would not necessarily be repeated on successive heating and cooling cycles. Because the magnetizations of the domains point in many different directions, their contributions to a magnetic field outside the bar tend to cancel, and so the cooled bar is a feeble magnet.

To reconstruct a powerful bar magnet, one needs to align the magnetizations of the many domains. This can be done by using the magnetic field of another "permanent" magnet or by employing the magnetic field produced by an electric current. Three processes operate when an external field is applied to the bar and gradually increased in magnitude.

1. Domain growth. A domain whose magnetization points along the external field (or nearly parallel to it) grows at its boundary. Individual moments in adjacent domains but near the boundary switch allegiance and become aligned with the external field (or nearly so).

2. Domain rotation. In some directions relative to the crystal lattice, alignment of magnetic moments is easier to achieve spontaneously (because of favorable mutual interactions) than in others; such directions are called *easy axes*. In a sufficiently strong external field, the moments in a domain rotate *en masse* to alignment along the easy crystallographic axis that is parallel to the external field (or most nearly parallel). The crystal structure itself inhibits such switching, but that is ultimately a blessing: when the external field is removed, the crystal structure holds the domain magnetizations in rough alignment and thus preserves the large over-all magnetization needed for a strong "permanent" magnet.

3. Coherent rotation. In a yet stronger external field, the magnetic moments in *all* domains are swung away from an easy axis and forced into closer alignment with the external field. This is a brute force or, better, a "brute torque" process.

The evolution of domains is a fascinating topic in its own right, but—for critical phenomena—it suffices to focus attention on the magnetization *within a single domain*, and we do that henceforth.

Recall that a paramagnetic system (such as the cesium titanium alum discussed in section 5.3) acquires a magnetization when placed in an external magnetic field $\mathbf{B}$, that is, a field produced by external sources (such as a coil carrying an electric current). Remove the external field, and the magnetization disappears, both on a macroscopic scale and also on the scales of 10^{12} or half-a-dozen atomic paramagnets. Nothing is left.

In contrast, hot iron that is cooled in the *absence* of an external field develops magnetization spontaneously (in domains) and retains it. The property of *spontaneous magnetization* characterizes *ferromagnetism* and distinguishes it from paramagnetism (and from diamagnetism).

Figure 16.4 displays the spontaneous magnetism of iron as a function of temperature. The temperature at which spontaneous magnetism ceases to exist is a critical temperature and is denoted by T_c. Above that temperature, iron behaves paramagnetically. The critical temperature for ferromagnetism is usually called the *Curie temperature*, after Pierre Curie, who studied the temperature dependence of magnetism extensively (before turning to the study of radioactivity with Marie Curie). By a happy

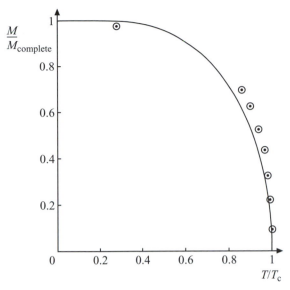

Figure 16.4 The spontaneous magnetization of single-crystal iron as a function of temperature. Plotted vertically is the magnetization divided by its value at complete alignment. The circles represent experimental data. The solid line is a theoretical curve based on developments in sections 16.3 and 16.4: the Ising model and mean field theory. [*Source*: H. H. Potter, *Proc. R. Soc. Lond.* **A146**, 362–87 (1934).]

coincidence, you may regard the subscript c in T_c as standing for either "Curie" or "critical." Table 16.1 lists some Curie temperatures.

For ferromagnetism, what needs to be explained theoretically? In broad outline, here are some major items:

1. the occurrence of spontaneous magnetization;

Table 16.1 *Some ferromagnetic materials and their Curie temperatures.*

Material	T_c (K)
Iron	1,043
Nickel	633
Fe & Ni alloy (50% each)	803
Gadolinium	293
Gadolinium chloride (GdCl$_3$)	2.2
Chromium bromide (CrBr$_3$)	37
Europium oxide (EuO)	77
Europium sulfide (EuS)	16.5

Source: D. H. Martin, *Magnetism in Solids* (MIT Press, Cambridge, MA, 1967).

2. the existence and numerical value of a critical temperature (the Curie temperature);
3. how the magnetization varies with temperature T and external field B over various temperature ranges: well below T_c, near T_c, and well above T_c.

16.2 Critical exponents

Of the various quantities worth studying near a critical point, spontaneous magnetization is the easiest to begin with. The sharp growth of M when the temperature descends below T_c suggests that one may be able to represent M by the temperature difference $T_c - T$ raised to some positive power less than 1:

$$M = \text{const} \times \left(\frac{T_c - T}{T_c} \right)^{\beta} \tag{16.1}$$

when $T \leqslant T_c$ and where the constant β lies in the interval $0 < \beta < 1$. (Provided β is less than 1, the derivative of M with respect to T will diverge as T approaches T_c from below. That gives a vertical slope, such as one sees in figure 16.4.) The division by T_c provides two benefits: (1) it compares the literal temperature difference $T_c - T$ to the natural magnitude scale, set by T_c itself, and (2) it gives a dimensionless quantity. A one-term expression like that in equation (16.1) can be expected to hold only close to T_c.

Experiments on ferromagnetic systems give values of β near $1/3$ but not literally that simple fraction (even with allowance for experimental error). The exponents lie in the range 0.33 to 0.42.

Other quantities associated with critical phenomena have one-term expressions like that for the magnetization. A few are displayed in table 16.2; the associated exponents are called *critical exponents*. The experimental exponents and most of the theoretical

Table 16.2 *Some critical exponents. The heat capacity C and magnetization M are measured in the absence of an external magnetic field, denoted by the subscript $B = 0$.*

Quantity	Temperature range	Proportional to	An experimental value
$C\vert_{B=0}$	$T > T_c$	$\left(\dfrac{T - T_c}{T_c} \right)^{-\alpha}$	$\alpha = 0.05$ for EuS
$M\vert_{B=0}$	$T < T_c$	$\left(\dfrac{T_c - T}{T_c} \right)^{\beta}$	$\beta = 0.368$ for $CrBr_3$
$\rho_{\text{liq}} - \rho_{\text{vap}}$	$T < T_c$	$\left(\dfrac{T_c - T}{T_c} \right)^{\beta}$	$\beta = 0.34$ for CO_2
$-\dfrac{1}{V} \left(\dfrac{\partial V}{\partial P} \right)_T$	$T > T_c$	$\left(\dfrac{T - T_c}{T_c} \right)^{-\gamma}$	$\gamma = 1.35$ for CO_2

Source: H. Eugene Stanley, *Introduction to Phase Transitions and Critical Phenomena* (Oxford University Press, New York, 1971).

ones are *not* simple fractions formed from small integers. Sometimes, as in the case of heat capacity, the exponents may differ between the interval just below T_c and the range just above it. (To be sure, there is now evidence, both experimental and theoretical, that the exponents have the same numerical value on the two sides of T_c for those phenomena—like heat capacity—that are qualitatively the same on both sides. The proportionality constants continue to differ between the two sides.) Moreover, it would only be fair to note that sometimes the behavior is singular in a way that cannot be described by an exponent. For example, a famous theoretical solution (Lars Onsager's solution for the two-dimensional Ising model) has a logarithmic singularity in the heat capacity: $C_V = \text{const} \times \ln(|T - T_c|/T_c)$. Complexity proliferates rapidly, and so the detailed discussion in this chapter is restricted to one exponent.

16.3 Ising model

In this section we construct a theoretical model whose goal is to explain ferromagnetism. To be sure, the aim is not to reproduce all the details, but the model should yield the qualitative and even the major quantitative aspects. Later we shall find that the symbols can be re-named and rearranged to provide a description of a fluid system as well.

The interaction energy

Spontaneous magnetization in a ferromagnetic domain arises when a majority of the electronic magnetic moments point in a single direction (more or less). In turn, that means that a majority of the electron spins point in a single direction (more or less). What interaction could make such parallel alignment energetically favorable?

Surprisingly, the answer lies in a combination of *electrostatics* and the Pauli exclusion principle. The following line of reasoning captures the essentials.

For simplicity, restrict attention to insulating ferromagnets. Examples are chromium bromide ($CrBr_3$), gadolinium trichloride ($GdCl_3$), and europium oxide (EuO). In these compounds, the metal ions are the magnetically relevant ions, and we focus attention on them. They are called the *magnetic ions*.

You may wonder, why the restriction to insulators? Ferromagnets that are electrical conductors, such as pure metallic iron, are more difficult to analyze because the conduction electrons move throughout the material. In contrast, the electrons in insulators are localized to the immediate vicinity of the ions or atoms.

Now consider two electrons, one from each of two adjacent magnetic ions. Note two items.

1. The electrons repel each other with the Coulomb electric force; associated with that repulsion is some positive potential energy.
2. The wave functions of the two electrons may overlap. Where they do, the Pauli exclusion principle correlates the relative positions of the electrons with their

relative spin orientations. If the spins are anti-parallel (and hence point differently), the electrons are permitted to get close together, and so the Coulomb potential energy is high. If the spins are parallel, then the two electrons may not be at the same location. Their joint wave function must vanish if we imagine that the electrons are at the same location. Moreover, continuity for the joint wave function makes mere closeness improbable; that leads to lower potential energy. In short, the Coulomb potential energy depends on the relative spin orientations.

Choose the zero of energy so that the spin-dependent interaction energy is

$-J$ if the spins are parallel,

$+J$ if the spins are anti-parallel,

(16.2)

where J is a positive constant with the dimensions of energy. (The letter J is often used for angular momentum, but not so in this chapter, where we follow the convention in ferromagnetism.)

The interaction energy $\pm J$ pertains to adjacent or—more technically—to nearest-neighbor magnetic ions in the lattice. In the simplest model (which we are constructing), pairs of ions that are more distant are taken to have negligible spin-dependent interaction. (Why? Because the electron wave functions go to zero quickly with increasing distance from the nominal atomic radius, and overlap is required if the Pauli principle is to be relevant.)

The "Coulomb" interaction of equation (16.2) is often called the *exchange interaction*. Here is the reason why. The Pauli exclusion principle requires—as a mathematical statement—that the joint wave function for a pair of electrons change sign if one mentally "exchanges" the two electrons (with respect to their positions and spin orientations). If the two electrons are assigned the same location and spin orientation and are then "exchanged," the wave function can "change sign" only if the wave function is actually zero. Thus the exchange property of the joint wave function is indirectly responsible for the lower electrostatic potential energy that is associated with the parallel alignment of nearest neighbor spins.

The summed energy

Specify further a uniaxial ferromagnetic system, that is, a system where spontaneous magnetization arises along only one axis relative to the crystal structure. [Cobalt provides an example of a uniaxial ferromagnet (but it is not an insulator).] Choose a direction along that axis and call it the positive z-axis. Then only spin orientations parallel or anti-parallel to the z-axis matter. The vector nature of electronic spin is reduced to a two-valued scalar property. Let

$\sigma_i = +1$ if spin i is parallel to $\hat{z}$,

$\sigma_i = -1$ if spin i is anti-parallel to $\hat{z}$,

(16.3)

where the subscript i denotes the ith spin and $\hat{\mathbf{z}}$ is a unit vector along the positive z-axis. The interaction energy of nearest neighbors i and j is then

$$-J\sigma_i\,\sigma_j. \tag{16.4}$$

(You can check the four possibilities: $\sigma_i = \pm 1$ and $\sigma_j = \pm 1$.)

If the system consists of a one-dimensional array of magnetic ions, as displayed in figure 16.5 (a), then the total spin-dependent interaction energy is given by the following sum:

$$E_{\text{int}} = -J(\sigma_1\sigma_2 + \sigma_2\sigma_3 + \sigma_3\sigma_4 + \cdots). \tag{16.5}$$

In general, for any regular array in 1, 2, or 3 dimensions, the interaction energy is a sum of terms $-J\sigma_i\sigma_j$ taken over all nearest neighbor pairs. The model thus constructed is called the *Ising model*. Wilhelm Lenz proposed such an interaction in a 1920 publication. Two years later, Lenz suggested to his graduate student, Ernst Ising, that he look for solutions to the model. Ising was able to solve the one-dimensional array exactly, and citations of his subsequent publication gave the model a name that stuck.

Actually, in the presence of an external magnetic field $\mathbf{B}$ directed along the positive z-axis, the full energy (that is relevant for ferromagnetism) is the sum of the interactions between nearest neighbor pairs plus the interaction with the external field. We write the relevant full energy as

$$E = -J \sum_{\text{n-n pairs}} \sigma_i\sigma_j + m_{\text{B}}B \sum_{\text{all } i} \sigma_i, \tag{16.6}$$

where "n-n pairs" means a sum over all nearest neighbor pairs of magnetic ions. The $+$ sign preceding the second term acknowledges that the electron's magnetic moment $\mathbf{m}_{\text{B}}$ is anti-parallel to its spin $\mathbf{s}$. Consequently, the interaction energy $-\mathbf{m}_{\text{B}} \cdot \mathbf{B}$ equals

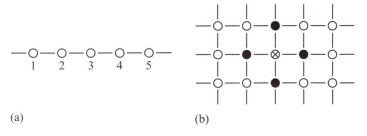

(a)   (b)

Figure 16.5 Two arrays of spins (associated with the magnetic ions). (a) One-dimensional array. (b) Two-dimensional square array. For the ion marked $\otimes$, the four nearest neighbors are marked with solid circles.

$-(-m_B\sigma)B$, that is, $+m_B B\sigma$. (Be aware that some authors ignore this fine algebraic point, which has no effect on the major conclusions about solutions.)

In constructing the energy displayed in equation (16.6), we assumed (implicitly) that only one electron is relevant in each magnetic ion and that the electron's orbital motion is not relevant to the interaction with the external magnetic field. Those assumptions are permissible for model building. One should note, however, that actual magnetic ions often provide several relevant electrons per ion and that the electrons' orbital motion often contributes to the magnetic moment per ion.

Before we explore some consequences of the Ising model, let us note a piece of physics that has been left out, quite deliberately. That item is the magnetic interaction of one magnetic moment with another, either nearby or distant. In section 14.3, we noted that a moment m_B produces a magnetic field of order 0.01 tesla at an interatomic distance of 5×10^{-10} meter. A dipole field falls off with distance r as $1/r^3$. Thus, at a nearest neighbor distance of 3×10^{-10} meter, the field would be approximately 0.05 tesla. Another magnetic moment m_B at that distance has a magnetic energy of order $m_B \times (0.05$ tesla$)$, which is approximately 3×10^{-6} eV. Even on an atomic scale, that is a small energy. Moreover, if the magnetic dipole interaction were responsible for ferromagnetism, we would expect the Curie temperature to be approximately that interaction energy divided by Boltzmann's constant, which amounts to 0.03 K. In comparison with measured Curie temperatures, that temperature is much too low. In fact, the "Coulomb" energy J is typically of order 3×10^{-2} eV, which is larger by a factor of 10^4 (although there are wide variations). The magnetic dipole–dipole interaction is too weak by far to explain spontaneous magnetization as it actually occurs.

16.4   Mean field theory

In the Ising model, the nearest neighbor interactions couple together all the magnetic ions, directly or indirectly. That extensive coupling may be able to align all the spins in a domain, but it has another consequence as well: predictions are difficult to calculate, usually extremely so. Is there some approximation that turns the coupled N-spin problem into merely a one-spin problem, which might be easy to solve?

Yes, there is. As our first method of solving the Ising model for the magnetization, we reason as follows.

When discussing the ith spin and its variable σ_i, we treat the nearest neighbor spins as though each had the average value $\langle\sigma\rangle$ for its spin variable. In principle, the numerical value of $\langle\sigma\rangle$ is computed by averaging over all the spin variables of the entire physical system. Moreover, the average $\langle\sigma\rangle$ is equal to the expectation value estimate for any single spin because all spins in the lattice are equivalent statistically. Therefore we can express $\langle\sigma\rangle$ in terms of $\langle\sigma\rangle$ itself and then solve for it. (If this is confusing, just read ahead; the details will make it clear.)

Let lower case z denote the number of nearest neighbors for the ith spin. Then we extract from equation (16.6) an effective energy for the ith spin:

$$E_i = -zJ\langle\sigma\rangle\sigma_i + m_\mathrm{B}B\sigma_i$$

$$= m_\mathrm{B}B_*\sigma_i, \tag{16.7}$$

where

$$B_* \equiv B - \frac{zJ}{m_\mathrm{B}}\langle\sigma\rangle. \tag{16.8}$$

The interaction with the nearest neighbors acts like an extra magnetic field of size $-(zJ/m_\mathrm{B})\langle\sigma\rangle$. The extra field is called the *mean field*. Pierre Weiss, a French expert on magnetism, developed an approximation like this in 1907. He had in mind magnetic interactions between magnetic dipoles (which we rejected as inadequate for ferromagnetism), and so his corresponding expression was literally an average or mean magnetic field. Calling his expression a "mean field" makes a lot of sense. Nowadays, any effective field that arises by Weiss's kind of approximate averaging is called a "mean field," and the corresponding approximate theory is called a *mean field theory*.

The great merit of the mean field approximation is that it enables us to treat the ith spin as though it were a single particle interacting with *constant* fields (not fields that fluctuate and that vary with the behavior of other particles).

The canonical probability distribution gives the probability $P(\sigma_i)$ for the two possible values of σ_i:

$$P(\sigma_i) = \frac{e^{-m_\mathrm{B}B_*\sigma_i/kT}}{2\cosh(m_\mathrm{B}B_*/kT)}. \tag{16.9}$$

The Boltzmann factor appears in the numerator, and the denominator provides the correct normalization.

The expectation value estimate $\langle\sigma_i\rangle$ follows as

$$\langle\sigma_i\rangle = \sum_{\sigma_i=\pm1} \sigma_i P(\sigma_i)$$

$$= -\tanh(m_\mathrm{B}B_*/kT). \tag{16.10}$$

Because all spins are statistically equivalent, the expectation value $\langle\sigma_i\rangle$ must have the common value $\langle\sigma\rangle$. Therefore we may turn equation (16.10) into a self-consistent equation for $\langle\sigma\rangle$:

$$\boxed{\langle\sigma\rangle = -\tanh\left[\frac{m_\mathrm{B}}{kT}\left(B - \frac{zJ}{m_\mathrm{B}}\langle\sigma\rangle\right)\right].} \tag{16.11}$$

To check this equation, set zJ equal to zero and compare with equation (5.13). In the present context, the quantity "$\langle$magnetic moment along **B**$\rangle$" is represented by $-m_\mathrm{B}\langle\sigma\rangle$, and so equation (16.11) recovers the earlier result in section 5.3.

Spontaneous magnetization

Does mean field theory predict spontaneous magnetization? To find out, set $B = 0$ in equation (16.11), which yields

$$\langle \sigma \rangle = \tanh \left(\frac{zJ}{kT} \langle \sigma \rangle \right), \tag{16.12}$$

and then solve for $\langle \sigma \rangle$. Two routes are worth following.

1. Graphical

Figure 16.6 graphs both the left-hand side and the right-hand side as functions of $\langle \sigma \rangle$ for positive $\langle \sigma \rangle$. Any intersection gives a solution to the equation. Because

$$\tanh x = x + \cdots \text{ when } |x| \ll 1,$$

the initial slope of the right-hand side's curve is zJ/kT. Provided that $zJ/kT > 1$, there will be an intersection at a nonzero value of $\langle \sigma \rangle$. The critical temperature T_c is given by

$$kT_c = zJ. \tag{16.13}$$

When $T > (zJ/k)$, only $\langle \sigma \rangle = 0$ is a solution, and so no spontaneous magnetization occurs in the higher temperature range.

When the temperature is below T_c, there is a negative solution for $\langle \sigma \rangle$ as well as a positive solution; the two solutions have the same magnitude. That duality reflects the basic symmetry of the context when no external field is present.

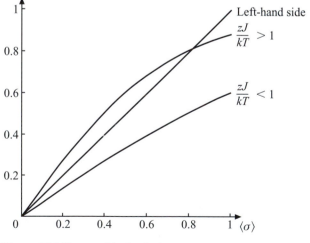

Figure 16.6 The graphical solution.

2. Closed form

To separate $\langle\sigma\rangle$ and T in equation (16.12), first regard the right-hand side as a function of $y \equiv \exp(zJ\langle\sigma\rangle/kT)$ and solve algebraically for y as a function of $\langle\sigma\rangle$. Then take the logarithm of y, finding

$$\frac{zJ}{kT} = \frac{1}{2\langle\sigma\rangle}\ln\left(\frac{1+\langle\sigma\rangle}{1-\langle\sigma\rangle}\right). \tag{16.14}$$

Figure 16.7 displays $\langle\sigma\rangle$ as a function of T/T_c (for positive $\langle\sigma\rangle$).

Critical exponent

The slope $d\langle\sigma\rangle/dT$ appears to become vertical as $T \to T_c$ from below. To find the detailed behavior near T_c, expand the logarithm in equation (16.14) through order $\langle\sigma\rangle^3$ around the value $\langle\sigma\rangle = 0$ and then solve for $\langle\sigma\rangle$. The result is

$$\langle\sigma\rangle = \sqrt{3}\left(\frac{T_c - T}{T_c}\right)^{1/2} \tag{16.15}$$

plus small corrections. Because $\langle\sigma\rangle$ and the magnetization M are proportional, mean field theory predicts that the critical exponent β equals $1/2$.

The one-term expression in equation (16.15) is shown in figure 16.7 for comparison with the precise solution extracted numerically from equation (16.14).

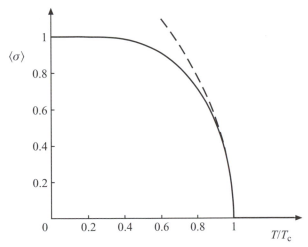

Figure 16.7 Magnetization in the Ising model according to mean field theory. The solid curve gives the positive and zero values of $\langle\sigma\rangle$, to which the magnetization is proportional. The dashed curve illustrates equation (16.15), the one-term expression that captures the essential behavior just below T_c.

Appraisal

How does one assess the combination of Ising model as a framework plus mean field theory as a solution route? Here is a list of successes and failures.

Successes

Together, the Ising model and mean field theory predict

1. spontaneous magnetization,
2. a critical temperature, given by $kT_c = zJ$,
3. a curve of $\langle \sigma \rangle$ (or M) versus T as given by figure 16.7, and
4. a non-integral critical exponent $\beta = 1/2$ for the behavior of the magnetization near T_c.

These are notable qualitative successes.

Failures

The failures are discussed under the same sequence numbers, 1 through 4.

1. Ernst Ising's exact solution for the one-dimensional array shows no spontaneous magnetization (except at absolute zero). The absence of a transition from paramagnetic behavior to ferromagnetic at some nonzero temperature was a disappointment for Ising. Lars Onsager's exact solution for the two-dimensional lattice does show a transition to spontaneous magnetization at nonzero temperature, and there is good theoretical evidence that the Ising model predicts such a transition in three dimensions. Mean field theory fails to capture the dependence on the spatial dimensionality of the spin array: 1, 2, or 3 dimensions. (All that mean field theory incorporates is the number of nearest neighbors. That number can be the same for lattices with different dimensions. For example, both a planar triangular lattice and a three-dimensional simple cubic lattice provide six nearest neighbors.) In short, the actual predictions of the Ising model depend on the number of spatial dimensions, but mean field theory fails to display that dependence.

2. When a nonzero critical temperature actually exists (at least theoretically), mean field theory yields a qualitatively correct estimate—but only that. Table 16.3 provides a quantitative comparison, and one sees that mean field theory can be inaccurate by as much as a factor of 2.

3. The curve of $\langle \sigma \rangle$ or M versus T is qualitatively correct. Quantitative accuracy, however, is lacking, as the discussion of item 4 will demonstrate.

4. Experimental exponents β lie in the range 0.33 to 0.42. The exact value for the two-dimensional Ising model is $\beta = 1/8$ (for all lattices). Numerical estimates for the three-dimensional simple cubic lattice appear to converge on $\beta = 0.325$ or near that value. Yet mean field theory persists in assigning $\beta = 1/2$ to all dimensions.

An over-all summary is this: mean field theory provides a qualitatively good approximation for the Ising model, but the solution is quantitatively inadequate.

Table 16.3 *Critical temperatures predicted from the Ising model.*

Lattice	Number of nearest neighbors	Mean field theory's T_c	Accurate T_c
1-dim	2	$2J/k$	No transition above zero
2-dim:			
honeycomb	3	$3J/k$	$0.506 \times (3J/k)$
square	4	$4J/k$	$0.567 \times (4J/k)$
3-dim:			
simple cubic	6	$6J/k$	$0.752 \times (6J/k)$
body-centered cubic	8	$8J/k$	$0.794 \times (8J/k)$
face-centered cubic	12	$12J/k$	$0.816 \times (12J/k)$

Source: Daniel C. Mattis, *The Theory of Magnetism* II (Springer, New York, 1985), p. 100.

Table 16.3 provides a clue to why mean field theory fails quantitatively. Note that, as the number of nearest neighbors increases, mean field theory gets better: its values for T_c are closer to the accurate values. The more nearest neighbors, the more nearly their collective behavior equals the average behavior of all spins. Mean field theory ignores local fluctuations. It can be quantitatively accurate only when the fluctuations are insignificant.

Yet, near a critical point, fluctuations are rampant. They occur on a wide spectrum of length scales, from nearest neighbors to the size of the physical system. One needs a calculational scheme that can take into account those fluctuations. The next section develops such a procedure.

16.5 Renormalization group

When the fundamental problem is to cope with many length scales, then a basic method is to treat one scale at a time and to repeat a standard procedure. Iteration is the key.

The term *renormalization group* refers to a body of techniques that implement the iterative insight. As applied in thermal physics, the name is not especially informative, and that is an historical accident. The renormalization group arose in quantum field theory in the 1950s as a way to cope with spurious infinities: to "renormalize" quantities to physically sensible values. Work in thermal physics by Leo P. Kadanoff in the 1960s and then a brilliant development by Kenneth G. Wilson in 1970–1971 gave us a practical calculational framework for critical phenomena. For his work, applicable to both thermal physics and quantum field theory, Wilson received the Nobel Prize for Physics in 1982.

In this section, we use the renormalization group to calculate the partition function for the Ising model, first in one dimension and then in two. At the end, we survey our procedures and extract some of the key methodological elements.

One-dimensional Ising model

Using the nearest neighbor interaction of equation (16.5), we write the partition function for the one-dimensional Ising model as

$$Z = \sum e^{(\sigma_1\sigma_2+\sigma_2\sigma_3+\sigma_3\sigma_4+\cdots)J/kT}. \tag{16.16}$$

The sum goes over the values ± 1 for each spin σ_i. The external field B has been set to zero, and figure 16.5 (a) displays the scene geometrically.

Now we prepare to sum over the ± 1 values for all the even-numbered spins. Thus we will sum over alternate spins (or "every other" spin), a procedure that can subsequently be iterated. Introduce the abbreviation

$$K \equiv J/kT, \tag{16.17}$$

where K is called the *coupling constant*. Then factor the exponential so that each factor contains only a single even-numbered spin:

$$Z(N, K) = \sum e^{K(\sigma_1\sigma_2+\sigma_2\sigma_3)} \times e^{K(\sigma_3\sigma_4+\sigma_4\sigma_5)} \times \cdots \tag{16.18}$$

where N denotes the total number of spins.

Summing over $\sigma_2 = \pm 1$ replaces, in effect, the first factor by two terms, and a similar effect is true for all other even-numbered spins:

$$Z(N, K) = \sum [e^{K(\sigma_1+\sigma_3)} + e^{-K(\sigma_1+\sigma_3)}] \times [e^{K(\sigma_3+\sigma_5)} + e^{-K(\sigma_3+\sigma_5)}] \times \cdots. \tag{16.19}$$

To avoid any difficulty with the end points, specify that the total number N of spins is even, wrap the linear chain of spins into a large circle, and let the Nth spin interact with the spins $i = N - 1$ and $i = 1$. When a system is large, boundary conditions have an insignificant effect, and a wrap-around boundary condition is convenient.

The remaining sum in (16.19) goes over the odd-numbered spins. The tactical problem is this: how to make that summation look similar to the original summation, so that we can again sum over alternate spins. Is there a constant K' and a function $f(K)$ such that the equation

$$f(K)e^{K'\sigma_1\sigma_3} = e^{K(\sigma_1+\sigma_3)} + e^{-K(\sigma_1+\sigma_3)} \tag{16.20}$$

holds?

Let's see. When $\sigma_1 = \sigma_3 = \pm 1$, the relationship

$$f(K)e^{K'} = e^{2K} + e^{-2K} \tag{16.21}$$

must hold. When $\sigma_1 = -\sigma_3 = \pm 1$, then

$$f(K)e^{-K'} = 2 \tag{16.22}$$

must hold. These are merely two constraints on two unknowns. The solutions are

$$f(K) = 2(\cosh 2K)^{1/2}, \tag{16.23}$$

$$K' = \tfrac{1}{2}\ln(\cosh 2K). \tag{16.24}$$

Thus equation (16.19) can be written as

$$Z(N, K) = f(K)^{N/2} \sum e^{K'\sigma_1\sigma_3} \times e^{K'\sigma_3\sigma_5} \times \cdots. \tag{16.25}$$

The remaining sum goes over the odd-labeled spins, which are $N/2$ in number. That sum is numerically the partition function for $N/2$ spins with the coupling constant K': $Z(N/2, K')$. So equation (16.25) can be written as the terse statement

$$Z(N, K) = f(K)^{N/2} Z(N/2, K'). \tag{16.26}$$

Is equation (16.26) suitable for iteration? Not really, because the partition functions refer to systems with different numbers of particles. To eliminate that awkward dependence on N, recall that $\ln Z$ must scale as the number of particles when the system is macroscopic. (For one line of reasoning, note that the Helmholtz free energy is an extensive quantity and equals $-kT \ln Z$.) Thus, when $N \gg 1$, the partition function must have the structure

$$\ln Z(N, K) = N\zeta(K) \tag{16.27}$$

for some function zeta that depends on K. (The choice "zeta" is intended to have mnemonic value, associated with the Z for partition function. The ζ here should not be confused with the Riemann zeta function.) Now take the logarithm of equation (16.26); use the forms (16.27) and (16.23); and find the relationship

$$\zeta(K') = 2\zeta(K) - \ln[2(\cosh 2K)^{1/2}]. \tag{16.28}$$

Equations (16.24) and (16.28) relate the partition function for the original spins to a partition function for a new system where alternate spins (the even-numbered spins) have been eliminated by summation. The shortest length scale in the original system— one lattice spacing—has been eliminated. In the new system, the remaining spins are separated by two lattice spacings. One would expect their coupling constant K' to be smaller than K. To confirm that expectation, write (16.24) as

$$e^{2K'} = e^{2K} \times \left(\frac{1 + e^{-4K}}{2}\right). \tag{16.29}$$

Because of the inequality $\exp(-4K) < 1$, the second factor on the right-hand side is less than one, and so the inequality $K' < K$ holds. The spins in the new system interact via the spins that have been summed over; so the coupling constant K' is a statistical average of $\pm K$, which is less than K in magnitude.

Equations (16.24) and (16.28) relate the set $\{K', \zeta(K')\}$ to the set $\{K, \zeta(K)\}$, and so they are called *recursion relations*. The equations do not provide ζ as a function of its argument. (If they did, the calculation would be finished.)

In principle, equations (16.24) and (16.28) can be iterated. Each iteration sums over alternate spins and eliminates a larger length scale. One can continue the process until the spins in the newest system are so far separated (in actual space) that their mutual interactions are negligible. [Expressed mathematically, the process iterates equation (16.29) and hence drives the effective coupling constant toward zero.] For a spin system with negligible interactions, the partition function is readily evaluated, and then one need only collect all the factors in order to express the original partition function.

To implement this scheme, it is convenient to work from the system with the largest length scale down to the desired system. Why? Because that route provides a numerically definite starting value for the partition function. (The physics logic goes from small length scales to large, but—in this instance, anyway—the numerical computation is easier from large to small.) Thus one needs K as a function of K' and $\zeta(K)$ as a function of $\zeta(K')$ and K'. Equation (16.24) may be solved for K:

$$K = \tfrac{1}{2}\ln[e^{2K'} + (e^{4K'} - 1)^{1/2}]. \tag{16.30}$$

Equation (16.28) becomes

$$\zeta(K) = \tfrac{1}{2} \times [\zeta(K') + K' + \ln 2]. \tag{16.31}$$

For N spins that are spatially fixed and non-interacting, the partition function is

$$Z(N, 0) = Z_1(0)^N = 2^N. \tag{16.32}$$

The first equality follows from the analysis in section 5.6 and problem 5.5. The partition function for one non-interacting spin, $Z_1(0)$, is simply the sum of two Boltzmann factors, each of the form e^0, and so $Z_1(0) = 2$. Take the logarithm of (16.32) and compare with equation (16.27) to find

$$\zeta(0) = \ln 2. \tag{16.33}$$

The iterative scheme cannot start at precisely $K' = 0$. Why? Because if $K' = 0$ is inserted into (16.30), the equation yields $K = 0$. No progress would be made. One says that zero coupling constant is a *fixed point* of the iterative transformation. (Setting $K' = 0$ exactly means "no interaction," and then no reshuffling of summations can generate an interaction.) To get off the ground, one needs to use small but nonzero K' and to approximate $\zeta(K')$. Thus let us take the values

$$K' = 0.01,$$
$$\zeta(0.01) \cong \ln 2. \tag{16.34}$$

Table 16.4 displays the consequences of iterating from the starting point (16.34). The exact value of $\zeta(K)$ is computed from the closed-form relationship developed in problem 16.4:

$$\lim_{N\to\infty} \frac{1}{N} \ln Z(N, K)_{\text{exact}} = \ln(e^K + e^{-K}). \tag{16.35}$$

The iterative approach works impressively well.

Table 16.4 *The function $\zeta(K)$ as calculated by iteration and its exact value.*

K	$\zeta_{iterated}$	ζ_{exact}
0.01	0.693 147	0.693 197
0.100 334	0.698 147	0.698 172
0.327 447	0.745 814	0.745 827
0.636 247	0.883 204	0.883 21
0.972 71	1.106 30	1.106 30
1.316 71	1.386 08	1.386 08
1.662 64	1.697 97	1.697 97
2.009 05	2.026 88	2.026 88
2.355 58	2.364 54	2.364 54
2.702 15	2.706 63	2.706 63
3.048 72	3.050 96	3.050 96

Source: Humphrey J. Maris and Leo P. Kadanoff, *Am. J. Phys.* **46**, 652–7 (1978).

Recall that the coupling constant K equals J/kT. Knowing the temperature dependence of the partition function $Z(N, J/kT)$ enables one to calculate the energy per spin, $\langle E \rangle/N$, and then the heat capacity C/N (both in zero external magnetic field). Both figure 16.8 and the exact expression show only smooth behavior for ζ or Z as a function of K. There is no indication of a phase transition (at some finite K). Indeed, the one-dimensional Ising model has neither spontaneous magnetization nor anomaly in heat capacity at any nonzero temperature.

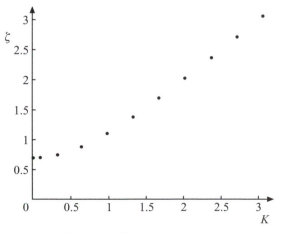

Figure 16.8 The run of $\zeta(K)$ versus K when computed by the renormalization group (for the one-dimensional Ising model).

Two-dimensional Ising model

To find a system that does exhibit a phase transition, we turn to the two-dimensional Ising model. In the exponent of its Boltzmann factor, the full partition function has a sum over all nearest neighbor pairs in a two-dimensional square lattice. The analog of summing over alternate spins is illustrated in figure 16.9. The step consists of summing over alternate spins in both x and y directions. The remaining spins form another square lattice (with lattice spacing larger by $\sqrt{2}$ and rotated by 45°).

After the Boltzmann factor has been summed over alternate spins, can the new exponent be written as a sum over all the new nearest neighbors? No. An algebraic attempt analogous to equations (16.18) to (16.26) generates not only nearest neighbor terms but also qualitatively different terms. Terms arise that represent direct interaction between the new next-nearest neighbor spins; those terms are proportional to $\sigma_i \sigma_j$ for the new next-nearest neighbors. [To visualize a next-nearest neighbor pair, choose four filled circles in figure 16.9 that form a square (with one open circle inside). Then focus attention on a pair of filled circles that lie on opposite corners, that is, one filled circle is half-way around the new square from the other.] There are also terms that describe the direct interaction of four spins (located on the corners of a square); those terms are proportional to $\sigma_i \sigma_j \sigma_k \sigma_l$. (The algebra is omitted here; the paper by Maris and Kadanoff, cited in the further reading section, provides details.) An exact recursion relation cannot be derived by this route.

The simplest approximation that retains a phase transition consists of this: (1) increase the coupling constant of the nearest neighbor interaction to approximate the additional aligning effect of the next-nearest neighbor interaction (and then omit the latter interaction from explicit inclusion) and (2) ignore the direct interaction of four spins (which has a numerically small coupling coefficient). These two steps produce analogs of equations (16.24) and (16.28):

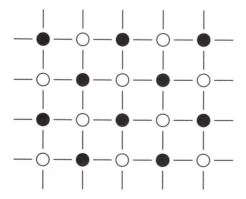

Figure 16.9 The square lattice of the two-dimensional Ising model. The alternate spins for the first summation are shown as open circles; filled circles represent the spins that remain to be summed over.

$$K' = \tfrac{3}{8}\ln(\cosh 4K), \tag{16.36}$$

$$\zeta(K') = 2\zeta(K) - \ln[2(\cosh 2K)^{1/2}(\cosh 4K)^{1/8}]. \tag{16.37}$$

These are (approximate) recursion relations for the two-dimensional Ising model. What do they imply?

To proceed as we did with the one-dimensional model, invert equations (16.36) and (16.37) so that one can start with the almost-free context of (16.34). Figure 16.10 (a) shows what iteration generates. The values of K and ζ converge on the values

$$K_c = 0.506\,98\ldots,$$
$$\zeta(K_c) = 1.0843\ldots, \tag{16.38}$$

where the subscript c denotes—suggestively—a critical value.

To get values of K greater than K_c, one must start the iteration above K_c. At very large coupling constant, most spins will be aligned. The partition function should be dominated by the Boltzmann factors that represent complete alignment (and that represent some nearby states as well). Each spin has four nearest neighbors, but each

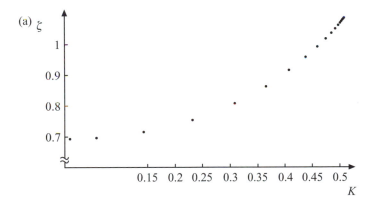

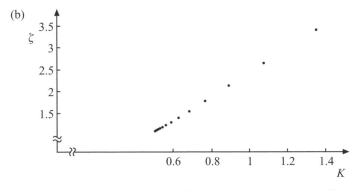

Figure 16.10 (a) Iteration from small K, the nearly-free context. (b) Iteration from large K, the context of almost total alignment.

interaction is shared by two spins. So there are $\frac{1}{2} \times 4N$ nearest neighbor terms in the energy, and all will be negative in the two states of complete alignment. Thus

$$Z(N, K) \cong 2 \times e^{K \times (1/2) \times 4N} \times \begin{pmatrix} \text{factor to include} \\ \text{nearby states} \end{pmatrix}. \tag{16.39}$$

The exponential dominates, and so $\zeta(K) \cong 2K$ when K is large relative to 1. [Figure 16.8, which refers to the one-dimensional model, shows that $\zeta(K)$ is approximately proportional to K already when K exceeds 2. The proportionality constant is $\frac{1}{2} \times 2 = 1$ because each spin has only two nearest neighbors in a one-dimensional system.]

Figure 16.10 (b) shows the consequence of starting with $K' = 10$ and $\zeta = 2K'$ and using the inverted recursion relations. Iteration takes K downward toward K_c and $\zeta(K_c)$.

Neither sequence of iterations takes K literally to K_c; rather, K_c is the limiting value for each sequence. As such, K_c has the property that inserting K_c into the recursion relation reproduces K_c. Thus K_c is a fixed point of the recursion relations. Moreover, because K_c is neither zero nor infinity, it is called a *non-trivial fixed point*.

Figure 16.11 combines the two halves of figure 16.10. It shows that $\zeta(K)$ is continuous at K_c. The smoothness of the join suggests that the first derivative, $d\zeta/dK$, is continuous, also; numerical analysis confirms that. Figure 16.12 displays the second derivative, $d^2\zeta/dK^2$, as computed numerically from the original data points. Aha! At last we see striking behavior at $K = K_c$.

The second derivative gives the heat capacity per spin, C/N, in the following way:

$$\frac{C}{N} = \frac{J^2}{kT^2} \frac{d^2\zeta}{dK^2}. \tag{16.40}$$

[For a derivation, start with equations (5.16) and (16.27), the relation $K = J/kT$, and

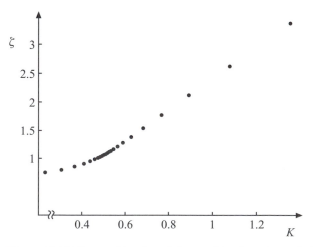

Figure 16.11 The combination of the two iterations shown in figure 16.10. To expand the juncture region, some distant points—at both low K and high K—have been omitted.

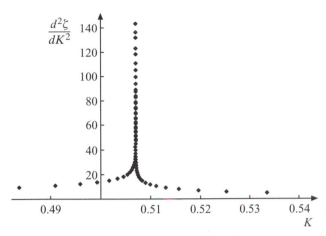

Figure 16.12 The second derivative, $d^2\zeta/dK^2$, has a singularity at K_c.

the chain rule for $d\zeta(K)/dT$.] Thus figure 16.12 implies that the heat capacity C/N becomes infinite as $K \to K_c$ from each side. This singularity suggests that a phase transition occurs at K_c.

Imagine fixing the interaction constant J and decreasing the temperature T from a high value. That will generate a smooth increase in the coupling constant K. For high T and small K, the spin system is paramagnetic and shows no magnetization (in the absence of an external magnetic field). As K_c is passed, the system undergoes a phase transition and develops a spontaneous magnetization. (Because we omitted all mention of an external field in the partition function, our expressions cannot be used to calculate even the spontaneous magnetization, but a separate calculation confirms its existence. Indeed, it suffices to use mean field theory far from the critical point, where that theory is adequate.)

Critical exponent

To characterize the singular behavior in the heat capacity, recall from table 16.2 that C frequently varies as $|T - T_c|^{-\alpha}$ for some critical exponent α. Because C is computed as a second derivative of $\zeta(K)$, suppose that

$$\zeta(K) = a|K - K_c|^{2-\alpha} + \text{(function analytic in } K), \tag{16.41}$$

where a and α are unknown constants. The amplitude a may have different values on opposite sides of K_c. The adjective "analytic" means that the function has a convergent Taylor series about the point $K = K_c$.

To determine the exponent α, we apply the recursion relations (16.36) and (16.37) near K_c. (Those relations move the system away from K_c on both sides.) First calculate K' symbolically:

$$K' = K'(K) = K'[K_c + (K - K_c)]$$

$$= K'(K_c) + \frac{dK'}{dK}\bigg|_{K=K_c} \times (K - K_c)$$

$$= K_c + \frac{dK'}{dK}\bigg|_{K=K_c} \times (K - K_c). \tag{16.42}$$

The step to the second line is a first-order Taylor expansion. The last line follows because K_c is a fixed point.

Equation (16.42) merely relates K and K'. Now we investigate what the structure in (16.41) implies in the recursion relation for ζ, namely (16.37). On the right-hand side, the logarithmic term is analytic in K; so

$$\text{r.h.s.} = 2 \times a|K - K_c|^{2-\alpha} + (\text{function analytic in } K). \tag{16.43}$$

The left-hand side is

$$\text{l.h.s.} = a|K' - K_c|^{2-\alpha} + (\text{function analytic in } K')$$

$$= a|K - K_c|^{2-\alpha} \times \left|\frac{dK'}{dK}\right|^{2-\alpha}_{K=K_c} + (\text{function analytic in } K), \tag{16.44}$$

upon using (16.42) to substitute for K'.

The terms in $|K - K_c|^{2-\alpha}$ on the two sides must have equal coefficients; so one deduces the relation

$$2 = \left|\frac{dK'}{dK}\right|^{2-\alpha}_{K=K_c}. \tag{16.45}$$

To solve for α, take the logarithm of both sides, finding

$$\alpha = 2 - \frac{\ln 2}{\ln\left|\dfrac{dK'}{dK}\right|_{K=K_c}}. \tag{16.46}$$

The critical exponent is determined by a first derivative of the recursion relations at the fixed point.

Appeal to the explicit recursion relation yields

$$\alpha = 2 - \frac{\ln 2}{\ln(\frac{3}{2}\tanh 4K_c)} = 0.1308. \tag{16.47}$$

How do the results for K_c and the critical exponent compare with Lars Onsager's exact results? Onsager derived the value $K_c = 0.4407$, and so the agreement on location is gratifyingly good. In the exact solution, the heat capacity has only a logarithmic singularity at K_c, varying there as $-\ln|T - T_c|$. Such behavior is milder than the power law that the approximation generates. Logarithmic singularities are rare among critical phenomena, and so one should not be dismayed that the approximation produced the much more common power law behavior.

Methods

From the preceding work, we can extract some methodological elements of the renormalization group.

1. Iterate (in general).
2. Incorporate different length scales successively.
3. Derive recursion relations, for example, the pair of equations (16.24) and (16.28) for the one-dimensional model.
4. Expect to approximate in item 3, for example, as in deriving the recursion relations (16.36) and (16.37) of the two-dimensional model.
5. Iterate the recursion relations and look for notable behavior.
6. Wherever a divergence occurs, study the local behavior by a first-order Taylor expansion. Thereby extract critical exponents.

The section and these items provide some insight into how the techniques of the renormalization group are applied. A definitive prescription does not exist—at least not yet. As Kenneth Wilson once remarked, "One cannot write a renormalization group cookbook."

16.6 First-order versus continuous

By now we have amassed sufficient examples to make a crucial distinction among phase transitions.

First consider heating a liquid at constant pressure; the pressure is specified to be *less* than the pressure P_c of the critical point. A glance at figure 12.1 shows that the point representing the system moves horizontally to the right through the liquid region, pauses at the vaporization curve (while liquid becomes vapor), and then moves further rightward in the vapor region. The chemical potential, $\mu(T, P)$, changes continuously, as illustrated in part (a) of figure 16.13. In particular, at the vaporization curve—a coexistence curve—the chemical potentials of liquid and vapor are equal.

In contrast, the slope, $(\partial \mu / \partial T)_P$, changes discontinuously at the vaporization curve, as shown in part (b) of the figure. We can deduce that property as follows.

(1) The Gibbs–Duhem relation, equation (12.23), gives the slope as

$$\left(\frac{\partial \mu}{\partial T}\right)_P = -s, \tag{16.48}$$

where s denotes the entropy per molecule.

(2) The latent heat of vaporization may be written [according to equation (12.8)] as

$$L_{\text{vap}} = T \times (s_{\text{vap}} - s_{\text{liq}}). \tag{16.49}$$

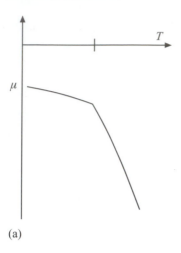

(a)

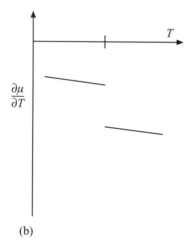

(b)

Figure 16.13 The behavior of the chemical potential and its first derivative under conditions of constant pressure and $P < P_c$. The tick mark denotes the temperature at which the system crosses the coexistence curve (from liquid to vapor).

The latent heat is positive; so the inequalities $s_{vap} > s_{liq} > 0$ hold. Thus the downward slope of the μ versus T graph steepens discontinuously as the system passes through the vaporization curve.

In 1933, the Austrian physicist Paul Ehrenfest proposed a classification of phase transitions: if any nth derivative of the chemical potential is discontinuous (and is the lowest discontinuous derivative), then the transition is called "nth order." According to Ehrenfest's scheme, the liquid to vapor transition—when the inequality $P < P_c$ holds—is a first-order transition.

Succinctly and in general, if the phase transition has a nonzero latent heat, then the transition is *first order*.

Now repeat the process: heat a liquid at constant pressure, but specify $P = P_c$ exactly. What can be qualitatively different? The latent heat of vaporization. Empirically, if one considers points on the vaporization curve successively closer to the critical point, the latent heat becomes smaller. After all, one is approaching the point where the liquid and vapor cease to differ. So, in the limit, no heating should be required to convert liquid into vapor. Liquid and vapor come to have the same entropy per molecule. Thus $L_{vap} = 0$ at the critical point, and so the slope $(\partial \mu / \partial T)_P$ is continuous on the vaporization curve (at the critical point).

The phase transition is no longer first order. Indeed, the quantities that become singular do so with power law singularities: as $|T - T_c|$ raised to some negative power. There are no discontinuities *per se*, and the Ehrenfest classification fails to apply. Rather, such transitions are best called *continuous phase transitions*. (To be sure, such transitions are sometimes called "second order," a relic of a mistaken identification in the 1930s and a usage to be shunned.)

For more detail about continuous phase transitions, consider the two-dimensional Ising model of the previous section. The general structure in equation (16.27) enables us to calculate the chemical potential as

$$\mu = \frac{\partial F}{\partial N} = \frac{\partial(-kT \ln Z)}{\partial N} = -kT\zeta(K). \tag{16.50}$$

The partial derivative is to be taken at constant T and constant external parameters; the latter is here the external magnetic field, which is zero. [When the external field is zero, the Gibbs and Helmholtz free energies coincide; so equation (16.50) follows also from the relation $G = \mu N$ and equation (16.27).] Figure 16.11 and numerical analysis show that both $\zeta(K)$ and $d\zeta/dK$ are continuous at $K = K_c$. Hence $\mu(T)$ and $d\mu/dT$ are continuous at $T = T_c$. The second derivative has a power law singularity but no discontinuity. Thus the two-dimensional Ising model (solved approximately) provides a continuous phase transition (from zero spontaneous magnetization to a nonzero value).

In general, critical phenomena are associated with continuous phase transitions.

16.7 Universality

Two preliminaries come before we can turn to the topic of "universality."

Order parameter

The first of those preliminaries is the notion of an order parameter. The basic idea is to identify a macroscopic quantity that changes from zero to nonzero during a continuous phase transition. Called the *order parameter*, that quantity gives a succinct characterization of the system's macroscopic state. Some examples will illustrate this.

For a magnetic system, the high-temperature, paramagnetic regime exhibits no spontaneous magnetization. In contrast, at low temperature, in the ferromagnetic region—below the Curie temperature T_c—spontaneous magnetization exists and has some nonzero value. The spontaneous magnetization provides an order parameter for a magnetic system.

Is there an order parameter for the liquid–vapor system? Yes. At temperatures below the critical point, liquid and vapor have different densities. Above T_c, there is only a single, undifferentiated fluid, spatially uniform in density. Thus the density difference, $\rho_{liq} - \rho_{vap}$, provides an order parameter for a liquid–vapor system.

The order parameter $\rho_{liq} - \rho_{vap}$ is obviously a scalar. In the Ising model, the spins are constrained to point parallel or anti-parallel to a certain fixed axis, and so the spontaneous magnetization is again a scalar. In a more realistic description of ferromagnetism, the spins may point in any direction. Then the spontaneous magnetization is a vector and has three components.

Later, the number of components that an order parameter possesses will play a significant role.

An order parameter can be defined for the superfluid transition in liquid ^{4}He, for the more complicated superfluid transition in liquid ^{3}He, for a mixture of partially miscible liquids, and for alloys.

Lattice gas

The second preliminary is the idea that the Ising model can help to explain the behavior of a fluid. Specify that a monatomic classical fluid consists of N_{atoms} particles. The atoms may be distributed among a vapor phase, a liquid phase, or—above T_c—merely a fluid phase. The classical analysis in section 13.1 can be extended from one atom to N_{atoms}. The partition function will have the structure

$$Z = \frac{1}{N_{atoms}!} \left(\frac{1}{\lambda_{th}^3} \right)^{N_{atoms}} \times \int_V \exp[-(\text{potential energy})/kT] d^3x_1 d^3x_2 \ldots . \qquad (16.51)$$

The factorial incorporates (approximately) the indistinguishability of identical atoms, as we learned in section 5.6. The integration over momenta produces the factor with the thermal de Broglie wavelength. The remaining integral is the difficult one. The integration goes over the conceivable positions of each atom within a box of volume V. If there were no interatomic forces, the potential energy in the Boltzmann factor would be zero, and the integral would be merely $V^{N_{atoms}}$. But, in fact, there are repulsive forces (of short range) and attractive forces (of intermediate range).

For a tractable approximation, divide the volume V into small cubical cells, each the size of an atom (more or less). Let there be N_{cells} of these tiny volumes. Each cell may be empty or occupied by one atom. This approximation is called a *lattice gas*. Table 16.5 begins to display the correspondences with the three-dimensional Ising model.

Table 16.5 *The correspondences between a lattice gas and the three-dimensional Ising model.*

Lattice gas	Ising model
Number of cells	Number of spins
Cell occupied	Spin up: $\sigma = +1$
Cell empty	Spin down: $\sigma = -1$
Potential energy of attractive forces	$-\varepsilon_0 \left(\dfrac{\sigma_i + 1}{2} \right) \left(\dfrac{\sigma_j + 1}{2} \right)$
Number of atoms	Number of up spins

Limiting occupancy to a maximum of one atom incorporates the short-range repulsive forces. To acknowledge the attractive forces, one says that, if adjacent cells are occupied, then the potential energy drops by $-\varepsilon_0$, where ε_0 is a positive constant. This potential energy corresponds to a nearest neighbor attractive interaction. For adjacent cells labeled i and j, the potential energy may be expressed as

$$-\varepsilon_0 \left(\frac{\sigma_i + 1}{2} \right) \left(\frac{\sigma_j + 1}{2} \right). \tag{16.52}$$

[According to table 16.5, $(\sigma_i + 1)/2$ equals 1 when cell i is occupied and equals zero when the cell is empty. Therefore the entire expression in (16.52) is $-\varepsilon_0$ when both cells are occupied and is zero otherwise.] When multiplied out, the interaction energy contains the product $\sigma_i \sigma_j$ that is characteristic of the Ising model. The terms linear in σ_i or σ_j are analogous to the interaction of a magnetic moment with an external field.

Indeed, the only significant difference between the lattice gas and the Ising model is a constraint. The number of occupied cells equals N_{atoms}. Because occupancy corresponds to spin up, one must impose the constraint that the number of up spins equals N_{atoms}. This can be done efficiently (in a good approximation), but we need not go into that detail. It suffices to know that a lattice gas and an Ising model can be matched up, item by item. That detail must carry over to their critical behavior. In particular, both will have a phase transition, and the critical exponents will be the same. For one system, the order parameter is the spontaneous magnetization; for the other, the density difference $\rho_{liq} - \rho_{vap}$. The language will differ, but the mathematics will be the same.

Universality classes

In 1970, a medley of investigations—both experimental and theoretical—led to the suggestion that diverse physical systems have *precisely* the same critical behavior. The phrase "same critical behavior" means that corresponding physical quantities (such as heat capacities or order parameters) have the same kinds of singularities and that the critical exponents are numerically the same.

These are three determinants of critical behavior:

1. the dimensionality of the physical space,
2. the number of components of the order parameter, and   (16.53)
3. the range of the interaction.

To start with, one characterizes the interaction range as either short or long. If short—for example, if nearest neighbor interactions exist—then the mere characterization "short range" usually suffices. If the range is long, then one needs to specify the exponent in a presumed power-law decrease with separation r.

Physical systems that share common values of items 1 to 3 belong to the same *universality class*. For example, a lattice gas and the three-dimensional Ising model are in the universality class specified by

1. space dimensionality $= 3$,
2. components of order parameter $= 1$, and
3. range of interaction $=$ short.

Moreover, the two models are faithful to the systems that they represent—at least near the critical point—and so real gases and actual uniaxial ferromagnets are in the universality class specified above. All have precisely the same critical exponents.

In short, all members of a universality class have the same critical exponents.

How universality arises

Return to figure 16.11. The figure was computed by iteration from starting values at low K and high K. Recall that $K \equiv J/kT$ and that ζ was defined in (16.27). That is the background. Now think of the points in figure 16.11 as providing $(1/N)\ln Z$ as a function of J/kT at fixed J. High temperature means low K and hence appears at the left-hand end. Cooling the system moves the system rightward. As figure 16.12 indicated, a phase transition occurs at a temperature T_c such that $J/kT_c = K_c = 0.506 \ldots$. We determined one critical exponent from the behavior of $\zeta(K)$ near T_c, and other exponents can be extracted there also. In short, the behavior of $\zeta(K)$ near T_c determines a system's critical behavior.

To go on, note that the specific lattice in the two-dimensional Ising model—square, honeycomb, or triangular—makes no difference to the critical exponents. How can that be? Recall that our original renormalization procedure took the system, iteration by iteration, from a point near K_c to low K (or to high K). Because K_c is a limit point, many iterations are required to move from the starting value of K to a value where ζ can be assessed reliably by approximation. Each iteration sums over alternate spins. The process averages over the short-distance properties of any specific lattice and removes their influence. Simply put, the largeness of the number of iterations washes out the short-distance details.

Analogously, systems as different as gases and ferromagnets can exhibit the same critical exponents because the iterations average over short-distance details.

To be sure, uniaxial ferromagnets and isotropic ferromagnets, for example, fall into different universality classes (because their order parameters have different numbers of components: 1 versus 3). The number of coupling constants (each akin to K for the simplest approximation to the Ising model) typically will differ, and the number will exceed 1. The way in which the coupling constants change upon iteration will differ. But, for each universality class, there is a characteristic pattern of change for the coupling constants under iteration. Such a pattern is the more complicated analog of the juncture region in figures 16.10 and 16.11. Moreover, certain rates of change are the analogs of $(dK'/dK)|_{K=K_c}$, and those rates determine the critical exponents. The pattern near the critical point will be independent of details like the lattice structure because the many iterations wash out such details.

One should not ask too much, however, of the argument that "multiple iterations wash out the short-distance details." Although the specific lattice structure does *not* affect the critical exponents, it does influence the critical temperature. A glance back at table 16.3 substantiates the claim. The table displays accurate values of T_c for three members of the universality class to which the three-dimensional Ising model belongs; the critical temperatures are all different. Similarly, T_c changes with change of lattice structure for the two members of the universality class to which the two-dimensional Ising model belongs.

One might ask next, why are so many iterations required in the calculation? The mathematical reason was noted several paragraphs back: a fixed point is a limit point of many iterations. But what is the "physical" reason?

For the answer, we return to critical opalescence, the dramatic feature of a liquid–vapor system near its critical point. Light is scattered by density fluctuations that occur on a wide range of length scales. Both at the critical point and near it, local fluctuations (from mean behavior) are correlated over an immense range of length scales. (The phrase "local fluctuations" means fluctuations in the number of atoms in a region of linear size equal to a few atomic diameters or—for a magnetic system—fluctuations in the values of a few nearby spins.) Such a wide spectrum of length scales occurs for all kinds of systems: liquid–vapor, magnetic, etc.

In a fluid, the long-range correlations produce the large-scale variations in density that scatter the light. The large length scales themselves suggest that small-scale details—such as lattice structure in a magnetic system—are irrelevant to critical behavior. Moreover, the presence of correlations on many length scales requires many iterations of a renormalization procedure before the separation of the remaining spins (in a magnetic system) is so large that one can reliably approximate the partition function and terminate the sequence of steps. Once again, the partition function near a critical point emerges only after a large number of iterations. Now we see a physical reason: local fluctuations (from mean behavior) are correlated over an immense range of length scales.

Admittedly, the description above is sketchy, but it should provide a glimpse into what is, in fact, a complicated analysis.

16.8 Essentials

1. A critical point is a point where the existence or coexistence of phases changes qualitatively. (Often a critical point marks the termination of an existence or coexistence curve.)

2. Near a critical point, fluctuations occur on a broad range of length scales.

3. Ferromagnetism is characterized by spontaneous magnetization (in domains and at temperatures below the Curie point).

4. Near a critical point, many physical quantities become singular (or vanish) with a power law behavior. The exponent in the power law is called a critical exponent.

5. The Ising model has $-J\sigma_i\sigma_j$ as the basic nearest neighbor interaction. Positive J favors parallel alignment of spins and hence can produce spontaneous magnetization. The interaction has its origin in a combination of electrostatics and the Pauli exclusion principle.

6. Mean field theory is an approximation scheme that reduces a coupled N-particle problem to a one-particle problem. For each particle, its interaction partners are replaced by the average behavior of all the particles. To determine the average behavior, a self-consistent equation is derived and solved. Mean field theory ignores local fluctuations, and so it can be quantitatively accurate only when such fluctuations are negligible.

7. The renormalization group uses iteration to incorporate different length scales successively. One derives recursion relations (which are usually only approximate), iterates them, and looks for fixed points, which represent a phase transition. A first-order Taylor expansion about a fixed point provides the critical exponents.

8. With few exceptions, phase transitions split into two sets:

   1. first-order transitions, which have a nonzero latent heat (usually) and a discontinuity in a first derivative of the chemical potential;
   2. continuous transitions, which have zero latent heat and power law singularities. (No discontinuities precede the appearance of the power law singularities.)

   Critical phenomena are associated with continuous phase transitions.

9. An order parameter is a (usually macroscopic) quantity that changes from zero to nonzero during a continuous phase transition. It gives a concise characterization of the system's macroscopic state. An order parameter may be a scalar, a vector, a complex number, or a more complicated geometric object.

10. Diverse physical systems have precisely the same critical exponents and fall into the same universality class. The determinants of class are the dimensionality of the physical space, the number of components of the order parameter, and the range of the interaction.

Critical behavior is associated with fixed points. Consequently, many iterations are required to calculate a partition function for a system near a critical point. The numerous iterations wash out specific details such as lattice structure and hence lead to the very existence of universality classes.

Further reading

The history of the Ising model (up to the middle 1960s) is ably narrated by Stephen G. Brush in "History of the Lenz–Ising model," *Rev. Mod. Phys.* **39**, 883–93 (1967).

Cyril Domb has contributed to our understanding of critical phenomena for over fifty years and knows whereof he speaks in *The Critical Point: A historical introduction to the modern theory of critical phenomena* (Taylor and Francis, Bristol, PA, 1996).

Clear writing and good insights are to be found in Nigel Goldenfeld, *Lectures on Phase Transitions and the Renormalization Group* (Addison-Wesley, Reading, MA, 1992). Another valuable perspective—although sometimes expressed too tersely—is provided by J. M. Yeomans, *Statistical Mechanics of Phase Transitions* (Oxford University Press, New York, 1992).

Section 16.5 is based on the article, "Teaching the renormalization group," by Humphrey J. Maris and Leo P. Kadanoff, *Am. J. Phys.* **46**, 652–7 (1978).

Another development, also based on the paper by Maris and Kadanoff, is provided by David Chandler, *Introduction to Modern Statistical Mechanics* (Oxford University Press, New York, 1987). Chandler makes some different points and hence complements section 16.5.

Kenneth G. Wilson described his work for a popular audience in "Problems in physics with many scales of length," *Sci. Am.* **241**, 158–79 (August 1979).

Problems

1. *Curie–Weiss law.*

(a) Solve approximately for $\langle \sigma \rangle$ from its mean field equation under the conditions $T \geqslant 1.2 T_c$ and $m_B B / k T_c \ll 1$. The ensuing expression is called the *Curie–Weiss law.*

(b) Compare your results with the analogous expression derived in section 5.3 for paramagnets that have no mutual interactions. Does the mean field interaction enhance or diminish the magnetization?

2. *Mean field revisited.* Here is another way to formulate the mean field theory of section 16.4. Write

$$\sigma_i = \langle \sigma \rangle + (\sigma_i - \langle \sigma \rangle) \equiv \langle \sigma \rangle + \delta \sigma_i;$$

use this form for each σ_i or σ_j in the energy expression (16.6); and then ignore terms in the product $\delta \sigma_i \delta \sigma_j$.

(a) Compute the partition function for a system of N spins, each having z nearest neighbors. Note that the final form can be written as

$$Z = \left(2 \cosh \frac{m_B B_*}{kT} \right)^N \times \exp \left(-\frac{z J N \langle \sigma \rangle^2}{2kT} \right).$$

(b) The *minima* in the Helmholtz free energy, $F = -kT \ln Z$, as a function of $\langle \sigma \rangle$ will give the most probable values of $\langle \sigma \rangle$. When you look for merely the *extrema* of F, what equation for $\langle \sigma \rangle$ emerges?

For parts (c) and (d), specify $B = 0$ and thereby study spontaneous magnetization.

(c) When $T < T_c$, does $\langle \sigma \rangle = 0$ correspond to a local minimum or maximum? What do you conclude about the "solution" $\langle \sigma \rangle = 0$?
(d) Graph F/NkT versus $\langle \sigma \rangle$ for both $T > T_c$ and $T < T_c$. What can you infer?
(e) Dropping the terms in $\delta \sigma_i \delta \sigma_j$ corresponds to ignoring correlations in the fluctuations of nearest-neighbor spins (relative to the mean behavior). Under which circumstances should this step be a good approximation? When a poor approximation?

3. *Susceptibility in mean field theory.* The derivative of the magnetization M with respect to the external field B is called the *susceptibility.* Often the limit $B \to 0$ is subsequently taken, and that limit yields the *initial* or *zero-field susceptibility.*

(a) Use mean field theory to study—near the critical point—the function

$$\chi(T) \equiv \lim_{B \to 0} \frac{\partial \langle \sigma \rangle}{\partial B},$$

which is proportional to the initial susceptibility per spin.
(b) In particular, determine the critical exponents above and below T_c.
(c) Also, determine the amplitudes, that is, the multiplicative coefficients of the power law in $|T - T_c|/T_c$.
(d) Provide a graph of $\chi(T)$ versus T for the interval around the critical temperature.

4. *Exact Ising model in one dimension.* Specify a linear chain of N spins, open at the ends, so that there are $N - 1$ nearest-neighbor interactions. Factor the partition function as

$$Z(N, K) = \sum e^{K\sigma_1\sigma_2} \times e^{K\sigma_2\sigma_3} \times \cdots .$$

The summation goes over the values ± 1 for each spin. Start with spin number 1. Regardless of what value σ_2 has, the sum over $\sigma_1 = \pm 1$ replaces the factor $\exp(K\sigma_1\sigma_2)$ by the factor $(e^K + e^{-K})$.

(a) Write out a justification for the claim in the preceding sentence. Then go on to determine $Z(N, K)$ completely.
(b) In which ways does your result justify the relationship (16.35), which was used for a closed chain?

(More about exact solutions to the one-dimensional model, given various boundary conditions, can be found in Goldenfeld's book, cited in the chapter references.)

5. *Fixed point in the two-dimensional Ising model.* Equation (16.36) determines how the coupling constant evolves under iteration. One can search for a fixed point, denoted $K_{\text{f.p.}}$, by asking whether that equation, now written as

$$K_{\text{f.p.}} = \tfrac{3}{8}\ln(\cosh 4K_{\text{f.p.}}),$$

has a solution.

(a) Search numerically for a solution.
(b) Then determine the numerical value of $\zeta(K_{\text{f.p.}})$.

6. Classify the phase transition associated with Bose–Einstein condensation (in an ideal three-dimensional gas).

7. *Latent heat and transition order.* Specify a ferromagnet at temperature $T < T_c$. Let the external magnetic field be $\mathbf{B}_{\text{ext}} = B\hat{\mathbf{z}}$, where B may be positive or negative.

(a) Consider the magnetization $\mathbf{M}$ in the two limits, $B \to 0$ through positive values and through negative values. Is $\mathbf{M}$ continuous? Or does it have a finite discontinuity? (You can analyze in terms of either the vector $\mathbf{M}$ or its component $\mathbf{M} \cdot \hat{\mathbf{z}}$ along the fixed direction $\hat{\mathbf{z}}$.)
(b) Classify the phase transition according to Ehrenfest's scheme.
(c) Do the two limits in part (a) have the same entropy per spin? Is there a nonzero latent heat (for the transition between the two "phases")? Is a nonzero latent heat a necessary or a sufficient condition for a first order transition?

8. *Ehrenfest's classification.* Paul Ehrenfest actually stated his classification system in terms of the Gibbs free energy G (per unit mass), rather than the chemical potential.

(a) If the physical system has only one species of particle, are formulations in terms of G and μ entirely equivalent? Defend your response.

(b) Now consider a binary mixture, such as a mixture of ^{3}He and ^{4}He liquids or zinc and copper atoms in an alloy like brass. Adopt the formulation in terms of the Gibbs free energy. What can you infer if the first derivative of G with respect to T is discontinuous?

Epilogue

All too easily, a book on thermal physics leaves the impression of a collection of applications: blackbody radiation, Debye theory, conduction electrons, van der Waals equation, and so on. Those applications are vital; they connect the theoretical and experimental worlds. Yet even more important is to come away from the book with a sense of its underlying theoretical structure. What is its equivalent of the equation $\mathbf{F} = m\mathbf{a}$ in a mechanics text? Or of Maxwell's equations in a book on electromagnetism?

In this book, the key theoretical ideas are the Second Law of Thermodynamics, the canonical probability distribution, the partition function, and the chemical potential. Of these, the Second Law of Thermodynamics comes first logically and is most nearly the central organizing principle. The majority of the applications, however, found us using the latter three items: $P(\Psi_j) = (1/Z)\exp(-E_j/kT)$, Z, or μ. By comparison with mechanics or electromagnetism, thermal physics suffers in that its most basic principle, the Second Law of Thermodynamics, is separated by layers of secondary theory from the physical applications. Perhaps that reflects the inherent difficulty of coping with 10^{20} particles all at once, or perhaps the root lies in the diversity of topics that can be addressed.

To emphasize the centrality of the Second Law, let me remind you that the evolution of entropy to a maximum (for an isolated system) led us to the general definition of temperature, $1/T = (\partial S/\partial E)_V$. That relation was a key ingredient in our derivation of the canonical probability distribution. The inequality $\Delta S \geq q/T$, which is another aspect of the Second Law, led us to the minimum property for the free energies. In turn, the minimum property led to the equality of chemical potentials (suitably weighted) in chemical and phase equilibria. Thus we can see the Second Law as the underlying principle in those equilibria.

Altogether, this epilogue should help you to see the forest despite the trees. A glance back to figure P1 in the preface, where the book's logical structure is outlined in flow-chart fashion, will help, too.

Appendix A Physical and Mathematical Data

Physical constants

Boltzmann's constant	$k = \begin{cases} 1.381 \times 10^{-23} \text{ J/K} \\ 0.8617 \times 10^{-4} \text{ eV/K} \end{cases}$
Planck's constant	$h = \begin{cases} 6.626 \times 10^{-34} \text{ J} \cdot \text{s} \\ 4.136 \times 10^{-15} \text{ eV} \cdot \text{s} \end{cases}$
Speed of light in vacuum	$c = 2.998 \times 10^8$ m/s
Electron charge (in magnitude)	$e = 1.602 \times 10^{-19}$ coulomb
Avogadro's number	$N_A = 6.022 \times 10^{23}$ items per mole
Bohr magneton	$m_B = \dfrac{e\hbar}{2m_e} = 9.274 \times 10^{-24}$ J/tesla
Stefan–Boltzmann constant	$\sigma_B = \dfrac{2\pi^5 k^4}{15\, c^2 h^3} = 5.67 \times 10^{-8}$ W/(m$^2 \cdot$ K^4)
Newtonian gravitational constant	$G = 6.673 \times 10^{-11}$ meter^3/(kg $\cdot$ s^2)
Mass of electron	$m_e = 0.9109 \times 10^{-30}$ kg
Mass of proton	$m_p = 1.673 \times 10^{-27}$ kg
Mass of hydrogen atom	$m_\text{H} = 1.674 \times 10^{-27}$ kg
Mass of helium atom (^{4}He)	$m_{^4\text{He}} = 6.649 \times 10^{-27}$ kg
Mass of diatomic nitrogen	$m_{\text{N}_2} = 4.653 \times 10^{-26}$ kg
Mass of the sun	$m_\text{sun} = 1.989 \times 10^{30}$ kg
Radius of the sun	$R_\text{sun} = 6.960 \times 10^8$ meters
Radius of the Earth's orbit (mean)	$r_\text{sun to Earth} = 1.496 \times 10^{11}$ meters

Conversions and equivalencies

1 electron volt (eV) corresponds to 1.602×10^{-19} joule.
1 atmosphere corresponds to 1.013×10^5 N/m^2.
1 calorie (20 °C) corresponds to 4.182 joules.

Convenient typical values

$\lambda_\text{blue-green} = 5 \times 10^{-7}$ m and $(h\nu)_\text{blue-green} = 2.5$ eV.
$\lambda_\text{orange} = 6 \times 10^{-7}$ m and $\nu_\text{orange} = 5 \times 10^{14}$ Hz.
$kT|_{T=300\text{K}} = 0.0259$ eV $= 1/40$ eV.
Room temperature $= 273 + 20 = 293$ K $\cong 300$ K.

Integrals of the form $\int e^{-ax^2} x^n \, dx$

Integrals of the form

$$I(n, a) \equiv \int_0^\infty e^{-ax^2} x^n \, dx, \tag{A1}$$

where a is a positive constant and n is zero or a positive integer, occur frequently in physics, particularly in classical thermal physics. Only for two values of n, namely, $n = 0$ and $n = 1$, need one work out the integrals in detail. All others follow by differentiation with respect to the parameter a, as follows:

$$\frac{\partial I(n, a)}{\partial a} = \int_0^\infty (-x^2) e^{-ax^2} x^n \, dx = -I(n + 2, a). \tag{A2}$$

Case of $n = 1$. When $n = 1$, the integrand contains the differential of the exponent (within a constant factor), and so the integral is easy:

$$I(1, a) = \int_0^\infty e^{-ax^2} x \, dx = -\frac{1}{2a} \int_0^\infty e^{-ax^2} d(-ax^2) = \frac{1}{2a}. \tag{A3}$$

Case of $n = 0$. The mathematician Pierre Simon, Marquis de Laplace, devised a neat trick for evaluating the integral

$$I(0, a) = \int_0^\infty e^{-ax^2} \, dx. \tag{A4}$$

Square both sides; replace the integration variable x in one integral by the variable y; and then construe the right-hand side as an integral over the positive quadrant of the x–y plane. Going to polar coordinates in that plane, where $r = \sqrt{x^2 + y^2}$, one splits the area into quarter-circle annular regions and finds

$$[I(0, a)]^2 = \int_0^\infty e^{-ar^2} \frac{1}{4} \times 2\pi r \, dr = \frac{\pi}{4a}. \tag{A5}$$

Then

$$I(0, a) = \sqrt{\frac{\pi}{4a}}. \tag{A6}$$

Table A1 displays the results for n running from 0 to 5.

Table A1 *Integrals of the form* $I(n, a) \equiv \int_0^\infty e^{-ax^2} x^n \, dx.$

$I(0, a) = \frac{1}{2}\sqrt{\pi} a^{-1/2}$	$I(1, a) = \frac{1}{2} a^{-1}$
$I(2, a) = \frac{1}{4}\sqrt{\pi} a^{-3/2}$	$I(3, a) = \frac{1}{2} a^{-2}$
$I(4, a) = \frac{3}{8}\sqrt{\pi} a^{-5/2}$	$I(5, a) = a^{-3}$

Integrals of the form $\int x^{p-1}/(e^x - 1)\, dx$

Provided $p > 1$, the integral cited above can be rearranged for evaluation by converting the awkward two-term denominator into a series of monomial expressions, as follows.

$$\int_0^\infty \frac{x^{p-1}}{e^x - 1}\, dx = \int_0^\infty e^{-x}(1 - e^{-x})^{-1} x^{p-1}\, dx$$

$$= \int_0^\infty e^{-x}\left[\sum_{m=0}^\infty (e^{-x})^m\right] x^{p-1}\, dx = \sum_{m=0}^\infty \int_0^\infty e^{-(m+1)x} x^{p-1}\, dx \qquad \text{(A7)}$$

$$= \left(\sum_{n=1}^\infty \frac{1}{n^p}\right) \times \int_0^\infty e^{-y} y^{p-1}\, dy = \zeta(p) \times \Gamma(p).$$

The step to a series uses a binomial expansion or recognizes the sum of a geometric series. The substitutions $n = m + 1$ and $y = nx$ reduce all the integrals to a single integral, the gamma function: $\Gamma(p)$. The remaining sum is the Riemann zeta function: $\zeta(p)$. For a generic value of p, neither function can be evaluated exactly or in closed form, but exact values are known for some of the p-values that concern us, as table A2 shows.

The logarithm of $1 + a$

There are several ways to justify the approximation $\ln(1 + a) \cong a$ when a is small (in magnitude) relative to 1.

First, one may turn to the integral expression for a natural logarithm and reason as follows:

Table A2 *Integrals of the form*
$\int_0^\infty x^{p-1}/(e^x - 1)\, dx = \zeta(p)\Gamma(p).$

p	$\Gamma(p)$	$\zeta(p)$	$\Gamma(p)\zeta(p)$
$\dfrac{3}{2}$	$\dfrac{\pi^{1/2}}{2}$	2.612	$\pi^{1/2}1.306$
$\dfrac{5}{2}$	$\dfrac{3\pi^{1/2}}{4}$	1.341	$\pi^{1/2}1.006$
3	2	1.202	2.404
4	6	$\dfrac{\pi^4}{90}$	$\dfrac{\pi^4}{15}$
6	120	$\dfrac{\pi^6}{945}$	$\dfrac{8\pi^6}{63}$

$$\ln(1 + a) = \int_1^{1+a} \frac{1}{y}\, dy \cong \int_1^{1+a} \frac{1}{1}\, dy = (1 + a) - 1 = a. \tag{A8}$$

The second step is permitted when $0 < |a| \ll 1$, so that even $1/(1 + a) \cong 1$.

Second, one may make a Taylor expansion of the logarithm about the value 1 for its argument:

$$\ln(1 + a) = \ln(1) + \left[\frac{d \ln x}{dx}\right]_{x=1} a + \cdots = 0 + \left[\frac{1}{x}\right]_{x=1} a + \cdots = a + \cdots. \tag{A9}$$

The derivative of $\ln x$ with respect to x is simply $1/x$, and so the second equality follows from the first.

Third, one may use a hand calculator to tabulate $\ln(1 + a)$ for small values of a and find the approximation as a numerical fact, provided $|a| < 0.01$ or so. Of course, the smaller the magnitude of a, the better the approximation. Table A3 illustrates these facts.

Approximating N!

The function $N!$ consists of N factors that range in size from 1 to N:

$$N! = 1 \times 2 \times 3 \times \cdots \times N. \tag{A10}$$

For a first approximation when N is large, take the "average" size of a factor and multiply that average together N times:

$$N! \cong \left(\frac{1 + N}{2}\right)^N \cong N^N 2^{-N}, \tag{A11}$$

provided $N \gg 1$. Using the arithmetic average might raise some doubts, but—we shall find—this approximation captures the essentials.

Our needs in thermal physics will be met by a good approximation to the logarithm of $N!$, and so we focus on $\ln N!$. The logarithm of a product is a sum of logarithms, and we can approximate the sum by an integral, as follows.

Table A3 *Numerical approximation* $\ln(1 + a)$ *when a is small relative to 1.*

x	$\ln x$	$\ln x$ rounded off
1.0	0	0
1.1	0.0953	0.1
1.01	0.009 95	0.01
1.001	0.000 9995	0.001
1.0001	0.000 099 995	0.0001

$$\ln N! = \ln(1 \times 2 \times 3 \times \cdots \times N) = \ln 1 + \ln 2 + \cdots + \ln N$$

$$\cong \int_1^N \ln x \, dx = (x \ln x - x)\big|_{x=1}^{x=N} \tag{A12}$$

$$\cong N \ln N - N.$$

Because the term $\ln 1$ in the first line is numerically zero, the sum has only $N - 1$ non-zero terms. It suffices to integrate $\ln x$ over the $N - 1$ unit intervals between $x = 1$ and $x = N$. A term of order 1 is ignored in the step to the last line. The final expression is an excellent approximation to $\ln N!$ when $N \gg 1$.

What about $N!$ itself? We carry on and rearrange equation (A12) as

$$\ln N! \cong N \ln N - N = N(\ln N - \ln e) = \ln(N^N e^{-N}). \tag{A13}$$

Logarithms are remarkably insensitive to modest changes in their arguments, and so we can *not* assert a near-equality for the two arguments. Nonetheless, to some lesser accuracy, we may equate the arguments and find

$$N! \cong N^N e^{-N}. \tag{A14}$$

Because $e \cong 2.7$, the present approximation sharpens our first approximation—but without making any radical change.

An even better approximation is provided by the expression

$$N! \cong \sqrt{2\pi N} \times N^N e^{-N}, \tag{A15}$$

the first term in *Stirling's approximation* (which is usually derived from an integral representation of $N!$).

Differentials

Consider a function f of two spatial variables, x and y: $f = f(x, y)$. By how much does the function f change in value when we shift attention from location (x, y) to location $(x + \Delta x, y + \Delta y)$? Figure A1 illustrates the context.

The shift in location is given by the vector

$$\Delta \mathbf{r} = \Delta x \, \hat{\mathbf{x}} + \Delta y \, \hat{\mathbf{y}}, \tag{A16}$$

where the circumflex (or "hat") denotes a unit vector along the corresponding direction. The gradient of the function f is

$$\mathrm{grad}\, f = \frac{\partial f}{\partial x} \hat{\mathbf{x}} + \frac{\partial f}{\partial y} \hat{\mathbf{y}}, \tag{A17}$$

when expressed in Cartesian coordinates. The vector $\mathrm{grad}\, f$ points in the direction that gives the maximal spatial rate of change of f. Moreover, the magnitude of $\mathrm{grad}\, f$ is equal to that maximal spatial rate of change. (In a direction perpendicular to $\mathrm{grad}\, f$, the function f does not change at all.)

To first order in the shift $\Delta \mathbf{r}$, the change Δf is given by the product of $|\mathrm{grad}\, f|$ times

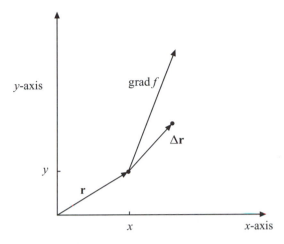

Figure A1 The vectors. Only the component of $\Delta\mathbf{r}$ along grad f contributes to the change in the value of f (to first order).

the component of $\Delta\mathbf{r}$ along grad f because only that component matters. Thus Δf is given by the scalar product of grad f and $\Delta\mathbf{r}$:

$$\Delta f \equiv f(\mathbf{r} + \Delta\mathbf{r}) - f(\mathbf{r})$$

$$= (\text{grad } f) \cdot \Delta\mathbf{r} + \text{higher order terms} \qquad (A18)$$

$$= \frac{\partial f}{\partial x}\Delta x + \frac{\partial f}{\partial y}\Delta y + \text{higher order terms}.$$

In effect, equation (A18) is the first portion in a Taylor's series (in two dimensions). Extension to three or more dimensions in Cartesian coordinates follows the same pattern: a sum of terms whose form is "partial derivative times change in the corresponding independent variable."

The general pattern holds true even if the independent variables are not spatial variables. Moreover, in the main text, all the Δ's are sufficiently small that only terms linear in Δ's need be retained.

Appendix B Examples of Estimating Occupation Numbers

Fermions

Suppose there are only three single-particle states (rather than the typical infinite number) and only two particles: $N = 2$. Table B1 shows the three full states Ψ_j and the corresponding sets of occupation numbers (given by the rows in the table). Now we use the second line of (8.7) to compute $\langle n_1 \rangle$ by summing over the admissible sets of occupation numbers. First take n_1 to be zero and sum over the admissible values of n_2 and n_3. Then take n_1 to be 1 and do likewise. Those steps produce the result

$$\langle n_1 \rangle = \frac{1}{Z(2)} \sum_{n_1} \sum_{n_2} \sum_{n_3} n_1 e^{-(n_1 \varepsilon_1 + n_2 \varepsilon_2 + n_3 \varepsilon_3)/kT}$$

subject to $n_1 + n_2 + n_3 = 2$

$$= \frac{1}{Z(2)} [0 e^{-(0+\varepsilon_2+\varepsilon_3)/kT} + 1 e^{-(\varepsilon_1+0+\varepsilon_3)/kT} + 1 e^{-(\varepsilon_1+\varepsilon_2+0)/kT}]. \tag{B1}$$

Looking at the second and last columns of table B1, we see that the outcome is exactly what we would have gotten by using the first line in equation (8.7). For large values of N, these direct methods are utterly impractical, and some special technique is needed.

Bosons

Again we specify $N = 2$ and only three single-particle states. Table B2 indicates that there are now six states Ψ_j and the corresponding sets of occupation numbers. To evaluate the boson analog of equation (8.7), sum over all sets of occupation numbers.

Table B1 *States and occupation numbers for two fermions, given only three single-particle states.*

Ψ_j	n_1	n_2	n_3	E_j
Ψ_1	0	1	1	$\varepsilon_2 + \varepsilon_3$
Ψ_2	1	0	1	$\varepsilon_1 + \varepsilon_3$
Ψ_3	1	1	0	$\varepsilon_1 + \varepsilon_2$

Table B2 *States and occupation numbers for two bosons, given only three single-particle states.*

Ψ_j	n_1	n_2	n_3	E_j
Ψ_1	0	0	2	$2\varepsilon_3$
Ψ_2	0	1	1	$\varepsilon_2 + \varepsilon_3$
Ψ_3	0	2	0	$2\varepsilon_2$
Ψ_4	1	0	1	$\varepsilon_1 + \varepsilon_3$
Ψ_5	1	1	0	$\varepsilon_1 + \varepsilon_2$
Ψ_6	2	0	0	$2\varepsilon_1$

First take n_1 to be zero and sum over the admissible values of n_2 and n_3. Next, take n_1 to be 1 and do likewise. Finally, take n_1 to be 2 and sum over the admissible values of the other n_αs:

$$\langle n_1 \rangle = \frac{1}{Z(2)} \sum_{n_1} \sum_{n_2} \sum_{n_3} n_1 e^{-(n_1\varepsilon_1 + n_2\varepsilon_2 + n_3\varepsilon_3)/kT}$$

subject to $n_1 + n_2 + n_3 = 2$

$$= \frac{1}{Z(2)} [0 + 0 + 0 + 1e^{-(\varepsilon_1 + 0 + \varepsilon_3)/kT} + 1e^{-(\varepsilon_1 + \varepsilon_2 + 0)/kT} + 2e^{-(2\varepsilon_1 + 0 + 0)/kT}]. \quad \text{(B2)}$$

Again, comparison with the second and last columns of table B2 shows that summing over sets of occupation numbers correctly evaluates $\langle n_1 \rangle$.

Appendix C The Framework of Probability Theory

Section 5.1 introduced the two schools of thought on what a probability means. This appendix carries on from that introduction and develops the framework of probability theory.

We noted that probabilities always arise in a context. The notation should reflect that fact. To express the idea, "the probability that a four appears, given that I roll a die once, is 1/6," we write succinctly

$$P(4 \text{ appears}|\text{roll once}) = \tfrac{1}{6}. \tag{C1}$$

The vertical line is read as "given that."

The two fundamental rules

Two fundamental rules govern the manipulation of probabilities. Dice and cards provide an easy way to introduce the rules by way of examples; then I will generalize the rules and later elaborate on why the generalizations are justified.

1. Negation ("not")

The probability that something does *not* occur is determined (numerically) by the probability that it does occur. For example, we can reason that

$$
\begin{aligned}
P(4 \text{ does not appear}|\text{roll once}) &= \frac{\text{tries when 4 does not appear}}{\text{all tries}} \\
&= \frac{\text{all tries} - \text{those in which 4 appears}}{\text{all tries}} \\
&= 1 - P(4 \text{ appears}|\text{roll once}). \tag{C2}
\end{aligned}
$$

The generalization is this:

$$P(\text{not } A|B) = 1 - P(A|B), \tag{C3}$$

where A and B denote verbal statements.

2. Conjunction ("and," written as "&")

The probability that both of two statements are true can be decomposed into separate probabilities. For an example, we ask, what is the probability of picking the queen of clubs when we pick one card from a well-shuffled deck? We might reason that there is only one queen of clubs in 52 cards, and so

$$P(\text{get queen of clubs}|\text{pick one card}) = \tfrac{1}{52}. \tag{C4}$$

Alternatively, we might look at figure C1 and note that a card's being the queen of clubs is equivalent to the card's being a club and being a queen simultaneously. Then we might reason that 1/4 of the cards are clubs and that, of the clubs themselves, only one in 13 is a queen. Thus we would compute the probability of picking the queen of clubs as $\tfrac{1}{4} \times \tfrac{1}{13} = \tfrac{1}{52}$, which agrees with our previous reasoning. Symbolically, we would write

$$P(\text{get queen of clubs}|\text{pick one card})$$
$$= P(\text{card is club \& card is queen}|\text{pick one card})$$
$$= P(\text{card is club}|\text{pick one card}) \times P(\text{card is queen}|\text{pick one card \& card is club})$$
$$= \tfrac{1}{4} \times \tfrac{1}{13} = \tfrac{1}{52}. \tag{C5}$$

The abstract generalization is

$$P(\text{A \& B}|\text{C}) = P(\text{A}|\text{C}) \times P(\text{B}|\text{C \& A}). \tag{C6}$$

Note that, for one of the probabilities on the right-hand side, the "given" condition is more than just the original statement C. The more informative and restrictive "given," C & A, can make a decisive difference. For an example, let A = "a 3 appears," B = "an even number appears," and C = "I roll a die once." Then $P(\text{A \& B}|\text{C}) = 0$ because 3 is not an even number, and so the conjunction is impossible. For use in the expansion (C6), one needs the probability

$$P(\text{B}|\text{C \& A}) = P(\text{an even number appears}|\text{I roll a die once \& a 3 appears})$$

$$= 0,$$

where the zero arises because an even number cannot appear if a 3 appears when I roll a die only once. Thus the right-hand side of equation (C6) has a zero factor, implying that the left-hand side equals zero, as it should. In contrast, the probability $P(\text{B}|\text{C})$, in

Clubs:	Ace	2 3 4 5 6 7 8 9 10 J (Q) K
Hearts:	Ace	2 ... Q
Diamonds:	Ace	2 ... Q
Spades:	Ace	2 ... Q

Figure C1 Picking the queen of clubs.

which the given is merely C, is $\frac{1}{2}$ and hence is nonzero. Using the latter probability by mistake would lead to inconsistency and error.

Equations (C3) and (C6) are the two fundamental rules of probability theory. To arrive at the rules, our development used the frequency version of what a probability means. Remarkably, the same rules emerge if one adopts the degree of belief interpretation from the very beginning. In a paper published in 1946, "Probability, frequency, and reasonable expectation," Richard T. Cox gave a compelling demonstration of that assertion. Cox's mathematical starting point was the following pair of modest assumptions:

(a) the probability $P(\text{not } A|B)$ is some function (presently unknown) of the probability $P(A|B)$.
(b) the probability $P(A \text{ \& } B|C)$ is some other function (also presently unknown) of the probabilities $P(A|C)$ and $P(B|C \text{ \& } A)$.

To determine the two unknown functions, Cox required that the functions yield rules that are consistent with ordinary logical reasoning. For example, because a double negative is equivalent to a positive statement, the functions must yield the relation $P(\text{not (not } A)|B) = P(A|B)$. After constructing several such constraints, Cox sought the most general functions that satisfy the constraints. The solution, Cox discovered, is unique. Aside from the freedom to reckon probabilities on different scales (0 to 100 percent, say, rather than 0 to 1), there is only one consistent set of rules for manipulating probabilities. Regardless of whether one construes a probability as a relative frequency or as a rational degree of belief, one must use precisely the rules (C3) and (C6).

The rule for "or"

The probability that either A or B or both statements are true can be decomposed into other, related probabilities. The rule for decomposition is derivable from the two fundamental rules, but we will follow a shorter route: an example and then generalization.

Suppose you are picking a single card from a well-shuffled deck. What is the probability that the card is a club or a queen? With the aid of figure C2, we reason that

Clubs:	Ace	2 3 4 5 6 7 8 9 10 J	Q	K
Hearts:	Ace	2 ...	Q	
Diamonds:	Ace	2 ...	Q	
Spades:	Ace	2 ...	Q	

Figure C2 Picking a queen or a club.

$$P(\text{club or queen}|\text{take 1 card from deck}) = \frac{13 \text{ clubs} + 4 \text{ queens} - 1 \text{ queen of clubs}}{52}$$

$$= P(\text{club}|\text{take 1 card}) + P(\text{queen}|\text{take 1 card}) - P(\text{club \& queen}|\text{take 1 card}). \tag{C7}$$

In the first line, we must subtract 1 queen of clubs because the queen of clubs appears among both the 13 clubs and the 4 queens; thus it has appeared *twice* already in the tally. Because the queen of clubs should appear only once (altogether), we must explicitly subtract 1 queen of clubs.

The generalization, known as the rule for *disjunction*, is

$$P(A \text{ or } B|C) = P(A|C) + P(B|C) - P(A \text{ \& } B|C). \tag{C8}$$

In the next subsection, we explore some consequences of this rule.

Mutually exclusive and exhaustive statements

If the statements A and B cannot both be true, given the context C, one says that the statements A and B are *mutually exclusive, given* C. The probability $P(A \text{ \& } B|C)$ is zero, and equation (C8) simplifies to

$$P(A \text{ or } B|C) = P(A|C) + P(B|C) \tag{C9}$$

when A and B are mutually exclusive, given C.

For an example, again let A = "a 3 appears," B = "an even number appears," and C = "I roll a die once." Then, upon going all the way back to equation (C8), we have

$$P(\text{a 3 appears or an even number appears}|\text{roll once})$$

$$= P(\text{a 3 appears}|\text{roll once}) + P(\text{an even number appears}|\text{roll once})$$

$$- P(\text{a 3 appears \& an even number appears}|\text{roll once})$$

$$= \tfrac{1}{6} + \tfrac{1}{2} - 0 = \tfrac{4}{6}. \tag{C10}$$

If at least one of the statements $A_1, A_2, \ldots, A_n$ that appear in a disjunction, A_1 or A_2 or $\ldots$ or A_n, must be true, given the context C, then the corresponding probability is 1:

$$P(A_1 \text{ or } A_2 \text{ or } \ldots \text{ or } A_n|C) = 1. \tag{C11}$$

Moreover, one says that the statements in the set $\{A_1, A_2, \ldots, A_n\}$ are *exhaustive, given* C.

If the statements A_1, A_2, ..., A_n are both mutually exclusive and exhaustive, given C, then equation (C11) and repeated use of equation (C8) imply

$$P(A_1 \text{ or } A_2 \text{ or } \ldots \text{ or } A_n|C) = \sum_{j=1}^{n} P(A_j|C) = 1. \tag{C12}$$

For example, if the statements have the form A_j = "the number j appears" and if the context C is that you roll a die once, then each $A_j = \frac{1}{6}$. Letting j run from 1 to 6 in equation (C12), we get the perhaps-familiar result that the sum of the probabilities yields 1. Note, however, that a sum equal to 1 requires (in general) that the statements be mutually exclusive and exhaustive in the given context.

For some further perspective on the result in equation (C12), remember that, given the statement C, the exhaustive property means that *at least one* of the statements A_j must be true. The mutual exclusion property means that *no more than one* of the statements may be true. Given the context, *one but only one* of the statements A_j is true, though we may not know which. That the probabilities sum up to unity, the number used to represent certainty, is then natural.

The expectation value estimate

Probabilities can provide us with estimates of physical quantities. Consider 40 paramagnetic particles that may, individually, have their magnetic moments parallel or anti-parallel to an external magnetic field. The net moment parallel to the field may be $netN = 40, 38, 36, \ldots$, or -40 times the value m_B of a single magnetic moment. (Section 5.3 sets the physical context more fully; the preceding sentences may suffice for here.) Figure C3 displays the set of probabilities $P(netN|C)$, where C denotes the

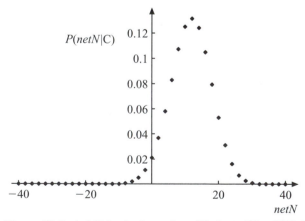

Figure C3 Probabilities in thermal equilibrium: $P(netN|C)$ for 40 particles. If the number of particles were much larger, the peak would be significantly narrower relative to the full range. (Note. The probabilities were computed from the canonical probability distribution, which is derived in section 5.2).

context sketched here and in section 5.3. Properly, to the left of the "given that" sign should appear the statement, "The net moment parallel to the field is *netN*," but using merely the variable *netN* provides brevity and is sufficient.

To estimate the value of *netN*, one could take the value that has the largest probability; inspection of figure C3 gives the estimate, $netN = 12$. Alternatively, one could consider each possible *netN*, weight its value by its probability of occurrence, and sum the terms:

$$\langle netN \rangle = \sum_{netN=-40}^{40} netN\, P(netN|C). \tag{C13}$$

Such an estimate is called an *expectation value* (or an *expectation value estimate*) and is denoted by angular brackets or an overbar. In the present situation, explicit summation yields $\langle netN \rangle = 11.65$. When thermal physics gives probabilities for a macroscopic property of a macroscopic system, the distribution is generally extremely narrow relative to its potential range. Then the expectation value estimate gives a reliable estimate and is often relatively easy to compute. For these reasons, we will use it extensively.

If the probabilities that enter an expectation value can be construed as frequencies in a long sequence of observations, then the expectation value itself can be construed as a simple arithmetic average over many observations. Such a situation might arise in an accelerator experiment in which many protons are shot at a beryllium target and the energy loss of each scattered proton is recorded. An expectation value calculated from quantum theory may be able to predict the average energy loss.

If, however, the situation is typical of a low temperature physics lab—perhaps a lab where the magnetic properties of liquid ^{3}He are under study—then an experiment is repeated only a few times. If one adopts the rational degree-of-belief interpretation, then the theoretical probabilities that we can calculate in thermal physics represent numerically the degree of conviction one should have in various possible values of a component of total magnetic moment, say. The expectation value estimate gives a weighted average over those possible values, where the weights are proportional to the strength of the conviction one should have, given the context (such as temperature and external magnetic field).

The narrowness of a probability distribution (for magnetic moment, say) arises because so many helium atoms contribute to the macroscopic property. What is large relative to 100 is not the number of experiments performed nor the number of helium samples that physicists literally work with. Rather, to repeat, it is the number of helium atoms in even one sample, perhaps 10^{20} atoms. In a sense, the law of averages operates *within* the macroscopic system. That enables us to make accurate predictions when we know a mere handful of experimental parameters.

Adherents of the frequency school will typically ask that one *imagine* a great many replicas of the actual experiment on helium. They will calculate a simple arithmetic average based on that collection of replicas and will then take the average as representative of what the single system in the lab will do. This approach is called the *ensemble* method. I will be candid and say that it seems artificial to me, but the practical conclusions are virtually identical to those drawn from the degree-of-belief interpretation.

A reduction procedure

In chapter 13, we derive a probability distribution for a gas molecule's velocity. Once we have that, we ought to be able to compute a probability distribution for the magnitude alone of the velocity, regardless of direction. To accomplish that reduction, we use a procedure that we derive now in some generality.

Suppose we have a set of probabilities $\{P(A_j \text{ \& } B|C)\}$, where the index j runs from 1 to n. What we really want, however, is merely the probability $P(B|C)$. To extract the latter, we sum over the index j and use equation (C6):

$$\sum_{j=1}^{n} P(A_j \text{ \& } B|C) = \sum_{j=1}^{n} P(B|C)P(A_j|C \text{ \& } B)$$

$$= P(B|C) \times \sum_{j=1}^{n} P(A_j|C \text{ \& } B)$$

$$= P(B|C). \tag{C14}$$

The step to the last line follows from equation (C12) *provided* the statements $A_1, A_2, \ldots, A_n$ are exhaustive and mutually exclusive, given the conjunctive statement C & B. [At the cost of a much more elaborate analysis, one can show that the conclusion in equation (C14) follows also if the statements $A_1, A_2, \ldots, A_n$ are exhaustive and mutually exclusive, given merely the original context C.]

The common sense interpretation of equation (C14) is that one "gets rid of" the statement A_j in $P(A_j \text{ \& } B|C)$ by summing over all the possibilities for A_j. In chapter 13, we sum (or, really, integrate) over all directions for the molecular velocity in order to compute a probability distribution for the speed alone.

References for further exploration of probability theory appear at the end of chapter 5.

Appendix D Qualitative Perspectives on the van der Waals Equation

The van der Waals equation, derived as equation (12.49), is reproduced here:

$$P = \frac{NkT}{V - Nb} - a\left(\frac{N}{V}\right)^2.$$ (D1)

This appendix offers further qualitative understanding of how the attractive and repulsive intermolecular forces affect the pressure.

Attraction

The attractive force reduces the pressure. A common line of reasoning proceeds as follows. When a molecule is deep in the interior of the gas, it is surrounded by a (more or less) spherically symmetric distribution of nearby molecules, and so it experiences no net force, on the average. When a molecule approaches the wall, however, numerous other molecules pull it away from the wall, but only a few pull it toward the wall (because few molecules lie between it and the wall). The molecule experiences a net pull away from the wall and toward the interior. The pull tends to diminish the momentum with which the molecule hits the wall and hence diminishes the pressure. The inward pull on a molecule will be proportional to the number of molecules behind it (which pull on it), and so the inward pull will be proportional to N/V. The number of collisions per second with the wall (by any molecules moving toward the wall) is proportional to N/V, as we reasoned in section 1.2. Thus the attractive force should reduce the pressure by a term proportional to $(N/V)^2$, which is precisely what emerged in equations (12.49) and (D1).

There is a flaw, however, in this common line of reasoning. The classical canonical probability distribution, as developed in chapter 13, implies that

$$\tfrac{1}{2}m\langle v^2\rangle = \tfrac{3}{2}kT$$ (D2)

everywhere in the gas, *regardless* of intermolecular forces. The estimated translational kinetic energy is $\tfrac{3}{2}kT$ both deep in the interior and very near the wall. Moreover, the distribution of molecular velocities is predicted to be isotropic everywhere. One needs to look at some aspect other than momentum for a tenable qualitative explanation.

Toward that end, we return to kinetic theory and to equation (1.6). For our purposes, that equation is best written as

$$P = \tfrac{2}{3} \times \tfrac{1}{2} m \langle v^2 \rangle \times \begin{pmatrix} \text{number density} \\ \text{very near the wall} \end{pmatrix}. \tag{D3}$$

The kinetic energy factor remains $\tfrac{3}{2}kT$, and so the effect of the attractive intermolecular forces must arise from a change in the number density very near the wall, a change relative to the over-all number density N/V.

Several paragraphs back, we noted that a molecule near the wall experiences a pull directed toward the interior. That pull will reduce the number density, as follows.

We can model the variation in number density with the "isothermal atmosphere" that we derived in chapter 7. Equation (7.7) can be written

$$\begin{pmatrix} \text{number density} \\ \text{at height } H \end{pmatrix} = \begin{pmatrix} \text{number density} \\ \text{at height zero} \end{pmatrix} \exp(-mgH/kT)$$

$$\cong \begin{pmatrix} \text{number density} \\ \text{at height zero} \end{pmatrix} \times \left(1 - \frac{mgH}{kT} \right), \tag{D4}$$

provided $mgH/kT \ll 1$. The gravitational force points from height H toward height zero; so height H corresponds to the wall, and height zero to the gaseous interior. What corresponds to the change in gravitational potential energy, mgH? It is a change in intermolecular potential energy associated with the attractive forces. Earlier, we expressed the potential energy of a single molecule as $-a \times (N/V)$ when the molecule is deep in the interior. When the molecule is close to the wall and fewer molecules attract it, its potential energy will have increased by an amount of order $a \times (N/V)$. Precisely how much is hard to say. Nonetheless, being guided by equation (D4), we can write

$$\begin{pmatrix} \text{number density} \\ \text{very near the wall} \end{pmatrix} \cong \frac{N}{V} \times \left[1 - \frac{\text{order of } a \times (N/V)}{kT} \right]. \tag{D5}$$

Substituting this form into equation (D3), we find

$$P \cong kT \frac{N}{V} \times \left[1 - \text{order of} \left(\frac{aN}{VkT} \right) \right]. \tag{D6}$$

If you mentally multiply out the factors, you will see that the pressure reduction term in the van der Waals equation, (D1), can indeed be understood as arising from a density reduction near the wall.

Repulsion

Now we ask how the repulsive forces can affect the number density. Represent the molecules by hard spheres of radius r_0. Figure D1 shows two wafer-like regions of space; the thickness of each is a small fraction of r_0. The first region is in the interior; the second has its upper edge at a distance $r_0 - \varepsilon$ from the wall (where ε is infinitesimal). Parts of these regions are excluded from occupancy by the center of any molecule because the exclusion spheres of nearby molecules extend into the region. For the interior region, the centers of such nearby molecules may lie on either side (or within the region. For the "near wall" region, molecular centers cannot fit between the wall and

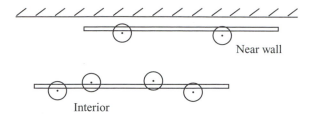

Figure D1 Wafer-like geometric volumes in the interior and near the wall. Seen edge-on, the volumes appear as strips. The circles represent molecular surfaces of radius r_0. The exclusion spheres (that is, spheres from which other molecular centers are excluded) have radii twice as large.

the region itself; only molecules with centers below (or within) the region can exclude other molecules. In essence, for the near wall region, only half as much volume is excluded as for the interior region. The number density in the near wall region should be higher than average; that enhances the collision rate and hence the pressure.

At first reading, you may wonder, how can a region of less excluded volume have higher number density? The key lies in the thinness of the regions. Almost all the "excluding" is done by molecules whose centers lie outside the regions.

For a quantitative assessment, we can reason as follows. According to section 12.9, a single molecule excludes the center of other molecules from a volume $8v_0$, where $v_0 \equiv (4\pi/3)r_0^3$ is the volume of a single molecule. In the interior, a geometric volume V_{geo} provides an accessible volume V_{acc} given by $V_{\text{acc}} = V_{\text{geo}} - 8v_0 \times (N/V)V_{\text{geo}}$. To understand this relationship, first take a large geometric volume—large in all directions—so that most of the excluded volume arises from molecules whose centers and exclusion spheres lie within the geometric volume. The number of such molecules is the number density, N/V, times V_{geo}. Each molecule excludes a volume $8v_0$. The full product gives the total excluded volume, and the difference between it and V_{geo} is the volume accessible to the center of another molecule.

By the homogeneity of the interior—on the average—the relationship between V_{acc} and V_{geo} applies to an interior volume of any size. In particular, it must apply to the wafer-like interior volume of figure D1.

Very near the wall (in the sense of figure D1), the excluded volume would be only half as large (because molecules can intrude from only one side, not from two). Thus the accessible volume near the wall is enhanced by the factor $1 + 4v_0 \times (N/V)$. In a dilute gas, the probability that a molecular center is within a given geometric region is proportional to the accessible volume. Thus the number density near the wall is enhanced by the factor $1 + 4v_0 \times (N/V)$. This factor implies that the coefficient b in van der Waals equation is approximately $4v_0$.

The approach in this appendix was espoused by Arnold Sommerfeld, *Lectures on Theoretical Physics, Vol. 5: Thermodynamics and Statistical Mechanics* (Academic Press, New York, 1964) and by Fritz Sauter, *Ann. Phys. (Leipzig)* **6**, 59–66 (1949). The reasoning for the effect of repulsion goes back to Ludwig Boltzmann, *Lectures on Gas Theory* (University of California Press, 1964), Part 2, Sections 3 to 5, first published in the last years of the nineteenth century.

Index

absorptivity, defined, 124
adiabatic, defined, 13
adiabatic relation, for
    classical ideal gas,
    14–16
angular brackets, defined,
    6, 95
Avogadro's number, 13

black, defined for surface,
    123
BEC, defined, 207
Bohr magneton, defined, 94
Boltzmann factor, defined,
    93
Boltzmann's constant, 6
Bose–Einstein
    condensation:
  defined, 202–3
  experiments, 205–9
  in a dilute gas, 207–9
  theory, 199–205
Bose temperature, defined,
    201–3
boson, defined, 168

canonical probability
    distribution, derived,
    91–4
Carnot cycle:
  defined, 51–2
  efficiency of, 54–5
Celsius, unit, 6
chemical equilibrium:
  conditions for, 246–7
  derived from kinetic
    theory, 244–6

derived from minimum
    property, 246–50
influences that determine,
    260–2
in semi-classical ideal
    gases, 247–50
chemical potential:
  at absolute zero, 226–8
  defined with Helmholtz
    free energy, 152–3
  early synopsis, 155
  equivalent expressions
    for, 226, 231, 239
  for adsorbed gas, 157–9
  for fermions at low
    temperature, 186–8
  for ideal gas, 153, 157
  for photon gas, 228–9
  for semi-classical ideal
    gas, 228
  functional dependence of,
    232–3
  from Gibbs free energy,
    231
  lemma for computing,
    156–7
  meaning of, 153–5
  understanding numerical
    values of, 226–30
  parallel between
    temperature and, 154
  when creation and
    annihilation occur,
    262–4
classical ideal gas: see ideal
    gas, classical
classical limit, of canonical

probability
    distribution, 306–14
Clausius–Clapeyron
    equation:
  derived, 280–1
  and vaporization curve,
    281–2
closed system, defined, 222
CM, defined, 8
conduction:
  as a heating process, 1–3
  equation of thermal,
    369–71
  thermal, studied, 367–75
convection, 3
cooling:
  by adiabatic compression,
    282–9
  by adiabatic
    demagnetization,
    331–7
  see also heating
coupling constant, defined,
    398
critical exponents, defined,
    388–9
critical opalescence, 383
critical phenomena, defined,
    384
critical point:
  defined in general, 384
  for liquid and vapor,
    211–12

Debye T^3 law, derived,
    134–7

Debye temperature, defined, 134
Debye theory, 130–9
degenerate, defined, 192
density of modes:
  defined, 117
  for electromagnetic modes, 118–9
  for sound waves, 133–4
density of states:
  defined, 75
  for entire system, 79–80
  for single spinless particle, 75–9
  for single particle with spin, 176–7
diatomic molecule, 250–7, 318–20
diffusion equation, derived, 369–71
domain, magnetic, defined, 385

efficiency:
  of Carnot cycle, 54–5
  maximum, 55–9
  of Otto cycle, 65
Einstein model, 131–2, 145
emissivity, defined, 125
energy:
  average translational kinetic, 7–8, 177–8
  internal, defined, 8
  modes of transfer of, 1–3
  per particle or mode, classically, 348–9
  zero point, 316–17
enthalpy:
  and latent heat, 274
  defined, 242, 274
  quantities derivable from, 242
entropy:
  additivity of, 42
  defined, 34
  disorder and, 44–5
  energy input by heating

and, 34–5, 41–4
evolution and, 128–30
etymology of, 34
heat capacity and, 62
heating and, 349–50
of ideal gas, 104, 107
of radiation, 122–3
paramagnetism and, 329–31
probabilities and, 327–9
variation with temperature, 39–41
rapid change and, 60–2
when canonical p.d. applies, 100–1
equilibrium constant:
  defined, 245, 249
  for semi-classical ideal gases, 249
equipartition theorem, derived, 314–16
exchange interaction, defined, 389–90
expectation value, defined, 95
expectation value estimate, defined, 95
extensive variable:
  defined, 105, 230
  examples: 230
external parameter, 2
extremum principles:
  for F and G, 233–4
  for S, F and G, 237

fermion, defined, 168
Fermi energy, defined, 184
Fermi energy, calculated, 184–5
Fermi function, defined, 183
Fermi temperature, defined, 185
ferromagnetism, 385–8
First Law of Thermodynamics, defined, 8–10

Fourier's equation, derived, 369–71
fusion curve: see melting curve

gas constant R, 13
Gibbs–Duhem relation, derived, 279–80
Gibbs free energy:
  chemical potential and, 231
  defined, 230–1
  quantities derivable from, 231
  generalization, 232
  minimum property, 233–4
Gibbs' phase rule, derived, 290–1

heat:
  as noun, 16–18
  as verb, 16–18
heat capacity:
  and $(\partial S / \partial T)_X$, 275–6
  at constant pressure:
    defined, 12
    for classical ideal gas, 12
  at constant volume:
    defined, 11
    for non-relativistic monatomic classical ideal gas, 11
  Debye theory of, 134–9
  defined, 11
  Dulong and Petit value of, 137
  heat as noun and, 16–17
  of diatomic molecules, 318–20
  of fermions at low temperature, 188–9
heat capacities, ratio of:
  defined, 15
  for diatomic molecules, 318–20

heat engine:
  defined, 55–6
  reversible, 56
heat equation, derived,
  369–71
heating:
  as inducing temperature
    change, 17
  as process, 16–18
  defined, 1–3
  entropy and, 349–50
  symbols q and Q, 9
heat reservoir, defined, 52
Helmholtz free energy:
  defined, 151–2
  expressed by partition
    function, 152
  minimum property,
    155–6, 233–4
  quantities derivable from,
    225–6
  why "free energy"?,
    234–5
heteronuclear, defined, 254
homonuclear, defined, 254

ideal gas:
  classical, defined, 12
  defined, 4
  degenerate: see quantum
    ideal gas
  nearly classical, 175–8
  quantum, 166–79,
    182–214
  treated semi-classically,
    101–7
ideal gas law, defined, 6–7
intensive variable:
  defined, 105, 230
  examples, 230
isentropic, defined, 288, 331
Ising model:
  defined, 389–92
  mean field theory and,
    392–7
  renormalization group
    and, 397–407

isotherm:
  defined, 292
  critical, 293
isothermal, defined, 14

Kelvin, biographical
  paragraph, 83
kelvin, unit, 6
kinetic theory, defined, 109

lambda:
  point, 205–6
  transition, 205–6
Langmuir model, 163–4
latent heat:
  discussed, 273–6
  entropy change and,
    275–6
  of vaporization, 273–4
  versus heat capacity,
    274–6
lattice gas, defined, 410–11
law of mass action:
  derived from kinetic
    theory, 244–6
  derived from minimum
    property, 246–50
Legendre transformation:
  defined, 225
  geometrically conceived,
    240–1

macroscopic regularity, 25
macrostate, defined, 27
magnetic field, "local", 336
magnetic ions, defined, 389
magnetic moment,
  reviewed, 94
magnetization:
  defined, 385
  easy axes, defined, 386
  spontaneous, defined, 386
  spontaneous, in mean
    field theory, 394–5
Maxwell relations, 240–2
Maxwell velocity

distribution, derived,
  310–11
mean field theory:
  appraised, 396–7
  defined, 392–3
mean free path:
  defined, 356
  refined, 375–6
  simplest expression for,
    358
melting curve, defined, 270
microstate, defined, 25
mole, defined, 13
multiplicity:
  defined, 27
  energy range in quantum
    version, 99–101
  energy transfer by
    heating and, 31–4
  entropy and, 34
  quantum version, 80,
    99–101

natural variables, defined,
  236
nearly classical:
  defined, 212
  ideal gas: see ideal gas,
    nearly classical
normal modes, defined, 116
normalization, 314
number density, defined, 5

occupation number:
  classical limit, 173–4
  defined, 169
  definition of estimated,
    170
  derivation of estimated,
    170–3
open system, defined, 222
order parameter, defined,
  409–10
Otto cycle (simplified):
  automobiles and, 65–7
  compression ratio and,
    66–7

Otto cycle (simplified):
   (*cont.*)
   defined, 57, 62–5
   efficiency of, 65
   entropy changes in, 63–5

paramagnetism:
   entropy and, 329–31
   Pauli, 192–4
   of conduction electrons,
      192–4
   of spatially fixed
      particles, 94–6,
      329–37
partition function:
   computing energy
      with, 97
   computing magnetic
      moment with, 98
   computing pressure with,
      97–8
   defined, 93
   for diatomic molecule,
      252–7
   for single particle (Z_1),
      defined, 102–4
   internal, defined, 248
   list of specific explicit
      forms, 108
   semi-classical limit of
      quantum, 174–5
   semi-classical, 103–4
perfect gas law: *see* ideal
      gas law
phase, defined, 270
phase equilibrium,
      conditions for,
      276–9
phase space, definition of
      classical, 306
phase transitions:
   continuous, 409
   first order, 407–9
phonons, and sound waves,
      130–9
Planck, determination of h
      and k by, 7, 127–8

Planck distribution, derived,
      119
Pomeranchuk refrigerator,
      285
Poiseuille's law, 366–7
pressure:
   according to kinetic
      theory, 4–6
   defined, 4
   extreme relativistic
      regime, 197–9
   of classical ideal gas, 4–7
   of electromagnetic
      radiation, 121
   of degenerate fermions,
      191
   of nearly classical ideal
      gas, 175–8
probabilities:
   entropy and, 327–9
   when temperature is
      fixed, 91–4
probability, defined, 89–91
probability distribution:
   defined, 310
   for speed, 311–12
   for velocity, 310–11
probability density, 309–10
probability theory,
      framework of,
      428–34

q and Q, defined, 9
quantum ideal gas, 166–79,
      182–214

radiation:
   absorptivity, 124
   as a heating process, 1–3
   cosmic background,
      127–8
   electromagnetic, 118–28
   emissivity, 125
   energy flux from black
      surface, 124
   energy flux from hole,
      123–4

energy per photon, 122
flux as function of
      frequency, 125–6
Planck distribution, 119
pressure, 121
total energy, 119–21
Wien's displacement law,
      112
radiative flux, defined, 123
random walk, and mean free
      path, 360–2
recursion relations, 399
renormalization group:
   1-dim. Ising model and,
      398–401
   2-dim. Ising model and,
      402–407
   defined, 397, 407
   fixed points and, 400, 404
   methods of, 407
reservoir: *see* heat reservoir
reversible:
   cycle, 56
   heat engine, 56
   process, defined, 67
reversibility:
   Carnot cycle and, 55–6
   conditions for, 67–9
   defined, 56, 67
   entropy change and,
      55–6, 67–9
   dissipative processes and,
      67–9
   maximum efficiency and,
      56–9
   Otto cycle and, 63–5
   slowness and, 67–9

Saha equation, 259–60
scaling, 105, 230
Second Law of
      Thermodynamics:
   defined, 29
   and desert analogy, 28–9
   in terms of entropy, 45–6
semi-classical
      approximation,

range of validity,
    105–7
semi-classical, defined, 175,
    212
specific heat, defined, 11–
    12
speed:
    most probable molecular,
        312–313
    mean molecular, 313
    root mean square
        molecular, 313
spin, degeneracy factor,
    252, 257
state function, defined, 18
statistical mechanics,
    defined, 109
Stefan–Boltzmann
    constant, defined,
    124
Stefan–Boltzmann law,
    derived, 123–4
stoichiometric coefficient,
    defined, 246–7
sublimation curve, defined,
    271
sublimation, defined, 270–1
superfluidity:
    in ^{3}He, 288–9
    in ^{4}He, 205–6

temperature:
    absolute, provisional
        definition of, 6
    average translational
        kinetic energy and,
        177–8, 348–9
    as deeper than average

kinetic energy, 349
Celsius, defined, 85
characteristic rotational,
    253–4
characteristic vibrational,
    252–3
critical, 271
Curie, defined, 386–7
Debye, 134
energy per particle or per
    mode and, 348–9
qualitative definition, 3
general quantitative
    definition, 80–2
individual versus
    relational, 85–6
Kelvin's definition of,
    83–4
negative absolute, 343–7
recapitulated, 347–9
Thomson's definition of,
    83–4
thermal:
    conductivity, coefficient
        of, 368
    contact, defined, 2–3
    de Broglie wavelength,
        defined, 103
    diffusivity, defined, 371
    equilibrium, defined, 4
    ionization, 257–60
    physics, defined, 18
thermodynamic potentials:
    defined, 236
    natural variables for,
        236–7
thermodynamics:
    defined, 109

etymology, 9
thermometer, defined, 3
Third Law of
    Thermodynamics,
    337–41
triple point:
    defined, 83, 271
    of water, 83
transport:
    of energy, 367–75
    of momentum, 362–7

universality:
    class, defined, 411–12
    determinants of, 412

van der Waals equation of
    state:
    developed, 293–300
    qualitative perspectives
        on, 435–8
vaporization curve:
    Clausius–Clapeyron
        equation and, 281–2
    defined, 271
    model for, 276–8
viscosity:
    studied, 362–7
    coefficient of, 364

w and W, defined, 9
white dwarf stars, 194–9
work:
    defined, 2
    pressure–volume, 10
    symbols w and W for, 9
working substance,
    defined, 55